Exposure to Engineered Nanomaterials in the Environment

Exposure to Engineered Nanomaterials in the Environment

Edited by

Nelson Marmiroli

University of Parma, Parma, Italy

National Interuniversity Consortium for Environmental Sciences (CINSA), Parma, Italy

Jason C. White

The Connecticut Agricultural Experiment Station, New Haven, CT, United States

Jing Song

Institute of Soil Science, Chinese Academy of Sciences, Nanjing, P.R. China

ELSEVIER

Elsevier
Radarweg 29, PO Box 211, 1000 AE Amsterdam, Netherlands
The Boulevard, Langford Lane, Kidlington, Oxford OX5 1GB, United Kingdom
50 Hampshire Street, 5th Floor, Cambridge, MA 02139, United States

Notices
Knowledge and best practice in this field are constantly changing. As new research and experience broaden our understanding, changes in research methods, professional practices, or medical treatment may become necessary.

Practitioners and researchers must always rely on their own experience and knowledge in evaluating and using any information, methods, compounds, or experiments described herein. In using such information or methods they should be mindful of their own safety and the safety of others, including parties for whom they have a professional responsibility.

To the fullest extent of the law, neither the Publisher nor the authors, contributors, or editors, assume any liability for any injury and/or damage to persons or property as a matter of products liability, negligence or otherwise, or from any use or operation of any methods, products, instructions, or ideas contained in the material herein.

British Library Cataloguing-in-Publication Data
A catalogue record for this book is available from the British Library

Library of Congress Cataloging-in-Publication Data
A catalog record for this book is available from the Library of Congress

ISBN: 978-0-12-814835-8

Publisher: Matthew Deans
Acquisition Editor: Simon Holt
Editorial Project Manager: Thomas Van Der Ploeg
Production Project Manager: Joy Christel Neumarin Honest Thangiah
Cover Designer: Greg Harris

Typeset by MPS Limited, Chennai, India

Contents

List of contributors

Ishaq O. Adisa
Environmental Science and Engineering PhD Program, The University of Texas at El Paso, El Paso, TX, United States; The Center for Nanotechnology and Agricultural Pathogen Suppression, New Haven, CT, United States

J. Enrique Alba-Rosales
Sciences and Engineering Division, University of Guanajuato, Leon, Mexico

Gustavo Basurto-Islas
Sciences and Engineering Division, University of Guanajuato, Leon, Mexico

Enrico Bergamaschi
Department of Public Health Science and Pediatrics, University of Turin, Turin, Italy

Dongxue Bi
Yantai Institute of Coastal Zone Research, University of Chinese Academy of Sciences, Yantai, P.R. China

Massimiliano G. Bianchi
Laboratory of General Pathology, Department of Medicine and Surgery, University of Parma, Parma, Italy

Geoffrey D. Bothun
Department of Chemical Engineering, University of Rhode Island, Kingston, RI, United States

Ovidio Bussolati
Laboratory of General Pathology, Department of Medicine and Surgery, University of Parma, Parma, Italy

Cristina Caffarra
Department of Medicine and Surgery, University of Parma, Parma, Italy

Davide Calestani
Institute of Materials for Electronics and Magnetism (IMEM)-CNR, Parma, Italy

Rigoberto Castro-Beltrán
Sciences and Engineering Division, University of Guanajuato, Leon, Mexico

Martina Chiu
Laboratory of General Pathology, Department of Medicine and Surgery, University of Parma, Parma, Italy

G. de la Rosa
Sciences and Engineering Division, University of Guanajuato, Leon, Mexico

Lirong Feng
Yantai Institute of Coastal Zone Research, University of Chinese Academy of Sciences, Yantai, P.R. China

Maricla Galetti
Center of Excellence for Toxicological Research, INAIL, University of Parma, Parma, Italy; Department of Medicine and Surgery, University of Parma, Parma, Italy

Concepción García-Castañeda
Sciences and Engineering Division, University of Guanajuato, Leon, Mexico

Jorge L. Gardea-Torresdey
Environmental Science and Engineering PhD Program, The University of Texas at El Paso, El Paso, TX, United States; Department of Chemistry and Biochemistry, The University of Texas at El Paso, El Paso, TX, United States; University of California Center for Environmental Implications of Nanotechnology (UC CEIN), The University of Texas at El Paso, El Paso, TX, United States; The Center for Nanotechnology and Agricultural Pathogen Suppression, New Haven, CT, United States

Amparo Guerrero Gerboles
Department of Medicine and Surgery, University of Parma, Parma, Italy

Jose A. Hernandez-Viezcas
Department of Chemistry and Biochemistry, The University of Texas at El Paso, El Paso, TX, United States; University of California Center for Environmental Implications of Nanotechnology (UC CEIN), The University of Texas at El Paso, El Paso, TX, United States

Patricia A. Holden
Bren School of Environmental Science & Management, Earth Research Institute, University of California Center for Environmental Implications of Nanotechnology (UC CEIN), University of California, Santa Barbara, Santa Barbara, California, United States

Arturo A. Keller
Bren School of Environmental Science & Management, University of California, Santa Barbara, CA, United States

Perla Y. López-Camacho
Sciences and Engineering Division, University of Guanajuato, Leon, Mexico; Natural Sciences Department, Metropolitan Autonomous University, CDMX, Mexico

Elena Maestri
SITEIA.PARMA, University of Parma, Parma, Italy; Department of Chemistry, Life Sciences and Environmental Sustainability, University of Parma, Parma, Italy

Marta Marmiroli
Department of Chemistry, Life Sciences and Environmental Sustainability, University of Parma, Parma, Italy

Nelson Marmiroli
Department of Chemistry, Life Sciences, and Environmental Sustainability, University of Parma, Parma, Italy; National Interuniversity Consortium for Environmental Sciences (CINSA), Parma, Italy

Michele Miragoli
Department of Medicine and Surgery, University of Parma, Parma, Italy; Institute of Genetic and Biomedical Research, IRGB-CNR, Milan, Italy

Monika Mortimer
Bren School of Environmental Science & Management, Earth Research Institute, University of California Center for Environmental Implications of Nanotechnology (UC CEIN), University of California, Santa Barbara, Santa Barbara, California, United States

Vinka Oyanedel-Craver
Department of Civil and Environmental Engineering, University of Rhode Island, Kingston, RI, United States

Luca Pagano
Department of Chemistry, Life Sciences and Environmental Sustainability, University of Parma, Parma, Italy

Keunhan Park
Department Mechanical Engineering, University of Utah, Salt Lake City, UT, United States

Nicol Parker
Bren School of Environmental Science & Management, University of California, Santa Barbara, CA, United States

Jose R. Peralta-Videa
Environmental Science and Engineering PhD Program, The University of Texas at El Paso, El Paso, TX, United States; Department of Chemistry and Biochemistry, The University of Texas at El Paso, El Paso, TX, United States; University of California Center for Environmental Implications of Nanotechnology (UC CEIN), The University of Texas at El Paso, El Paso, TX, United States

Venkata L. Reddy Pullagurala
Environmental Science and Engineering PhD Program, The University of Texas at El Paso, El Paso, TX, United States; University of California Center for Environmental Implications of Nanotechnology (UC CEIN), The University of Texas at El Paso, El Paso, TX, United States

Swati Rawat
Environmental Science and Engineering PhD Program, The University of Texas at El Paso, El Paso, TX, United States; University of California Center for Environmental Implications of Nanotechnology (UC CEIN), The University of Texas at El Paso, El Paso, TX, United States

Ana M. Rincon
European Food Safety Authority, Parma, Italy

Brett H. Robinson
Department of Chemistry, University of Canterbury, Christchurch, New Zealand

Stefano Rossi
Department of Medicine and Surgery, University of Parma, Parma, Italy

Roberta Ruotolo
Department of Chemistry, Life Sciences and Environmental Sustainability, University of Parma, Parma, Italy

Jing Song
Institute of Soil Science, Chinese Academy of Sciences, Nanjing, P.R. China

Giuseppe Taurino
Laboratory of General Pathology, Department of Medicine and Surgery, University of Parma, Parma, Italy

Benny K.G Theng
Manaaki Whenua—Landcare Research, Palmerston North, New Zealand

Edgar Vázquez-Núñez
Sciences and Engineering Division, University of Guanajuato, Leon, Mexico

Jason C. White
Connecticut Agricultural Experiment Station, New Haven, CT, United States; Department of Analytical Chemistry, The Connecticut Agricultural Experiment Station, New Haven, CT, United States; The Center for Nanotechnology and Agricultural Pathogen Suppression, New Haven, CT, United States

Guodong Yuan
School of Environmental and Chemical Engineering, Zhaoqing University, Zhaoqing, P.R. China

Nubia Zuverza-Mena
Department of Analytical Chemistry, The Connecticut Agricultural Experiment Station, New Haven, CT, United States

Short biographies

Prof. Nelson Marmiroli is a full professor in the Department of Chemistry, Life Sciences, and Environmental Sustainability of the University of Parma. Prof. Marmiroli coordinates a team that includes full professors, associate professors, research associates, and technicians. His current research topics are focused on the application of environmental biotechnologies for sustainability; phytoremediation, bioremediation, emerging contaminants (nanomaterials), interaction of plants with pollutants, genetic and molecular bases of genotype–environment interactions in different organisms (proteomic, genomic, transcriptomic analyses); molecular traceability of food supply chains for food safety and authenticity protection, coexistence of genetically modified plants with nonmodified plants.

Dr. Jing Song received his PhD degree in "soil chemistry" in 2002 from the Institute of Soil Science, Chinese Academy of Sciences (ISSCAS) and became an associate researcher in 2005. He served as deputy director of the Soil and Environment Bioremediation Research Centre (SEBC, ISSCAS), and deputy director of the CAS Key Laboratory of Soil Environment and Pollution Remediation (KLSEPR, CAS). Dr. Song's main research interests include bioavailability of soil pollutants, derivation of risk-based soil environmental standards, and research/development on the physiochemical remediation technologies for polluted soil/site. He has been active in joint research projects with Rothamsted Research, United Kingdom and Wageningen University and Research Center, the Netherlands. Dr. Song is a member of the Soil Environment Committee, Chinese Society of Soil Sciences, Senior Associate Editor for heavy metals of the *International Journal of Phytoremediation*, and a member of the Board of Directors of the International Society of Phytotechnology.

Dr. Jason C. White is the vice director and chief analytical chemist at the Connecticut Agricultural Experiment Station (CAES). Dr. White received his PhD in "environmental toxicology" from Cornell University in 1997. After 1 year as a postdoctoral associate at CAES, Dr. White joined the Department of Soil and Water in 1998. In 2009 he assumed the Department Head position in Analytical Chemistry, and in 2013, he was also appointed as CAES a vice director. The CAES Analytical Chemistry Department provides sample analysis to all other state agencies and also participates in the FDA Food Emergency Response Network (FERN) Chemistry Cooperative Agreement Program (cCAP). Dr. White also has research programs in several separate areas: nanomaterial contamination of agricultural crops, nano-enabled agriculture, and the phytoremediation of soils contaminated with persistent organic pollutants. Dr. White also has adjunct faculty appointments at the University of Texas—El Paso, University of Massachusetts, and Post University.

Foreword

The development and use of nanotechnologies and nanostructured materials are growing at an extremely rapid pace, both within the scientific community and without, in a diverse range of fields and disciplines. There is obviously significant interest in new methods of synthesis, which gave rise to specifically tailored materials for agricultural, medical, mechanical, or electrical application, but this has also led to simultaneous research on the problems of safety and sustainability associated with the dispersion of nanomaterials and nanoparticles in the environment. The fact that engineered nanomaterials (ENMs) have been included in the risk assessment requirements at the same level as for any new chemicals makes their development subject to some specific conditions. Among these, certainly safety for both environment and human health comes as of paramount concern. Safety issues are often complex; indeed in many cases the actual risk determined scientifically and that perceived by the public at large may be totally different. A typical example of this dichotomy has been the case of genetically modified organisms (GMOs) or transgenic plants, where widespread global application has been severely hampered by exaggerated risk perception and poor risk communication, on both sides.

Therefore those involved in the field of nanotechnologies and ENMs must work diligently to avoid the same pitfalls experience by GMOs. To this end a reliable and transparent risk assessment evaluation of the different ENMs must be pursued as independently of the cost and the time needed to comply with the actual legislation as possible. Industrialists and applicants have to be engaged with and trust the scientists and their independent opinions as this will be critical to accurate risk perception by the general public. End users and stakeholders have to be taken "inside" the evaluation process to ensure the convergence, independence, and validity of the results. A lack of transparency gives the impression of hiding one's head under the sand, and this only increases distrust and recrimination.

The second aspect that requires significant consideration concerns the sustainability either of producing and using ENMs. Sustainable ENMs production is important as it certainly impacts the unintentional dispersion of any bulk reagents, and a lack of consideration here will lead to excessive use of energy/resources, and more important, potentially compromised occupational health and safety. In addition, the sustainable use of ENMs will also provide feedback iteratively to intentional "safe by design" strategies that ultimately convey benefit from product cradle to grave.

There is a large convergence occurring in this field that includes an incredibly diverse range of disciplines and interests; in many ways making this a "vision frontier" for new technologies. But converging technologies cannot stand alone; there must be a robust "Nanosociety Ecosystem" where all efforts and programs are transparently evaluated prior their final deployment and destination. And since some fields more advanced in their application than others, manufacturers, regulators, and consumers can all assist in "real-time" assessment of true costs and benefits of a given approach.

However, like a new and emerging field, technologies linked to ENMs cannot progress if there is not significant interest and effort dedicated to developing capacity building programs and self-sustainability. As such, the target audience of this text includes the communities of undergraduate/graduate students, researchers, as well as policy and decision makers. The intentional organization and division of the respective chapters reflect this goal and specifically seek to avoid being the collection of highly specialized reviews.

The level of understanding is maintained at a high level, with a progression within the different chapters from basic information to those more necessarily specialized and advanced. The track within the different chapters follows also the same logic of progression, with the overall intent of making the convergence of disciplines a general strategy for a holistic approach to explain a rapidly development field of science with the vision of a common goal, the sustainable application of nanotechnologies.

Topics within the text span from environment to human health and from agriculture to food production; all are presented with the intent of making the topics understandable "at first sight," including the use of supportive illustrations and examples. To this end the book is divided in 4 parts and 14 chapters.

The first section includes two chapters on the synthesis, manufacturing, analysis, and characterization of ENMs; all of which are essential to understanding the "safe by design" approach.

The second section deals with the presence of ENMs in the environment and seeks to understand the transfer and interactions with organisms, in natural ecosystems, agroecosystems, urban environments, the workplace, and in everyday life.

The third section involves the application of ENMs to medicine, with the analysis of toxicological information, methods and approaches, and also the use of ENMs in therapy and diagnostics, including risk assessment and impacts on human health.

The fourth section summarizes the impacts of ENMs on society, with an analysis of ethical issues, consumer perception, education efforts, and regulation.

All chapters provide the readers with key points as part of the final message, and a list of references for further reading. The chapters in their final forms are the result of a peer review process, involving both internal and external evaluators; the editors are greatly indebted to these colleagues for the time and expertise.

The editors wish to thank the Sustainable Nanotechnology Organization for serving as robust and interactive venue where many of the authors have met, exchanged ideas, been challenged, and ultimately have produced what we feel is a significant contribution to the field we all work in.

The editors wish also to thank Dr. Thomas Van der Ploeg from Elsevier Editorial Office for following and stimulating the preparation of the book.

A special thank you to Dr. Marta Marmiroli and Mirca Lazzaretti for artwork.

Last, all editors and authors wish to express our heartfelt gratitude to Prof. Elena Maestri, whose efforts were instrumental in the preparation and completion of this book and who skillfully managed all contacts with the authors, editors, and the editorial office throughout the entire process.

Nelson Marmiroli[1,2], Jing Song[3] and Jason C. White[4]

[1]*Department of Chemistry, Life Sciences, and Environmental Sustainability, University of Parma, Parma, Italy*

[2]*National Interuniversity Consortium for Environmental Sciences (CINSA), Parma, Italy*

[3]*Institute of Soil Science, Chinese Academy of Sciences, Nanjing, P.R. China*

[4]*Connecticut Agricultural Experiment Station, New Haven, CT, United States*

Preface: Novel technologies from the nanoscale require new responsibilities

This volume presents a remarkable selection of research contributions concerning the safety of nanomaterials, from design approaches to exposure through natural and man-made systems, to societal implications and acceptance. The evolution of nanotechnology safety studies parallels nanoscale science and engineering research in general, which has expanding from nanoparticle precursors to nanomaterial systems.

The current progress of nanotechnology is characterized by integration of knowledge at the nanoscale and creation of larger nanostructure-based material and architectures, devices, and systems. The average annual growth rate of nanotechnology applications was about 25% in the first decade of this century and accelerated after 2010, and societal implications grew with them. In 2013 world market of products and services incorporating nanotechnology as a condition for their competitiveness reached $1 trillion, according to Lux Research. New generations of nanotechnology and corresponding governance needs are timed with the introduction of new fundamental concepts and with the successive increases in the degree of control, integration, complexity, and risk. Responsible governance focuses attention on the confluence of nanotechnology with other domains, its role on social sustainability, and synergistic effect from integration with other emerging technologies ("Converging Knowledge, Technology and Society: Beyond Nano-Bio-Info-Cognitive Technologies," Springer 2013, available on www.wtec.org/NBIC2/).

After an initial focus on nanomaterials toxicity itself, scientific interest in exposure to consumer and industrial nanoproducts has risen and led to essential improvements in evaluating risk. There have been significant research and regulatory activities in addressing the environmental, health, and safety implications of nanotechnology applications, especially for the passive as well as chemically and biologically active nanoparticles. The field continues to grow at the confluence with biology, biomedicine, harnessing large data, neurotechnology, and artificial intelligence. Responsible development of nanotechnology increasingly addresses ethical, legal, and other societal issues and includes ecological and human development aspects.

In the second part of this decade, nanotechnology has continued to expand to new knowledge and application domains, as illustrated in this volume. The editors have selected representative international contributions focused on exposure to engineered nanomaterials. The first section of the book is on "safety by design" synthesis and characterization (Chapter 2: Characterization of the physical and chemical properties of engineered nanomaterials and Chapter 3: Fate of engineered nanomaterials in natural environments and impacts on ecosystems). A second group of topics has a focus on exposure to engineered nanomaterials in natural, productive, biomedical, and consumer-driven ecosystems (Chapters 4–11). The volume concludes with overview studies about the social, perception, and regulatory implications of engineered nanomaterials (Chapters 12–14).

This volume brings together some of the most vigorous minds in nanotechnology and exposure studies. We encourage all academic and industry experts to closely read these state-of-the-art contributions on nanomaterials exposure in natural, working, and other living environments.

Mihail C. Roco
U.S. National Science Foundation and National Nanotechnology Initiative, Alexandria, VA, United States
Alexandria, VA, December 20, 2018

SECTION

I

Synthesis and characterization of Engineered Nanomaterials, towards a "safe by design" approach

CHAPTER

Synthesis and production of engineered nanomaterials for laboratory and industrial use

1

G. de la Rosa[1], Concepción García-Castañeda[1], Edgar Vázquez-Núñez[1], Perla Y. López-Camacho[1,2], Gustavo Basurto-Islas[1], Rigoberto Castro-Beltrán[1] and J. Enrique Alba-Rosales[1]

[1]Sciences and Engineering Division, University of Guanajuato, Leon, Mexico

[2]Natural Sciences Department, Metropolitan Autonomous University, CDMX, Mexico

1.1 Introduction

Life is sustained at different nanoscale levels, processes such as the transmission of chemical information ensure the perpetuation of life. This means that humans have interacted with nanomaterials (NMs) since appeared on earth. However, it was until Richard Feynman's famous talk "there is plenty of room at the bottom" (Feynman, 1960) that technology at the atomic scale and later at the nanoscale came into consideration. Nowadays, NMs are considered the basis of new technologies. For this reason, their synthesis and production are of the outmost interest. This chapter highlights the importance of engineered NMs (ENMs) in everyday life as well as in achieving substantial scientific and technological progress. A summary on the different methodologies used to synthesize NMs and produce nanostructures is presented. Selected ENMs uses and applications are also included. As closing remarks, a case study analyzing the perspectives of ENMs in nonlinear plasmonics is incorporated. A conceptual map showing the parts of this chapter is provided in Fig. 1.1.

1.2 Synthesis of nanomaterials

ENMs are prepared by a plethora of different methods that can be classified by different ways. One of the classifications, top-down/bottom-up, is based on the "size" of the starting material. Top-down methods start with bulk materials and are shattered or etched until reaching sizes in the nanorange. These are mainly

Exposure to Engineered Nanomaterials in the Environment. DOI: https://doi.org/10.1016/B978-0-12-814835-8.00001-7

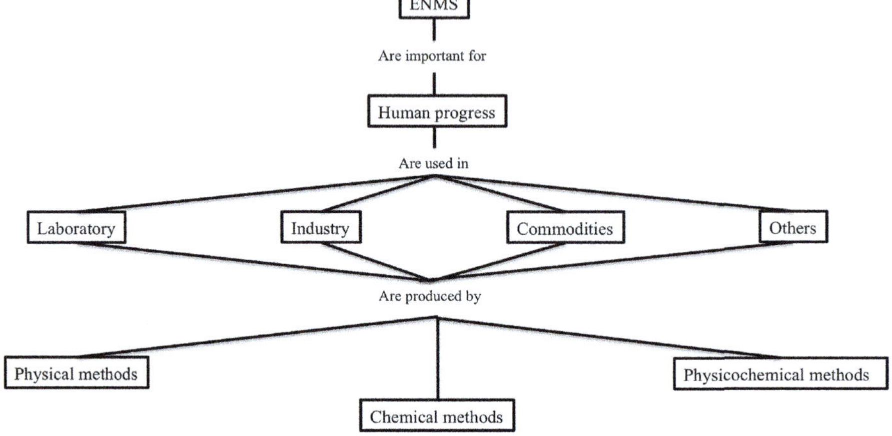

FIGURE 1.1

Conceptual map.

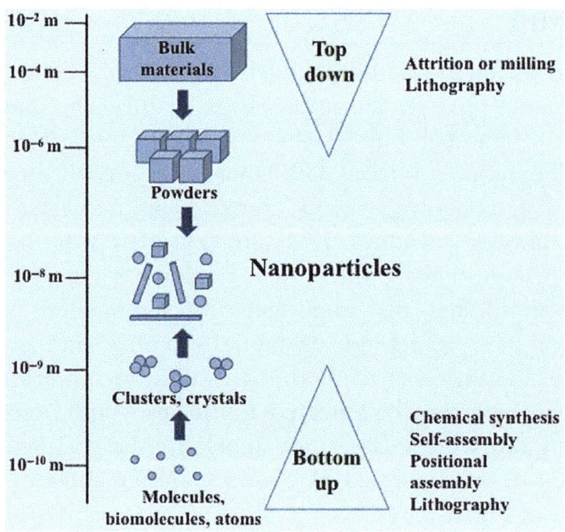

FIGURE 1.2

Engineered nanomaterials can be fabricated starting with bulk materials (top-down) or by chemical reactions (bottom-up).

based in physical phenomena including attrition, milling, and laser ablation. Bottom-up strategies are based in physical and chemical processes to build up from atoms or molecules to crystals and NMs. These include chemical synthesis, self-assembly, positional assembly, and lithography (Fig. 1.2). Chemical synthesis is mainly carried through sol–gel, oxidation–reduction, and biological processes.

To produced zero-dimension nanoparticles (NPs), dimensional synthesis uses homogeneous nucleation, heterogeneous nucleation, and kinetic confinement.

1.3 Use of nanomaterials in the laboratory

1.3.1 In research and analytical chemistry

In the laboratory, ENMs are mainly used for research, biomedical, and analytical purposes. These materials are being investigated with the aim of understanding their behavior and finding further applications. With regards to behavior, physical, chemical, physicochemical, and quantum size–related properties are studied in different contexts. With the new knowledge, novel applications can be found or improved. In addition, this information helps understanding other phenomena such as nanotoxicology. In the analytical chemistry arena, specifically in dealing with instrumental methods, ENMs help improving sensitivity, accuracy, and specificity (Liu and Zhao, 2009). For this reason, more and more techniques take advantage of these NMs and new analytical devices are being developed.

1.3.2 In biomedical applications

1.3.2.1 Metal-based nanomaterials

NMs provide an extensive platform for potential biomedical applications including immunoassays, imaging, biosensors, cellular labeling, therapeutics, drug delivery, and hyperthermia. Some NMs are metal-based NPs, such as nanogold, silver, copper, and nanometallic oxides (titanium dioxide, zinc oxide, iron oxide, and quantum dots). However, their low dispersion in water solution and the safety for therapeutics are still of concern. In particular cases high toxicity and low control of the NP residues in the organism may represent a hazard.

The biomedical research on NMs has focused not only on the therapeutic effect but also on the potential toxicity to the host. Different factors are related to this toxicity and they include NP chemistry, shape, size, self-aggregation, and electromagnetic properties. It has been stated that since NPs are formed by several atoms, toxicity is mainly based on their particular arrangement (Buzea et al., 2007). Their nanosize allows these materials to enter into living organisms. In humans their transport through the circulatory system and crossing the physiological barriers causes severe damages in tissues and organs at cellular levels disrupting their normal functions. Magnetic NP for biomedical applications is usually between 10 and 100 nm; the area has been associated to a high stability, ligand attachment properties, and easy degradation in kidneys or other organs (Gupta and Gupta, 2005; Karimi et al., 2013).

The size also determines the magnetic properties. Superparamagnetic NPs are the most prevalent for biomedical applications, because they do not retain magnetism after the removal of the magnetic field stimulation. On the contrary, nonsuperparamagnetic form agglomerates increasing their toxicity (Ma and Liu, 2007).

Coating is an important strategy for increasing the biocompatibility of engineered NPs and preventing their direct exposure to the host. Most of the magnetic NPs (Co, Ni, FePt) and metal oxides are toxic. Their hydrophobic surface avoids dispersion in water-forming cluster and losing their magnetic properties (Ma and Liu, 2007). Some polymers have been used to diminish the agglomeration, decreasing the particle size during the synthesis which causes the reduction in crystallinity (Sun et al., 2008). Other strategies for coating include hydrophobic NPs with hydrophilic polymers–forming stable colloids and ligand exchange (Jun et al., 2005), covalent bonding (Veiseh et al., 2005), and micelle encapsulation (Yu et al., 2007). Polymers have been widely studied because their properties allow the metal NPs to interact with host with low damage. In this context research has focused mainly on polyethylene glycol, chitosan, dextran, and poly(ethylamine). The coating assembly onto NP surface shows different arrangements: end-grafted polymers, surface absorption polymers, copolymers, and lipids (Karimi et al., 2013). Although coating plays an important role for biocompatibility, it may affect the performance of the NPs depending on the core and function.

Fig. 1.3 portraits several effects of the coating process in ENMs that may impact their performance in biomedicine and clinical applications. The biomacromolecules surface recognition by the NPs such as protein receptors, DNA, and lipids provides a potential tool for the control of cellular mechanisms including regulation of transcription, enzymatic inhibition, blocking or promoting both signaling and protein aggregation as therapeutics in pathologies, as well as

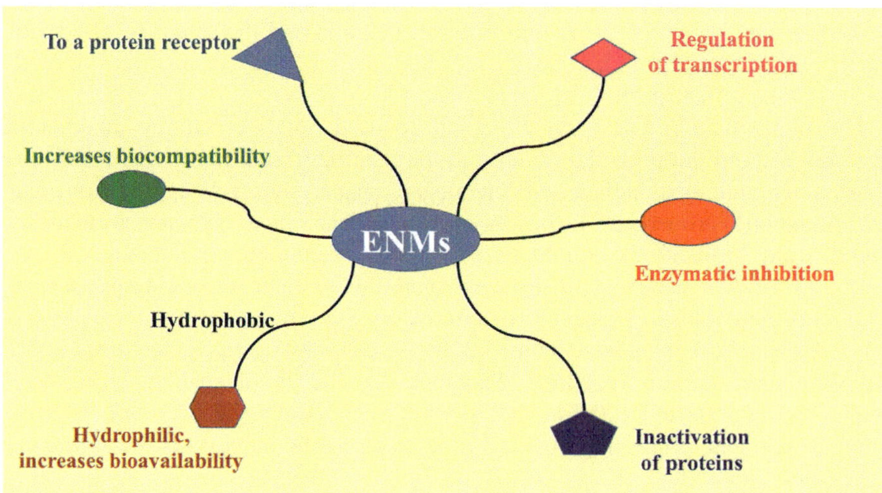

FIGURE 1.3

In biomedical applications, coating approach is used to provide specific characteristics to engineered nanomaterials.

nanocarriers linking NPs to proteins and DNA by direct covalent binding and noncovalent interactions between the NP and biomolecules (De et al., 2008).

Synthesis of superparamagnetic iron oxide NPs (SPIONs) is relatively simple. These materials display high biocompatibility and biodegradability. Also, no hysteresis is produced and this avoids agglomeration (Singh and Sahoo, 2014). In biomedicine SPIONs are one of the most frequently used NPs in magnetic resonance imaging as contrast agents (Xie et al., 2015). Furthermore, SPIONs have displayed capacity for magnetic drug targeting or drug delivery, suggesting they are promising therapeutic molecules (del Burgo et al., 2014).

Engineered NPs used to target cancer tissues contain surface biomolecules for cancer cells recognition. This specificity causes less damage in the organism as compared to radiation and chemotherapy that also may affect normal cells. Furthermore, several strategies are combined: targeting-specific cancer cells, internalization, carrying chemotherapeutics drugs, and further drug release (Veiseh et al., 2010). Nanocarriers for cancer incorporate drugs to the NPs such as doxorubicin, paclitaxel, methotrexate, and epirubicin (McBain et al., 2008; Sun et al., 2014); adsorption and linking molecules are the physical and chemical strategies used to functionalize the NPs (Mohammed et al., 2017). Radioimmunotherapy involves engineered NPs-linked antibodies targeting-specific cells and carrying low doses of radioactive isotopes for cancer treatment; the accumulation of radio in the target tumor induces cell death (Mohammed et al., 2017). Hyperthermia is probably the most promising protocol for cancer treatment. In this case NPs are locally injected in the target tissue. An external high-frequency alternating magnetic field is used to heat the NPs. Since temperature is crucial for cell survival, heating the NPs results in an increase in temperature of above 42°C that may cause necrosis, apoptosis, and activation of immune system (Shinkai et al., 1995). An alternative is magnetic fluid hyperthermia, which includes an injection of a magnetic fluid into the target followed by a frequency alternating magnetic field, resulting in the generation of heat (Laurent et al., 2011).

Gene therapy is the therapeutic delivery of exogenous nucleic acid into a host, manipulating the endogenous genes. Liposomes (low efficiency) and virus (serious safety concerns including acute immune response and immunogenicity) have been used for delivering, the challenge is to improve these mechanisms using nonviral carriers, NPs offer a solution as delivery systems with successful cellular uptake and efficient gene transfer because of the nanometer size ranges. Moreover, NPs have displayed stability of therapeutic nucleic acids against enzymatic degradation. In this case NPs are stable and biocompatible depending on functionalization, and efficiently transport of exogenous genetic material into the tissue-specific site. However, some difficulties still remain as transfection efficiency is low and, despite its biocompatibility, at high concentrations they show toxicity. Furthermore, the negatively charged molecules in serum and on cell the surface complicate the goal of achieving efficient transfection (Mohammed et al., 2017).

1.3.2.2 Carbon-based nanomaterials in biomedical applications

Normally, carbon-based NMs exist, as they are formed of carbon, one of the most abundant elements in nature. They have been extensively investigated since 1985 when fullerene was first described (Kroto et al., 1985). Solubility of these materials has shown low compatibility in biological systems because of their hydrophobicity; different strategies have been developed to achieve soluble materials in aqueous dispersions, in order to study them for biological applications. Some strategies include preparing two-phase colloidal solutions or synthesizing derivatives and polymers with hydrophilic substances such as amino acids, carboxylic acids, polyhydroxyl groups, or amphiphilic polymers (Gigault and Budzinski, 2016; Rašović, 2017; Kausar, 2017). Toxicity of carbon NMs is controversial even though many toxicological studies have been reported. Some research groups have reported no apparent toxicity of NMs and some others show significant toxicity in cell cultures and in vivo animal models (Zhang et al., 2014).

Carbon-based NMs, including fullerenes, graphene, carbon nanotubes, carbon nanodots, and nanodiamonds, have particular electronic, thermal, and optical properties, and in recent years they have been studied for biomedical applications including inhibition of human immunodeficiency virus-1 (HIV-1) protease, antioxidant, and cancer therapeutic, besides its application on diagnosis and alternative therapies as photoacoustic imaging and photothermal therapy, brachytherapy, chemotherapy, and dual-modal imaging (Bednarikova et al., 2016). In biomedical applications carbon-based NMs have been investigated as drug or gene delivery vehicles, or as drugs per se.

Physicochemical properties of NMs including size, surface area, solubility, chemical composition, shape, agglomeration state, crystal structure, surface energy, surface morphology, and surface charge have been extensively studied (Gatoo et al., 2014). Briefly, C_{60} fullerene is an icosahedrally shaped nanosized molecule, composed of 60 carbon atoms (C_{60}) with 20 hexagons and 12 pentagons with a crystallographic diameter of 0.7 nm and van der Waals diameter of 1 nm (Lebedeva et al., 2015). Nanotubes are graphitic tubules based on a hexagonal lattice of sp^2 carbon atoms; because of their physicochemical properties, they can potentially have bioimaging and tracing functions coupled with drug delivery. Graphene consists of a single atomically thin sheet of hexagonally bound sp^2 carbon atoms with large surface area and possibility of easy functionalization. Nanodiamonds are composed of carbon sp^3 structures in the core, with sp^2 and defect carbons on the surface (Bhattacharya et al., 2016).

NMs, particularly fullerenes, have been described as inhibitors of HIV-1 protease, an essential enzyme to yield infectious virions. In 1993 Friedman proposed the fitting of fullerene C_{60} with HIV-1 protease through steric and electrostatic interactions, using model building and simple physical chemical analysis (Friedman et al., 1993). Later, many researches have proposed several fullerene derivatives as inhibitors, looking to improve the affinity for the active site. Bosi et al. (2003) reported derivatives bearing solubilizing chains showing activity against HIV-1 type in the low micromolar range. Numerous fullerene derivatives have been developed displaying successful anti-HIV activity by targeting other important HIV enzymes, such as reverse transcriptase (Yasuno et al., 2015; Martinez et al., 2016).

Antioxidants are molecules that inhibit or quench free-radical reactions and delay or inhibit cellular damage. In human organisms there is an endogenous antioxidant defense system, and many exogenous molecules bear this capacity too, including NMs (Nimse and Pal, 2015). From the different types of carbon-based NMs group, fullerenes are the most efficient free-radical scavengers, because of the presence of several double bonds in the cage and functional groups attached to it, what makes it possible to react with free radicals as superoxide (O_2^-), hydroxyl radicals (OH^-), and hydrogen peroxide (H_2O_2) (Goodarzi et al., 2017). Fullerene is nonpolar and is not soluble in water; so many attempts to prepare more polar derivatives have been made. In general, polar derivative fullerenes are more favorable for radical scavenging (Sachkova et al., 2017; Hsieh et al., 2017).

Applications on cancer therapy have been described for carbon-based NMs, mainly in drug delivery systems. For instance, Panchuk et al. (2015) described a complex, fullerene C_{60} and doxorubicin, that inhibits tumor growth in a lung carcinoma model. Raza et al. (2016) demonstrated that conjugation of docetaxel with multiwalled carbon nanotubes enhances the anticancer activity of the drug; they present the pharmacokinetic profile of docetaxel improved by the conjugation.

Some reports have shown antimicrobial properties for carbon-based NMs such as fullerenes, carbon nanotubes, and graphene oxide NPs, with little resistance and tolerable cytotoxic effect on mammalian cells. It has been proposed that fullerene displays antimicrobial properties because of its ability to affect metabolic processes. In the case of nanotubes and graphene oxide NPs, antimicrobial mechanism occurs through physical damage and cell membrane disruption (Maleki Dizaj et al., 2015; Ji et al., 2016).

1.4 Use of nanomaterials in industrial and commercial applications

ENMs increase the productivity of industry by introducing changes in operational conditions, that is, temperature, pressure, reaction process, etc. Nanotechnology, as a revolutionary technology, has been capable to restructure the global economy with new initiatives based on the convergence of sciences. Industrial scale-up has effects in the strength of the labor force, transferring knowledge to industrial processes and commercial applications (Invernizzi, 2011).

NMs, as raw materials, imply new challenges to the nanotechnology industry. This is in part due to the fact that, in recent years, the "commoditization" for ENMs has emerged as a trend to standardize a variety of products. Maintaining the minimum standards of quality allows manufacturing large volumes and in turn more competitive prices can be offered (McGovern, 2010; Integrated Nanoscience and Commodity Exchange, 2012).

The commercialization of NMs requires an exhaustive assessment to determine their environmental, health, and safety implications (National Nanotechnology Initiative, 2015). The rate at which novel NMs are produced and incorporated into new products is a concern since it may outpace the rate at which the potential impact in humans and the environment can be assessed (Olson and Gurian, 2012).

In 2005 the Woodrow Wilson Institute started a Project on Emerging Nanotechnologies to create the first nanotechnology consumer products inventory (CPI). These products were grouped into eight categories according to their applications:

1. Health and fitness
2. Home and garden
3. Automotive
4. Cross-cutting
5. Electronics
6. Food and beverage
7. Appliances
8. Goods for children

United States leads the production with 60% followed by the Asian and European markets. In this inventory more than 50% corresponded to a health and fitness category where the main product in 2006 was clothing with 34 products in 2006 (Woodrow Wilson Institute, 2013). According to a new inventory (Vance et al., 2015), over 1800 commercial products were marketed in the world, and it is estimated that by the year of 2020, this figure will reach 3400. Fig. 1.4 shows the increase of the commercial products over the time since the inventory started. This list is periodically updated by the CPI.

With regards to chemical composition, metal and metal oxides have the highest percentage in products (37%), followed by carbonaceous, silicon, and other NMs. Some commodities may contain more than one NM in the composition. Unfortunately, at least 49% of commercial products do not describe their ENMs composition. In order to properly assess the life cycle and potential risks of products containing ENMs, obtaining precise data on their chemical composition and main exposure routes (dermal, ingestion, and inhalation) is extremely important.

Most of the NMs employed in industry are those based on silver (50%), titanium (10%), and zinc (10%); others include Pt, Pd, SiO_2, Fe_2O_3, Fe_3O_4, and Al_2O_3, which are classified as miscellaneous NPs (Piccinno et al., 2012). Herein, a diversity of applications is discussed.

1.4.1 Silver nanomaterials

The main countries producing silver NMs are Korea, the United States, and Germany (Inshakova and Inshakov, 2017); these countries produce and provide

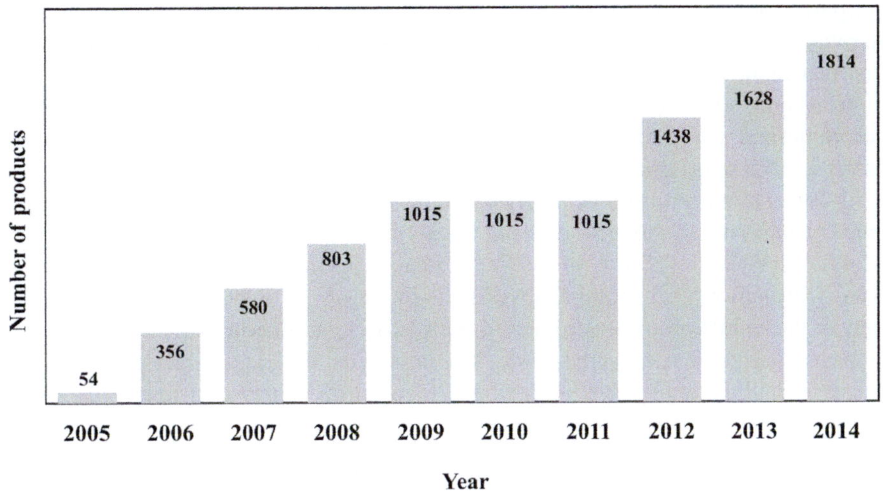

FIGURE 1.4

Number of total commercial products registered in the CPI from 2005 to 2014. *CPI*, Consumer products inventory.

Based on Vance M.E., Kuiken, T., Vejerano, E.P., McGinnis, S.P., Hochella, M.F., Rejeski, D., et al., 2015. Nanotechnology in the real world: redeveloping the nanomaterial consumer products inventory. Beilstein J. Nanotechnol. 6, 1769–1780.

silver NMs that are used in around 400 registered products all around the world. These ENMs have been widely used as antimicrobial agents, and for this reason, they are incorporated into health diagnosis equipment, food industry, and electronic devices used in medicine (You et al., 2012). They are also incorporated in the cosmetic and personal care industry, appliances, clothing and textiles, cross-cutting, food and drink, home and garden (Wijnhoven et al., 2010).

Some negative effects to the environment have been reported, for instance, Blaser et al. (2008) demonstrated the presence of Ag-NPs and NMs in air, due to the functionalization process of bags; they explained this phenomenon using a model; Franci et al. (2015), on the other hand, reported the effect of functionalization with Ag NMs and NPs on bacterial population.

Food industry uses Ag-NPs in the packaging process to prevent microbial contamination and prolong the commercial market life. The sonochemical coating is the most common technique to prepare materials covered with NPs, providing protection against both Gram-positive and -negative of bacteria (Darroudi et al., 2012). The biomedical sector must satisfy and guarantee the absence of pathogens and microbes, which naturally are present in the environment. In the last few years colloidal Ag-NPs have been tested at pilot scale in water treatment systems (Kumar et al., 2014).

Apart from the food sector, the Ag-NPs have applications in the electrochemical field, due to their properties. Ag-NPs can work as nanosensors; they offer fast response and can detect low concentrations of analytes (Anker et al., 2010). Manno et al. (2008) electrodeposited Ag-NPs onto alumina plates gold micropatterned electrode that showed a high sensitivity to hydrogen peroxide (Hahm and Lieber, 2004).

Ag-NPs have been used as catalysts in the luminol–hydrogen peroxide chemiluminescent system, showing a better catalytic activity than that for Au and Pt colloids (Guo et al., 2008). Moreover, Liu and Zhao (2009) used Ag-NPs supported with halloysite nanotubes (Ag/HNTs) with an Ag content of about 11% to catalyze the reduction of 4-nitrophenol with $NaBH_4$ in alkaline aqueous solution.

In the last years the synthesis of NPs intended to be in contact with food or for use in medical applications have been focused on using eco-friendly methods. Green synthesis methods include the use of microorganisms such as (1) fungi, yeasts, bacteria, and actinomycetes, or (2) plant extracts, (3) virus DNA, and (4) diatoms (Rafique et al., 2017).

Ag green synthetized NPs have been tested as cardiovascular implants (Grunkemeier et al., 2006); for reducing inflammation in the heart; as catheters, in order to reduce the generation of biofilms due to the antimicrobial activity (Chou et al., 2008). Zheng et al. (2010) and Liu et al. (2012) reported that bacterial contamination associated to implants and joint substitution are as high as 1.0%–4.0%; they are hard to treat and display high morbidity. Ag-NPs have been incorporated into plain poly bone reinforce, for safer connections of joint prostheses as an approach to decrease bacterial resistance.

In general, the applications of these NPs produced by green methods can be summarized as follows medical applications, environmental applications (air disinfection; Yoon et al., 2008, surface disinfection; Kumar et al., 2008), antiviral activity (showing effect against HIV-1, Herpes simplex virus, hepatitis B, Monkey pox virus; Ahmed et al., 2010; Lara et al., 2010; Rogers et al., 2008), and antifungal virus (showing fungicidal effect against *Candida albicans*, *Candida krusei*, *Trichophyton mentagophytes*; Coker et al., 2011; Kim et al., 2009; Manjumeena et al., 2014).

1.4.2 Titanium nanomaterials

Due to their versatility, properties, and characteristics, titanium NMs could be widely used in diverse applications, that is, UV protection, photocatalysis, photovoltaics, and electrochromics and photochromics (Kapilashrami et al., 2014; Mahlambi et al., 2015). Their stability, nontoxicity, and low cost made them more valuable than other NMs. Due to their hydrophobicity and hydrophilicity, Ti NMs have been used in surfaces by placing Ti nanorods on mirrors and eyeglasses. Also, two-phase separation processes systems have taken advantage of their superhydrophilicity and hydrophobicity (Feng et al., 2004). They are also used in self-cleaning coatings of devices (Parkin and Palgrave, 2005). In the clean energies sector the Ti NMs can have important applications due to their capacity to enhance the absorption of light by sensitizing the surfaces with them (Pelaez et al., 2012). In this field the energy production is not only focused on solar resources but also on hydrogen generation (Li et al., 2015).

The TiO_2 NMs are well known for their efficient and environmental performance as photocatalysts, having a great use in degradation of water pollutants. Their effect and efficiency in reducing pathogens (i.e., *Escherichia coli*) has also been reported (Liga et al., 2011). So far, three generations of NMs based on Ti have been developed, which have suffered modifications due to the doping the surface of this material with other elements (Wen et al., 2015). According to Environmental Protection Agency (2012), the first generation of nanostructures comprises those passive NMs not intended for a particular use; the second generation is related to active nanostructures designed for specific purposes, while the third one is related to 3D nanosystems as well as systems of nanosystems. The crystallinity of Ti has helped in improving the transfer and conversion of solar energy into electricity; in the photovoltaic industry, the Ti NMs have been mixed with other materials in order to increase their performance; Han et al. (2005) fabricated a hybrid TiO_2 nanocrystalline electrode, which the efficiency in the energy conversion process.

In experiments related to the photovoltaics processes, Ti/Au-NPs were used to split water and generate H_2 and O_2 (Gomes Silva et al., 2010); Salvador (2001) described the thermodynamics and photochemistry of the water splitting, providing valuable information that allowed later the introduction of doped active surfaces with Pt and NiO for better reaction efficiency. It is clear that the cheaper the energy conversion, the more attractive these processes become for the industry. For this reason, studies are being with NMs that may potentially carry out the split reaction under natural conditions, this means, without the need of external conditions. The introduction of TiO_2 nanotubes into the energy conversion sector has solved this. The conversion with nanotubes doped with other materials has been carried out using wavelengths that could not be possible executed with solely TiO_2 nanotubes (Wang et al., 2012).

In the electronic sector, the electrochromic devices have emerged as one of the most explored items that take advantage of the electrochromic properties of Ti (Thakur et al., 2012). Electrochromism refers to changes in optical surfaces due to the reduction or oxidation of elements, in this case Ti. Electrochromism has been produced in electrodes containing nanocrystalline TiO_2 and Li electrolytes (Chen et al., 2010b), also in nanocrystalline TiO_2 structures modified with viologens and anthrachinons with a surface anchoring group (Wagemaker et al., 2001).

Another important area is hydrogen storage using Ti NM. Lim et al. (2010) found that nanotubes of TiO_2 can store up to 2% H_2 (per total weight) at room temperature and 6 MPa. The gas can be later unconfined by releasing pressure. Some other configurations have been tested including multilayered walls of nanotubes of TiO_2 in a temperature range between $-195°C$ and $200°C$ and pressures from 0 to 6 bar (Schlapbach and Züttel, 2011).

Lately, the nanocrystalline structure of TiO_2 has been studied as sensors for gases (Moon et al., 2010), these sensors were tested to detect hydrogen at room temperature showing other benefits in their feature, for example, they can be auto cleaned (Mor et al., 2003). TiO_2 NMs are under continuous technological development. These materials seem to have a key role toward the environmental conservations and for the developing of new renewable energy technologies in a sustainable way.

1.4.3 Zinc nanomaterials

Zinc oxide (ZnO) is an inorganic compound and antimicrobial agent present in soils and generally recognized as safe (GRAS) (Rasmussen et al., 2010). It has been introduced in different industrial sectors, mainly in the food and textile industries. Zn- and Cu-NPs have been incorporated in synthetic fibers and commercial textiles, that is, carpet fabrics, water filters, materials for use in healthcare facilities, apparel, home furnishing, etc. (Espitia et al., 2012; Nayak and Padhye, 2014). One of the advantages of using Zn-NPs is that they have a minimum effect on color or texture, clarity, or processing.

As it was explained before, the Zn-NPs can provide antimicrobial activity to surfaces of materials; many coating formulations have included the ZnO-NPs into materials in order to provide long-term antimicrobial performance, such is the case of the healthcare industry (Smijs and Pavel, 2011), industrial and institutional cleaning service (Schlapbach and Züttel, 2011), and painting and coating (Mostafaei and Nasirpouri, 2014). Currently, the presence of Zn-NPs in formulations of coatings is widely spread. For instance, urethanes, acrylics, and vinyl acetates have shown good performance in UV protection tests in environmental conditions (Beltrame and Bos, 2008). Cosmetic industry also takes advantage of UV protection provided by ZnO- and TiO_2-NPs (Nash, 2006; Stark et al., 2015).

Zn-NPs may play a role in central nervous system by mediating neuronal excitability or even by releasing neurotransmitters. Some reports have indicated that Zn-NPs are biocompatible (Osmond and McCall, 2010; Rasmussen et al., 2010).

In agriculture it has been demonstrated that Zn-NPs have the potential to increase the crops yield and accelerate the growth of plants (Tolaymat et al., 2017), these NPs have been also used as germination promoters for seeds, for example, in peanuts (Prasad et al., 2012).

On the other hand, the use of plan extracts and microorganisms as biotemplated media for preparing Zn-NPs has opened possibilities for its use in the biomedical sector, for instance as antimicrobial and antibacterial agents (Poole, 2017; El-Batal et al., 2018), displaying selectivity. Different reports indicated that these NPs are nontoxic to cultivated human dermal fibroblasts; however, they exhibit toxicity toward metastatic tumor cells (Ivask et al., 2015) and vascular endothelial cells (Gojova et al., 2007) while inducing apoptosis in neural stem cells.

The presence of Zn-NPs represents an emergent and powerful tool for drug delivery and bioimaging industrial areas. In the first, Yuan et al. (2010) used the ZnO quantum dots as a drug delivery system for targeting doxorubicin to HeLa cells and encapsulating the NPs with chitosan to improve their stability; in gene delivery, Nie et al. (2006) demonstrated that tetrapod-like nanostructures were able to work as vehicles of this genetic material. Regarding the imaging sector and due to their semiconductor properties, the Zn-NPs have a high potential to replace the traditional Cd-related species, which traditionally have been used in biology and optical fields. Among the applications, these NPs were used in models like human skin and rat liver cells (Roberts et al., 2008), skin (Song et al., 2011), blood cells of zebrafish (Urban et al., 2012), human nasopharyngeal epidermal carcinoma cells (Kachynski et al., 2008), HeLa cells (Zhang et al., 2012), and B16F10 cells (Wu et al., 2011).

1.4.4 Platinum and palladium nanomaterials

In the energy industrial sector, the development of fuel cells is a concern; these devices can operate on the basis of hydrogen or small organic molecules (ethanol, formic acid, methanol) (Kulesza et al., 2013). Platinum (Pt) and palladium (Pd) are used to catalyze the reactions in these fuel cells and also are widely used in the petroleum and automotive sectors due to their high catalytic activity and stability (Bell et al., 2008). The latest efforts are based on the developing of new structures (at nanometric level) in order to increase the catalytic activity of Pt. This is because Pt is a very expensive metal and in economic terms represents an important investment that can drive nonprofitable processes (Bashyam and Zelenay, 2006).

Pt-based NMs are also used in the oxidation of CO. This step is important in the fuel cell operation because CO is an intermediate/product in organic compounds oxidation (methanol, ethanol, and formic acid) (Wang et al., 2004). In addition, it is present in the process of generation of hydrogen from hydrocarbons (Chiarello et al., 2010). Recently, some other applications in the pharmaceutical sector have been tested. For example, Pt nanotubes are being tested to catalyze the glucose oxidation reaction and finally to detect the glucose content of diabetic patients (Wang et al., 2008b).

Pd-NPs have been used as catalyzers in different reactions, including the hydrogenation of olefins in supercritical CO_2 media (water-in-CO_2 microemulsion) (Ohde et al., 2002), as well as in the conversion of the group nitro (NO_2) to amine (NH_2). This last reaction may have important commercial implications. In the same sense nanofiber-supported Pt- and Pd-NPs were used to catalyze a chemoselective hydrogenation of nitroarenes (Boymans et al., 2013); the reaction proceeds under mild conditions in high turnover numbers [up to 2400 mol (substrate)/mol (metal)], but contrarily to a conventional reaction, lower temperatures were used (50°C instead of 170°C).

1.4.5 SiO_2 nanoparticles

In order to develop new nanodevices and novel smart materials, nanotechnology has been focused in creating new structures that can provide new properties (Baumberg, 2006). The construction of new colloidal dyads and triads, strings, and other architectures have been performed in liquid–liquid interfaces (Böker et al., 2007). Si-NPs form part of this diverse groups of materials that have undergone these manipulations; most of the applications are in the electronic field (Sanchez et al., 2011), due to their optical and electrical properties.

Electric power industry has used these properties from the nanosilica materials (Cao et al., 2004) by including nanocomposites in insulators or semiconductors into the sector. Some of the applications include energy conversion, power delivery, energy storage, and power consumption (Oliver and Stone, 1995; Mayoux, 2000; Kaufhold et al., 2002). The nanodielectrics can enhance the reliability of current systems and can improve the efficiency, which is an advantage for renewable energy resources (Dang et al., 2013). The silica and alumina NMs have shown a great performance as insulators (Huang et al., 2011).

Si-NPs are also present in silica polymer composites used as fillers, where they have improved the homogeneity in the mixing and final product (Rong et al., 2006). Surface modification of silica by using silane-coupling agents (Jesionowski and Krysztafkiewicz, 2001) has allowed the grafting of conjugation of nanostructured silica with polymers or proteins for diverse applications, that is, biocatalysts and bioseparations (Yang et al., 2004). Thus they have been used in the immobilization of proteins, enzymes, and other bioactive agents in analytical biochemistry, medicine, and biotechnology (Haukanes and Kvam, 1993; Kiselev et al., 1999).

1.4.6 Fe_2O_3 and Fe_3O_4

Fe_2O_3-NPs have been included in the construction, environment, sensing, catalysis, water splitting, lubricant energy storage, and data storage sectors (Biswas and Wu, 2005). The use of Fe_2O_3-NPs in concrete formulations and its effect during the mixing has showed improvements on the strength of the material (Nazari et al., 2010).

Photocatalysis and sensing activities are areas where Fe_2O_3-NPs have been tested and evaluated. Maji et al. (2012) synthesized photocatalytically active mesoporous α-Fe_2O_3-NPs, and these were tested in the decomposition of aqueous Rose Bengal dye, which was completely destroyed by light irradiation. Other potential applications include gas sensing (Wang et al., 2008b), lithium-ion battery production (Chen et al., 2005), catalysis, and pollutant removal (Li et al., 2006b).

1.4.7 Al_2O_3 and cerium oxide nanomaterials

Nano-Al oxides containing Li are used in the energy storage sector. Using an initio molecular dynamic simulations, Jung and Han (2013) reported that

$Li_{3.4}Al_2O_3$ is a more favorable compound of lithiated alumina, these results may have implications on the fabrication of high-performance electrodes.

Chemical mechanical planarization (CMP), catalysis, and UV-coating and painting are among the most relevant application of nano-CeO_x (Dejhosseini et al., 2013; Sababi et al., 2014; Sampurno et al., 2009). CMP is the process that can remove topography from silicon oxide, metal, and polysilicon surfaces (Krishnan et al., 2009), it means the smoothing and planning surfaces with the combination of chemical and mechanical forces. The application is most importantly focused on microelectronics materials (Feng et al., 2006) to enhance intrinsic properties such as conductivity (Chris and Sandhu, 1993), electron transfer (Steigerwald et al., 2008), and polishing of structures (McClain and DeSimone, 2003).

1.5 Perspectives: case study on nonlinear plasmonics

Photonics is defined as the technology in charge of generating, using, and detecting photons. The subwavelength control of light through nanophotonic structures offers novel opportunities on photonics (Koenderink et al., 2015), plasmonics (Shi et al., 2013; Castro-Beltrán et al., 2017), nonlinear plasmonics (Kauranen and Zayats, 2012), and biosensing applications (Hunt and Armani, 2010). Nanophotonics refers to the study and manipulation of light at the nanoscale by novel 2D and 3D nanoarchitectures, which improves conventional functionality with respect to standard photonic platforms mainly because of their dimensions and inherent multifunctionality (Koenderink et al., 2015). Important nanophotonic functionalities can be achieved by considering hybrid plasmonic–photonic platforms, and metal–dielectric interfaces are good candidates for this (Castro-Beltrán et al., 2017; Hunt and Armani, 2010; Kauranen and Zayats, 2012; Zhu et al., 2017). When an electromagnetic wave interacts with the metal–dielectric interface, hybrid waves of photons and collective oscillations of conduction electrons gives rise to a resonant field confinement which are normally referred to as localized surface plasmons (LSPs) and surface plasmon polaritons (SPPs). LSPs and SPPs are respectively referred for metal NPs and metal–dielectric interfaces. Optical phenomena that are based on intensity-dependent effects find this field a unique opportunity to increase their effective optical responses (Kippenberg et al., 2011; Shen et al., 2018). The resonant confinement can be engineered by controlling both the size and the shape of the metallic surface on which the electromagnetic wave arrives. This extraordinary property has allowed many potential applications in photonics and biochemical sensing. In addition, manipulation of ordered plasmonic nanostructures has a key role on high-fidelity color rendering and spectral filtering applications (Gu et al., 2015; Kristensen et al., 2016).

Plasmon–polaritons interactions constitute another significant branch of opportunity for nonlinear plasmonics (Kauranen and Zayats, 2012). The confinement of light at subwavelength scales has shown that the diversity of nonlinear optical (NLO) phenomena can be easily achieved and tuned at these nanoscales. This growing area represents novel opportunities for hybrid photonic design platforms because photon–plasmons interactions, submicrometer integrated photonic structures, and metallic NMs provide more versatile devices for dynamically controlling the NLO effects: second- and third-order nonlinearities. In terms of NLO effects the phenomenon of frequency generation, which can be governed by second and third-order nonlinearities, has an important role in telecommunication and spectroscopy applications (Kippenberg et al., 2011). A photonic material with high NLO susceptibility [$\chi^{(2)}$ and $\chi^{(3)}$] and good optomechanical properties is a good candidate for these applications. In samples with dimensions larger than the excitation wavelength the nonlinear frequency generation process requires a phase matching condition in all its sources, which restricts the free access to frequency conversion applications (Castro-Beltrán et al., 2017). In addition, second-order effects require a material symmetry (Singh et al., 2013). At the nanoscale, these restrictions diminish and plasmonic structures are strongly positioned including the emerging field of nonlinear plasmonic.

It is known that localized resonant effects (achieved by resonant plasmon excitations) significantly enhance the optical field in the vicinity of the localized plasmons (Olesiak-Banska et al., 2012); however, dielectric–metallic structures can further support hybrid resonant modes that can be stronger and more concentrated than LSPs and therefore present stronger field enhancement (Castro-Beltrán et al., 2017; Yang et al., 2011).

In the fields of hybrid photonic–plasmonic platforms, integrated optics (Armani et al., 2003), and optical micrometer, devices coated with metallic NPs boost their inherent optical responses. Submicrometer (sub-μ) optical devices have proven to be useful for on-chip integration in a large number of applications ranging from environmental monitoring (Hunt and Armani, 2010) to microlasers (Shi et al., 2013). In sub-μ and on-chip devices, a versatile photonic structure represents better application opportunities, and microcavities are extraordinary photonic structures that can diversify their capabilities (Armani et al., 2003). One of the main concerns on these resonant platforms is the roughness after device fabrication (Armani et al., 2003; Castro-Beltran et al., 2014). There are several protocols to fabricate microresonators, and these are mainly referred to on-chip devices or fiber optics. In all cases, after fabrication, cavities must present smoothness in the sidewalls or in the periphery where light is supported and continually interacts. Quality factor (Q) is one of the main optical properties of cavities. If the roughness is high, the quality-factor collapses affecting potential applications. Microresonators with moderate and high Q represent good photonic platforms to achieve dielectric–metallic interactions. Here, the main purpose is to potentiate the optical properties of cavities through metallic NP interactions decorating the resonator surface. A prime example of these interactions for nonlinear plasmonics is presented in the design and applications of novel microlasers (μ-lasers) devices (Shi et al., 2013; Castro-Beltrán et al., 2017). The new μ-lasers platforms take advantage of the shared photonic and plasmonic properties, supported by microcavities and Au-NPs to potentiate the physical interaction between the two research areas: nanotechnology and integrated optics. The novel performance of these hybrid devices has impacted on the novel integrated laser development through two different phenomena: two-photon upconversion and four-wave mixing (FWM) based on Kerr effects. In the two-photon upconversion effect the hybrid microcavity is optically pumped in the longitudinal plasmonic band of the Au-NPs decorating the cavity surface to achieve visible (580 nm) optical lasing with threshold below the 50 μW, Fig. 1.5.

On the other hand, frequency combs (based on third-order FWM process) (Castro-Beltrán et al., 2017) display unique applications for spectroscopy and telecommunications from infrared spectroscopy, quantum information to optical telecommunication (Kippenberg et al., 2011). When the excitation wavelength is far from the main plasmonic band of the particles, the hybrid cavity can support a hybrid mode between the dielectric–metallic interface, creating a stronger localized mode than the conventional whispering gallery mode, Fig. 1.6. This hybrid mode has the potential to excite inner materials with stronger NLO coefficients, which represents novel opportunities for nonlinear plasmonic applications.

The use of metallic NPs expands the performance of the μ-lasers. Future work should include studies in fundamental and applied science in the fields of NP geometries, surface decoration of the cavities to boost the hybrid performance of the full photonic microplatform, and controlling the hybrid mode generation.

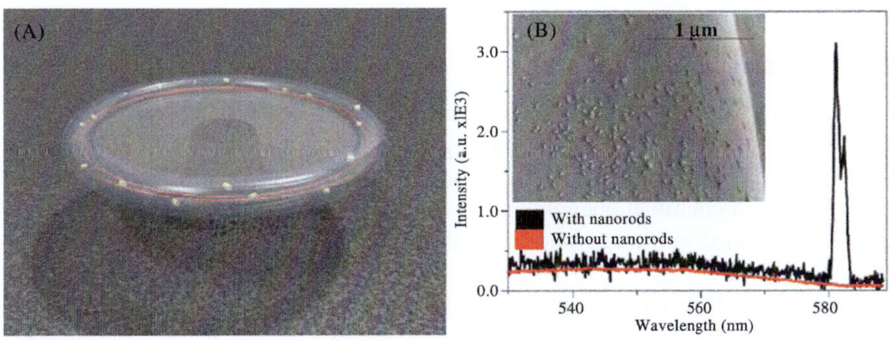

FIGURE 1.5

(A) Artistic render of a toroidal optical cavity coated with gold nanorods. (B) Main optical lasing results (~580 nm) obtained with the hybrid platforms. The scanning electron microscope image of the cavity coated with gold nanoparticles is presented as an inset figure.

Reprinted with permission from Shi, C., Soltani, S., Armani, A.M., 2013. Gold nanorod plasmonic upconversion microlaser. Nano Lett. 13 (12), 5827–5831. Copyright 2013, ACS.

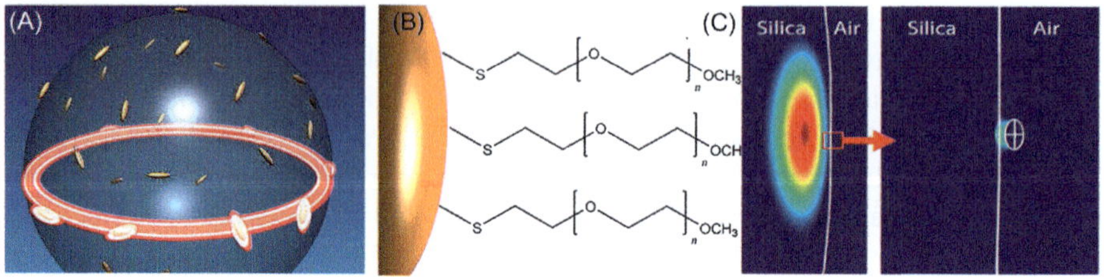

FIGURE 1.6

(A) Whispering gallery mode resonator coated with polyethylene glycol-functionalized gold nanoparticles. (B) Artistic representation of functionalized gold nanorod. (C) Finite element modeling where both whispering galley mode and the photonic–plasmonic mode hybridization are created.

Adapted with permission from Castro-Beltrán, R., Diep, V.M., Soltani, S., Gungor, E., Armani, A.M., 2017. Plasmonically enhanced Kerr frequency combs. ACS Photonics 4 (11), 2828–2834, Copyright 2017, ACS.

Highlights

- Nanotechnology has considerably improved the life of humankind.
- ENMs can be used in almost every area of human activity.
- Metallic and carbon-based ENMs are produced by several companies around the world.
- Improving ENMs synthesis is essential.

Key Points

- Nanotechnology is the manipulation of matter at nanoscale.
- Synthesis of ENMs is achieved starting with atoms or molecules (bottom-up) or with bulk materials (top-down).
- Physical and chemical phenomena are key points in ENMs synthesis.
- ENMs are used in the laboratory for research, analytical chemistry, and biomedicine.
- In industry, electronics, cosmetics, and the energy sector are among the many areas where ENMs are being applied.
- Photonics is one area where ENMs have shown great potential.

Acknowledgment

The authors acknowledge Universidad de Guanajuato.

References

Ahamed, M., AlSalhi, M.S., Siddiqui, M.K.J., 2010. Silver nanoparticle applications and human health. Clin. Chim. Acta 411 (23–24), 1841–1848.

Anker, J.N., Hall, W.P., Lyandres, O., Shah, N.C., Zhao, J., Van Duyne, R.P., 2010. Biosensing with plasmonic nanosensors. Nanosci. Technol. 308–319.

Armani, D.K., Kippenberg, T.J., Spillane, S.M., Vahala, K.J., 2003. Ultra-high-Q toroid microcavity on a chip. Nature 421, 925.

Bashyam, R., Zelenay, P., 2006. A class of non-precious metal composite catalysts for fuel cells. Nature 443 (7107), 63.

Baumberg, J.J., 2006. Breaking the mould: casting on the nanometre scale. Nat. Mater. 5 (1), 2.

Bednarikova, Z., Huy, P.D., Mocanu, M.M., Fedunova, D., Li, M.S., Gazova, Z., 2016. Fullerenol $C_{60}(OH)_{16}$ prevents amyloid fibrillization of $A\beta_{40}$-in vitro and in silico approach. Phys. Chem. Chem. Phys. 18 (28), 18855–18867.

Bell, A.T., Gates, B.C., Ray, D., Thompson, M.R., 2008. Basic Research Needs: Catalysis for Energy (No. PNNL-17214). Pacific Northwest National Laboratory (PNNL), Richland, WA.

Beltrame, R.E., Bos, M.A., 2008. U.S. Patent Application No. 10/583,615.

Bhattacharya, K., Mukherjee, S.P., Gallud, A., Burkert, S.C., Bistarelli, S., Bellucci, S., et al., 2016. Biological interactions of carbon-based nanomaterials: from coronation to degradation. Nanomedicine 12 (2), 333–351.

Biswas, P., Wu, C.Y., 2005. Nanoparticles and the environment. J Air Waste Manag. Assoc. 55 (6), 708–746.

Blaser, S.A., Scheringer, M., MacLeod, M., Hungerbühler, K., 2008. Estimation of cumulative aquatic exposure and risk due to silver: contribution of nano-functionalized plastics and textiles. Sci. Total Environ. 390 (2–3), 396–409.

Böker, A., He, J., Emrick, T., Russell, T.P., 2007. Self-assembly of nanoparticles at interfaces. Soft Matter 3 (10), 1231–1248.

Bosi, S., Da Ros, T., Spalluto, G., Balzarini, J., Prato, M., 2003. Synthesis and anti-HIV properties of new water-soluble bis-functionalized [60]fullerene derivatives. Bioorg. Med. Chem. Lett. 13 (24), 4437–4440.

Boymans, E., Boland, S., Witte, P.T., Müller, C., Vogt, D., 2013. Chemoselective hydrogenation of functionalized nitroarenes using supported Mo promoted Pt nanoparticles. ChemCatChem 5 (2), 431–434.

Buzea, C., Pacheco, I.I., Robbie, K., 2007. Nanomaterials and nanoparticles: sources and toxicity. Biointerphases 2 (4), MR17–MR71.

Cao, Y., Irwin, P.C., Younsi, K., 2004. The future of nanodielectrics in the electrical power industry. IEEE Trans. Dielectr. Electr. Insul. 11 (5), 797–807.

Castro-Beltran, R., Huby, N., Loas, G., Lhermite, H., Pluchon, D., Bêche, B., 2014. Improvement of efficient coupling and optical resonances by using taper-waveguides coupled to cascade of UV210 polymer micro-resonators. J. Micromech. Microeng. 24 (12), 125006.

Castro-Beltrán, R., Diep, V.M., Soltani, S., Gungor, E., Armani, A.M., 2017. Plasmonically enhanced Kerr frequency combs. ACS Photonics 4 (11), 2828–2834.

Chen, J., Xu, L., Li, W., Gou, X., 2005. α-Fe_2O_3 nanotubes in gas sensor and lithium-ion battery applications. Adv. Mater. 17 (5), 582–586.

Chen, S.F., Li, J.P., Qian, K., Xu, W.P., Lu, Y., Huang, W.X., et al., 2010b. Large scale photochemical synthesis of M@TiO_2 nanocomposites (M = Ag, Pd, Au, Pt) and their optical properties, CO oxidation performance, and antibacterial effect. Nano Res. 3 (4), 244–255.

Chiarello, G.L., Aguirre, M.H., Selli, E., 2010. Hydrogen production by photocatalytic steam reforming of methanol on noble metal-modified TiO_2. J. Catal. 273 (2), 182–190.

Chou, C.W., Hsu, S.H., Wang, P.H., 2008. Biostability and biocompatibility of poly(ether) urethane containing gold or silver nanoparticles in a porcine model. J. Biomed. Mater. Res. A 84 (3), 785–794.

Chris, C.Y., Sandhu, G.S., 1993. U.S. Patent No. 5,240,552. U.S. Patent and Trademark Office, Washington, DC.

Coker, R.J., Hunter, B.M., Rudge, J.W., Liverani, M., Hanvoravongchai, P., 2011. Emerging infectious diseases in southeast Asia: regional challenges to control. Lancet 377 (9765), 599–609.

Dang, Z.M., Yuan, J.K., Yao, S.H., Liao, R.J., 2013. Flexible nanodielectric materials with high permittivity for power energy storage. Adv. Mater. 25 (44), 6334–6365.

Darroudi, M., Zak, A.K., Muhamad, M.R., Huang, N.M., Hakimi, M., 2012. Green synthesis of colloidal silver nanoparticles by sonochemical method. Mater. Lett. 66 (1), 117–120.

De, M., Ghosh, P.S., Rotello, V.M., 2008. Applications of nanoparticles in biology. Adv. Mater. 20 (22), 4225–4241.

Dejhosseini, M., Aida, T., Watanabe, M., Takami, S., Hojo, D., Aoki, N., et al., 2013. Catalytic cracking reaction of heavy oil in the presence of cerium oxide nanoparticles in supercritical water. Energy Fuels 27 (8), 4624–4631.

del Burgo, L.S., Hernandez, R.M., Orive, G., Pedraz, J.L., 2014. Nanotherapeutic approaches for brain cancer management. Nanomedicine 10 (5), 905–919.

El-Batal, A.I., Mosalam, F.M., Ghorab, M.M., Hanora, A., Elbarbary, A.M., 2018. Antimicrobial, antioxidant and anticancer activities of zinc nanoparticles prepared by natural polysaccharides and gamma radiation. Int. J. Biol. Macromol. 107, 2298–2311.

Environmental Protection Agency, 2012. Nanotechnology White Paper, Science Policy Council. p. 13. Available from: <https://www.epa.gov/sites/production/files/2015-01/documents/nanotechnology_whitepaper.pdf>.

Espitia, P.J.P., Soares, N.D.F.F., dos Reis Coimbra, J.S., de Andrade, N.J., Cruz, R.S., Medeiros, E.A.A., 2012. Zinc oxide nanoparticles: synthesis, antimicrobial activity and food packaging applications. Food Bioprocess Technol. 5 (5), 1447–1464.

Feng, L., Zhang, Z., Mai, Z., Ma, Y., Liu, B., Jiang, L., et al., 2004. A super-hydrophobic and super-oleophilic coating mesh film for the separation of oil and water. Angew. Chem. Int. Ed. 43 (15), 2012–2014.

Feng, X., Sayle, D.C., Wang, Z.L., Paras, M.S., Santora, B., Sutorik, A.C., et al., 2006. Converting ceria polyhedral nanoparticles into single-crystal nanospheres. Science 312 (5779), 1504–1508.

Feynman, R., 1960. There is plenty of room at the bottom. Caltech Eng. Sci. 23 (5), 22–36.

Franci, G., Falanga, A., Galdiero, S., Palomba, L., Rai, M., Morelli, G., et al., 2015. Silver nanoparticles as potential antibacterial agents. Molecules 20 (5), 8856–8874.

Friedman, S.H., Decamp, D.L., Sijbesma, R.P., Srdanov, G., Wudl, F., Kenyon, G.L., 1993. Inhibition of the HIV-1 protease by fullerene derivatives—model-building studies and experimental-verification. J. Am. Chem. Soc. 115 (15), 6506–6509.

Gatoo, M.A., Naseem, S., Arfat, M.Y., Dar, A.M., Qasim, K., Zubair, S., 2014. Physicochemical properties of nanomaterials: implication in associated toxic manifestations. Biomed. Res. Int. 2014, 498420.

Gigault, J., Budzinski, H., 2016. Selection of an appropriate aqueous nano-fullerene (nC60) preparation protocol for studying its environmental fate and behavior. TrAC, Trends Anal. Chem. 80, 1–11.

Gojova, A., Guo, B., Kota, R.S., Rutledge, J.C., Kennedy, I.M., Barakat, A.I., 2007. Induction of inflammation in vascular endothelial cells by metal oxide nanoparticles: effect of particle composition. Environ. Health Perspect. 115 (3), 403.

Gomes Silva, C., Juárez, R., Marino, T., Molinari, R., García, H., 2010. Influence of excitation wavelength (UV or visible light) on the photocatalytic activity of titania containing gold nanoparticles for the generation of hydrogen or oxygen from water. J. Am. Chem. Soc. 133 (3), 595–602.

Goodarzi, S., Da Ros, T., Conde, J., Sefat, F., Mozafari, M., 2017. Fullerene: biomedical engineers get to revisit an old friend. Mater. Today 20 (8), 460–480.

Grunkemeier, G.L., Jin, R., Starr, A., 2006. Prosthetic heart valves: objective performance criteria versus randomized clinical trial. Ann. Thorac. Surg. 82 (3), 776–780.

Gu, Y., Zhang, L., Yang, J.K.W., Yeo, S.P., Qiu, C.-W., 2015. Color generation via subwavelength plasmonic nanostructures. Nanoscale 7 (15), 6409–6419.

Guo, J.Z., Cui, H., Zhou, W., Wang, W., 2008. Ag nanoparticle-catalyzed chemiluminescent reaction between luminol and hydrogen peroxide. J. Photochem. Photobiol., A 193 (2–3), 89–96.

Gupta, A.K., Gupta, M., 2005. Synthesis and surface engineering of iron oxide nanoparticles for biomedical applications. Biomaterials 26 (18), 3995–4021.

Hahm, J.I., Lieber, C.M., 2004. Direct ultrasensitive electrical detection of DNA and DNA sequence variations using nanowire nanosensors. Nano Lett. 4 (1), 51–54.

Han, H.W., Liu, W., Zhang, J., Zhao, X.Z., 2005. A hybrid poly(ethylene oxide)/poly (vinylidene fluoride)/TiO_2 nanoparticle solid-state redox electrolyte for dye-sensitized nanocrystalline solar cells. Adv. Funct. Mater. 15 (12), 1940–1944.

Haukanes, B.I., Kvam, C., 1993. Application of magnetic beads in bioassays. Nat. Biotechnol. 11 (1), 60.

Hsieh, F.Y., Zhilenkov, A.V., Voronov, I.I., Khakina, E.A., Mischenko, D.V., Troshin, P.A., et al., 2017. Water-soluble fullerene derivatives as brain medicine: surface chemistry determines if they are neuroprotective and antitumor. ACS Appl. Mater. Interfaces 9 (13), 11482–11492.

Huang, X., Jiang, P., Tanaka, T., 2011. A review of dielectric polymer composites with high thermal conductivity. IEEE Electr. Insul. Mag. 27 (4).

Hunt, H.K., Armani, A.M., 2010. Label-free biological and chemical sensors. Nanoscale 2 (9), 1544–1559.

Inshakova, E., Inshakov, O., 2017. World market for nanomaterials: structure and trends, MATEC Web of Conferences, vol. 129. EDPSciences, p. 02013.

Integrated Nanoscience and Commodity Exchange (INSCX), 2012. Nanotechnology: Producer/Downstream User Guide to INSCX™ Exchange. Available from: <www.inscx.com/wp/about/>.

Invernizzi, N., 2011. Nanotechnology between the lab and the shop floor: what are the effects on labor? J. Nanopart. Res. 13, 2249–2268.

Ivask, A., Titma, T., Visnapuu, M., Vija, H., Kakinen, A., Sihtmae, M., et al., 2015. Toxicity of 11 metal oxide nanoparticles to three mammalian cell types in vitro. Curr. Top. Med. Chem. 15 (18), 1914–1929.

Jesionowski, T., Krysztafkiewicz, A., 2001. Influence of silane coupling agents on surface properties of precipitated silicas. Appl. Surf. Sci. 172 (1–2), 18–32.
Ji, H., Sun, H., Qu, X., 2016. Antibacterial applications of graphene-based nanomaterials: recent achievements and challenges. Adv. Drug Deliv. Rev. 105 (Pt B), 176–189.
Jun, Y.W., Huh, Y.M., Choi, J.S., Lee, J.H., Song, H.T., Kim, S., et al., 2005. Nanoscale size effect of magnetic nanocrystals and their utilization for cancer diagnosis via magnetic resonance imaging. J. Am. Chem. Soc. 127 (16), 5732–5733.
Jung, S.C., Han, Y.K., 2013. How do Li atoms pass through the Al_2O_3 coating layer during lithiation in Li-ion batteries? J. Phys. Chem. Lett. 4 (16), 2681–2685.
Kachynski, A.V., Kuzmin, A.N., Nyk, M., Roy, I., Prasad, P.N., 2008. Zinc oxide nanocrystals for nonresonant nonlinear optical microscopy in biology and medicine. J. Phys. Chem. C 112 (29), 10721–10724.
Kapilashrami, M., Zhang, Y., Liu, Y.S., Hagfeldt, A., Guo, J., 2014. Probing the optical property and electronic structure of TiO_2 nanomaterials for renewable energy applications. Chem. Rev. 114 (19), 9662–9707.
Karimi, Z., Karimi, L., Shokrollahi, H., 2013. Nano-magnetic particles used in biomedicine: core and coating materials. Mater. Sci. Eng., C 33 (5), 2465–2475.
Kaufhold, M., Schafer, K., Bauer, K., Bethge, A., Risse, J., 2002. Interface phenomena in stator winding insulation-challenges in design, diagnosis, and service experience. IEEE Electr. Insul. Mag. 18 (2), 27–36.
Kauranen, M., Zayats, A.V., 2012. Nonlinear plasmonics. Nat. Photonics 6, 737.
Kausar, A., 2017. Advances in polymer/fullerene nanocomposite: a review on essential features and applications. Polym. Plast. Technol. Eng. 56 (6), 594–605.
Kim, K.J., Sung, W.S., Suh, B.K., Moon, S.K., Choi, J.S., Kim, J.G., et al., 2009. Antifungal activity and mode of action of silver nano-particles on *Candida albicans*. Biometals 22 (2), 235–242.
Kippenberg, T.J., Holzwarth, R., Diddams, S.A., 2011. Microresonator-based optical frequency combs. Science 332 (6029), 555–559.
Kiselev, M.V., Gladilin, A.K., Melik-Nubarov, N.S., Sveshnikov, P.G., Miethe, P., Levashov, A.V., 1999. Determination of cyclosporin A in 20% ethanol by a magnetic beads-based immunofluorescence assay. Anal. Biochem. 269 (2), 393–398.
Koenderink, A.F., Alù, A., Polman, A., 2015. Nanophotonics: shrinking light-based technology. Science 348 (6234), 516–521.
Krishnan, M., Nalaskowski, J.W., Cook, L.M., 2009. Chemical mechanical planarization: slurry chemistry, materials, and mechanisms. Chem. Rev. 110 (1), 178–204.
Kristensen, A., Yang, J.K.W., Bozhevolnyi, S.I., Link, S., Nordlander, P., Halas, N.J., et al., 2016. Plasmonic colour generation. Nat. Rev. Mater. 2, 16088.
Kroto, H.W., Heath, J.R., Obrien, S.C., Curl, R.F., Smalley, R.E., 1985. C-60 Buckminsterfullerene. Nature 318 (6042), 162–163.
Kulesza, P.J., Pieta, I.S., Rutkowska, I.A., Wadas, A., Marks, D., Klak, K., et al., 2013. Electrocatalytic oxidation of small organic molecules in acid medium: enhancement of activity of noble metal nanoparticles and their alloys by supporting or modifying them with metal oxides. Electrochim. Acta 110, 474–483.
Kumar, A., Vemula, P.K., Ajayan, P.M., John, G., 2008. Silver-nanoparticle-embedded antimicrobial paints based on vegetable oil. Nat. Mater. 7 (3), 236.
Kumar, S., Ahlawat, W., Bhanjana, G., Heydarifard, S., Nazhad, M.M., Dilbaghi, N., 2014. Nanotechnology-based water treatment strategies. J. Nanosci. Nanotechnol. 14 (2), 1838–1858.

Lara, H.H., Ayala-Nuñez, N.V., Ixtepan-Turrent, L., Rodriguez-Padilla, C., 2010. Mode of antiviral action of silver nanoparticles against HIV-1. J. Nanobiotechnol. 8 (1), 1.

Laurent, S., Dutz, S., Hafeli, U.O., Mahmoudi, M., 2011. Magnetic fluid hyperthermia: focus on superparamagnetic iron oxide nanoparticles. Adv. Colloid Interface Sci. 166 (1–2), 8–23.

Lebedeva, M.A., Chamberlain, T.W., Khlobystov, A.N., 2015. Harnessing the synergistic and complementary properties of fullerene and transition-metal compounds for nanomaterial applications. Chem. Rev. 115 (20), 11301–11351.

Li, L., Fan, M., Brown, R.C., Van Leeuwen, J., Wang, J., Wang, W., et al., 2006. Synthesis, properties, and environmental applications of nanoscale iron-based materials: a review. Crit. Rev. Environ. Sci. Technol. 36 (5), 405–431.

Li, L., Yan, J., Wang, T., Zhao, Z.J., Zhang, J., Gong, J., et al., 2015. Sub-10 nm rutile titanium dioxide nanoparticles for efficient visible-light-driven photocatalytic hydrogen production. Nat. Commun. 6, 5881.

Liga, M.V., Bryant, E.L., Colvin, V.L., Li, Q., 2011. Virus inactivation by silver doped titanium dioxide nanoparticles for drinking water treatment. Water Res. 45 (2), 535–544.

Lim, K.L., Kazemian, H., Yaakob, Z., Daud, W.W., 2010. Solid-state materials and methods for hydrogen storage: a critical review. Chem. Eng. Technol. 33 (2), 213–226.

Liu, P., Zhao, M., 2009. Silver nanoparticle supported on halloysite nanotubes catalyzed reduction of 4-nitrophenol (4-NP). Appl. Surf. Sci. 255 (7), 3989–3993.

Liu, Y., Zheng, Z., Zara, J.N., Hsu, C., Soofer, D.E., Lee, K.S., et al., 2012. The antimicrobial and osteoinductive properties of silver nanoparticle/poly(DL-lactic-*co*-glycolic acid)-coated stainless steel. Biomaterials 33 (34), 8745–8756.

Ma, Z.Y., Liu, H.Z., 2007. Synthesis and surface modification of magnetic particles for application in biotechnology and biomedicine. China Particuol. 5 (1–2), 1–10.

Mahlambi, M.M., Ngila, C.J., Mamba, B.B., 2015. Recent developments in environmental photocatalytic degradation of organic pollutants: the case of titanium dioxide nanoparticles—a review. J. Nanomater. 2015, 5.

Maji, S.K., Mukherjee, N., Mondal, A., Adhikary, B., 2012. Synthesis, characterization and photocatalytic activity of α-Fe_2O_3 nanoparticles. Polyhedron 33 (1), 145–149.

Maleki Dizaj, S., Mennati, A., Jafari, S., Khezri, K., Adibkia, K., 2015. Antimicrobial activity of carbon-based nanoparticles. Adv. Pharm. Bull. 5 (1), 19–23.

Manjumeena, R., Duraibabu, D., Sudha, J., Kalaichelvan, P.T., 2014. Biogenic nanosilver incorporated reverse osmosis membrane for antibacterial and antifungal activities against selected pathogenic strains: an enhanced eco-friendly water disinfection approach. J. Environ. Sci. Health. Part A 49 (10), 1125–1133.

Manno, D., Filippo, E., Di Giulio, M., Serra, A., 2008. Synthesis and characterization of starch-stabilized Ag nanostructures for sensors applications. J. Non-Cryst. Solids 354 (52–54), 5515–5520.

Martinez, Z.S., Castro, E., Seong, C.S., Cerón, M.R., Echegoyen, L., Llano, M., 2016. Fullerene derivatives strongly inhibit HIV-1 replication by affecting virus maturation without impairing protease activity. Antimicrob. Agents Chemother. 60 (10), 5731–5741.

Mayoux, C., 2000. Degradation of insulating materials under electrical stress. IEEE Trans. Dielectr. Electr. Insul. 7 (5), 590–601.

McBain, S.C., Griesenbach, U., Xenariou, S., Keramane, A., Batich, C.D., Alton, E.W., et al., 2008. Magnetic nanoparticles as gene delivery agents: enhanced transfection in the presence of oscillating magnet arrays. Nanotechnology 19 (40), 405102.

McClain, J.B., DeSimone, J.M., 2003. U.S. Patent No. 6,623,355. U.S. Patent and Trademark Office, Washington, DC.

McGovern, C., 2010. Commoditization of nanomaterials. Nanotechnol. Perceptions 6, 155–178.

Mohammed, L., Gomaa, H.G., Ragab, D., Zhu, J., 2017. Magnetic nanoparticles for environmental and biomedical applications: a review. Particuology 30, 14.

Moon, J., Park, J.A., Lee, S.J., Zyung, T., Kim, I.D., 2010. Pd-doped TiO_2 nanofiber networks for gas sensor applications. Sens. Actuators B: Chem. 149 (1), 301–305.

Mor, G.K., Varghese, O.K., Paulose, M., Grimes, C.A., 2003. A self-cleaning, room temperature titania-nanotube hydrogen gas sensor. Sens. Lett. 1 (1), 42–46.

Mostafaei, A., Nasirpouri, F., 2014. Epoxy/polyaniline–ZnO nanorods hybrid nanocomposite coatings: synthesis, characterization and corrosion protection performance of conducting paints. Prog. Org. Coat. 77 (1), 146–159.

Nash, J.F., 2006. Human safety and efficacy of ultraviolet filters and sunscreen products. Dermatol. Clin. 24 (1), 35–51.

National Nanotechnology Initiative, 2015. Quantifying exposure to engineered nanomaterials (QEEN) from manufactured products, addressing environmental, health, and safety implications. In: Workshop Proceedings. Arlington, VA. Available from: <www.nano.gov/node/1584>.

Nayak, R., Padhye, R., 2014. Antimicrobial finishes for textiles. Functional Finishes for Textiles: Improving Comfort, Performance and Protection. Elsevier, Amsterdam, p. 361.

Nazari, A., Riahi, S., Riahi, S., Shamekhi, S.F., Khademno, A., 2010. The effects of incorporation Fe_2O_3 nanoparticles on tensile and flexural strength of concrete. J. Am. Sci. 6 (4), 90–93.

Nie, L., Gao, L., Feng, P., Zhang, J., Fu, X., Liu, Y., et al., 2006. Three-dimensional functionalized tetrapod-like ZnO nanostructures for plasmid DNA delivery. Small 2 (5), 621–625.

Nimse, S.B., Pal, D., 2015. Free radicals, natural antioxidants, and their reaction mechanisms. RSC Adv. 5 (35), 27986–28006.

Ohde, H., Wai, C.M., Kim, H., Kim, J., Ohde, M., 2002. Hydrogenation of olefins in supercritical CO_2 catalyzed by palladium nanoparticles in a water-in-CO_2 microemulsion. J. Am. Chem. Soc. 124 (17), 4540–4541.

Olesiak-Banska, J.M., Gordel, R., Kolkowski, K., Matczyszyn, Samoc, M., 2012. Third-order nonlinear optical properties of colloidal gold nanorods. J. Phys. Chem. C 116 (25), 13731–13737.

Oliver, J.A., Stone, G.C., 1995. Implications for the application of adjustable speed drive electronics to motor stator winding insulation. IEEE Electr. Insul. Mag. 11 (4), 32–36.

Olson, M.S., Gurian, P.L., 2012. Risk assessment strategies as nanomaterials transition into commercial applications. J. Nanopart. Res. 14, 786–792.

Osmond, M.J., McCall, M.J., 2010. Zinc oxide nanoparticles in modern sunscreens: an analysis of potential exposure and hazard. Nanotoxicology 4 (1), 15–41.

Panchuk, R.R., Prylutska, S.V., Chumakl, V.V., Skorokhyd, N.R., Lehka, L.V., Evstigneev, M.P., et al., 2015. Application of C60 fullerene-doxorubicin complex for tumor cell treatment in vitro and in vivo. J. Biomed. Nanotechnol. 11 (7), 1139–1152.

Parkin, I.P., Palgrave, R.G., 2005. Self-cleaning coatings. J. Mater. Chem. 15 (17), 1689–1695.

Pelaez, M., Nolan, N.T., Pillai, S.C., Seery, M.K., Falaras, P., Kontos, A.G., et al., 2012. A review on the visible light active titanium dioxide photocatalysts for environmental applications. Appl. Catal. B: Environ. 125, 331–349.

Piccinno, F., Gottschalk, F., Seeger, S., Nowack, B., 2012. Industrial production quantities and uses of ten engineered nanomaterials in Europe and the world. J. Nanopart. Res. 14 (9), 1109.

Poole, K., 2017. At the nexus of antibiotics and metals: the impact of Cu and Zn on antibiotic activity and resistance. Trends Microbiol. 25 (10), 820–832.

Prasad, T.N.V.K.V., Sudhakar, P., Sreenivasulu, Y., Latha, P., Munaswamy, V., Reddy, K. R., et al., 2012. Effect of nanoscale zinc oxide particles on the germination, growth and yield of peanut. J. Plant Nutr. 35 (6), 905–927.

Rafique, M., Sadaf, I., Rafique, M.S., Tahir, M.B., 2017. A review on green synthesis of silver nanoparticles and their applications. Artif. Cells Nanomed. Biotechnol. 45 (7), 1272–1291.

Rasmussen, J.W., Martinez, E., Louka, P., Wingett, D.G., 2010. Zinc oxide nanoparticles for selective destruction of tumor cells and potential for drug delivery applications. Expert Opin. Drug Deliv. 7 (9), 1063–1077.

Rašović, I., 2017. Water-soluble fullerenes for medical applications. Mater. Sci. Technol. 33 (7), 777–794.

Raza, K., Kumar, D., Kiran, C., Kumar, M., Guru, S.K., Kumar, P., et al., 2016. Conjugation of docetaxel with multiwalled carbon nanotubes and codelivery with piperine: implications on pharmacokinetic profile and anticancer activity. Mol. Pharm. 13 (7), 2423–2432.

Roberts, M.S., Roberts, M.J., Robertson, T.A., Sanchez, W., Thörling, C., Zou, Y., et al., 2008. In vitro and in vivo imaging of xenobiotic transport in human skin and in the rat liver. J. Biophotonics 1 (6), 478–493.

Rogers, J.V., Parkinson, C.V., Choi, Y.W., Speshock, J.L., Hussain, S.M., 2008. A preliminary assessment of silver nanoparticle inhibition of monkeypox virus plaque formation. Nanoscale Res. Lett. 3 (4), 129.

Rong, M.Z., Zhang, M.Q., Ruan, W.H., 2006. Surface modification of nanoscale fillers for improving properties of polymer nanocomposites: a review. Mater. Sci. Technol. 22 (7), 787–796.

Sababi, M., Pan, J., Augustsson, P.E., Sundell, P.E., Claesson, P.M., 2014. Influence of polyaniline and ceria nanoparticle additives on corrosion protection of a UV-cure coating on carbon steel. Corros. Sci. 84, 189–197.

Sachkova, A.S., Kovel, E.S., Churilov, G.N., Guseynov, O.A., Bondar, A.A., Dubinina, I. A., et al., 2017. On mechanism of antioxidant effect of fullerenols. Biochem. Biophys. Rep. 9, 1–8.

Salvador, P., 2001. Semiconductors' photoelectrochemistry: a kinetic and thermodynamic analysis in the light of equilibrium and nonequilibrium models. J. Phys. Chem. B 105 (26), 6128–6141.

Sampurno, Y., Sudargho, F., Zhuang, Y., Ashizawa, T., Morishima, H., Philipossian, A., 2009. Effect of cerium oxide particle sizes in oxide chemical mechanical planarization. Electrochem. Solid-State Lett. 12 (6), H191–H194.

Sanchez, C., Belleville, P., Popall, M., Nicole, L., 2011. Applications of advanced hybrid organic–inorganic nanomaterials: from laboratory to market. Chem. Soc. Rev. 40 (2), 696–753.

Schlapbach, L., Züttel, A., 2011. Hydrogen-storage materials for mobile applications. Materials for Sustainable Energy: A Collection of Peer-Reviewed Research and Review Articles from Nature Publishing Group, 265–270.

Shen, X., Castro-Beltran, R., Diep, V.M., Soltani, S., Armani, A.M., 2018. Low-threshold parametric oscillation in organically modified microcavities. Sci. Adv. 4 (1), eaao4507.

Shi, C., Soltani, S., Armani, A.M., 2013. Gold nanorod plasmonic upconversion microlaser. Nano Lett. 13 (12), 5827–5831.

Shinkai, M., Suzuki, M., Iijima, S., Kobayashi, T., 1995. Antibody-conjugated magnetoliposomes for targeting cancer cells and their application in hyperthermia. Biotechnol. Appl. Biochem. 21 (Pt 2), 125–137.

Singh, A., Lehoux, A., Remita, H., Zyss, J., Ledoux-Rak, I., 2013. Second harmonic response of gold nanorods: a strong enhancement with the aspect ratio. J. Phys. Chem. Lett. 4 (22), 3958–3961.

Singh, A., Sahoo, S.K., 2014. Magnetic nanoparticles: a novel platform for cancer theranostics. Drug Discov. Today 19 (4), 474–481.

Smijs, T.G., Pavel, S., 2011. Titanium dioxide and zinc oxide nanoparticles in sunscreens: focus on their safety and effectiveness. Nanotechnol. Sci. Appl. 4, 95.

Song, Z., Kelf, T.A., Sanchez, W.H., Roberts, M.S., Rička, J., Frenz, M., et al., 2011. Characterization of optical properties of ZnO nanoparticles for quantitative imaging of transdermal transport. Biomed. Optics Express 2 (12), 3321–3333.

Stark, W.J., Stoessel, P.R., Wohlleben, W., Hafner, A., 2015. Industrial applications of nanoparticles. Chem. Soc. Rev. 44 (16), 5793–5805.

Steigerwald, J.M., Murarka, S.P., Gutmann, R.J., 2008. Chemical Mechanical Planarization of Microelectronic Materials. John Wiley & Sons.

Sun, C., Lee, J.S., Zhang, M., 2008. Magnetic nanoparticles in MR imaging and drug delivery. Adv. Drug Deliv. Rev. 60 (11), 1252–1265.

Sun, N.F., Liu, Z.A., Huang, W.B., Tian, A.L., Hu, S.Y., 2014. The research of nanoparticles as gene vector for tumor gene therapy. Crit. Rev. Oncol. Hematol. 89 (3), 352–357.

Thakur, V.K., Ding, G., Ma, J., Lee, P.S., Lu, X., 2012. Hybrid materials and polymer electrolytes for electrochromic device applications. Adv. Mater. 24 (30), 4071–4096.

Tolaymat, T., Genaidy, A., Abdelraheem, W., Dionysiou, D., Andersen, C., 2017. The effects of metallic engineered nanoparticles upon plant systems: an analytic examination of scientific evidence. Sci. Total Environ. 579, 93–106.

Urban, B.E., Neogi, P., Senthilkumar, K., Rajpurohit, S.K., Jagadeeshwaran, P., Kim, S., et al., 2012. Bioimaging using the optimized nonlinear optical properties of ZnO nanoparticles. IEEE J. Sel. Top. Quantum Electron. 18 (4), 1451–1456.

Vance, M.E., Todd, K., Vejerano, E.P., McGinnis, S.P., Hochella, M.F., Rejeski, D., et al., 2015. Nanotechnology in the real world: redeveloping the nanomaterial consumer products inventory. Beilstein J. Nanotechnol. 6, 1769–1780.

Veiseh, O., Gunn, J.W., Zhang, M., 2010. Design and fabrication of magnetic nanoparticles for targeted drug delivery and imaging. Adv. Drug Deliv. Rev. 62 (3), 284–304.

Veiseh, O., Sun, C., Gunn, J., Kohler, N., Gabikian, P., Lee, D., et al., 2005. Optical and MRI multifunctional nanoprobe for targeting gliomas. Nano Lett. 5 (6), 1003–1008.

Wagemaker, M., van de Krol, R., Kentgens, A.P., Van Well, A.A., Mulder, F.M., 2001. Two phase morphology limits lithium diffusion in TiO_2 (anatase): a 7Li MAS NMR study. J. Am. Chem. Soc. 123 (46), 11454–11461.

Wang, H., Jusys, Z., Behm, R.J., 2004. Ethanol electrooxidation on a carbon-supported Pt catalyst: reaction kinetics and product yields. J. Phys. Chem. B 108 (50), 19413–19424.

Wang, Q., Yang, X., Wang, X., Huang, M., Hou, J., 2012. Synthesis of N-doped TiO_2 mesosponge by solvothermal transformation of anodic TiO_2 nanotubes and enhanced photoelectrochemical performance. Electrochim. Acta 62, 158–162.

Wang, Y., Wang, Y., Cao, J., Kong, F., Xia, H., Zhang, J., et al., 2008b. Low-temperature H_2S sensors based on Ag-doped α-Fe_2O_3 nanoparticles. Sens. Actuators B: Chem. 131 (1), 183–189.

Wen, J., Li, X., Liu, W., Fang, Y., Xie, J., Xu, Y., 2015. Photocatalysis fundamentals and surface modification of TiO_2 nanomaterials. Chin. J. Catal. 36 (12), 2049–2070.

Wijnhoven, S.W.P., Oomen, A.G., Sips, A.J.A.M., Bourgeois, F.C., Te Dorsthorst, G.J.P. M., Kooi, M.W., et al., 2010. Development of an inventory for consumer products containing nanomaterials. In: 070307/2010/580587/SER/D3.

Woodrow Wilson Institute, 2013. The Nanotechnology Consumer Products Inventory: Project on Emerging Nanotechnologies. Available from: <www.nanotechproject.org/cpi>.

Wu, W., Shen, J., Banerjee, P., Zhou, S., 2011. A multifuntional nanoplatform based on responsive fluorescent plasmonic ZnO-Au@PEG hybrid nanogels. Adv. Funct. Mater. 21 (15), 2830–2839.

Xie, S.B., Zhang, B.L., Wang, L., Wang, J., Li, X., Yang, G., et al., 2015. Superparamagnetic iron oxide nanoparticles coated with different polymers and their MRI contrast effects in the mouse brains. Appl. Surf. Sci. 326, 32–38.

Yang, H.H., Zhang, S.Q., Chen, X.L., Zhuang, Z.X., Xu, J.G., Wang, X.R., 2004. Magnetite containing spherical silica nanoparticles for biocatalysis and bioseparations. Anal. Chem. 76 (5), 1316–1321.

Yang, X., Liu, Y., Oulton, R.F., Yin, X., Zhang, X., 2011. Optical forces in hybrid plasmonic waveguides. Nano Lett. 11 (2), 321–328.

Yasuno, T., Ohe, T., Takahashi, K., Nakamura, S., Mashino, T., 2015. The human immunodeficiency virus-reverse transcriptase inhibition activity of novel pyridine/pyridinium-type fullerene derivatives. Bioorg. Med. Chem. Lett. 25 (16), 3226–3229.

Yoon, K.Y., Byeon, J.H., Park, C.W., Hwang, J., 2008. Antimicrobial effect of silver particles on bacterial contamination of activated carbon fibers. Environ. Sci. Technol. 42 (4), 1251–1255.

You, C., Han, C., Wang, X., Zheng, Y., Li, Q., Hu, X., et al., 2012. The progress of silver nanoparticles in the antibacterial mechanism, clinical application and cytotoxicity. Mol. Biol. Rep. 39 (9), 9193–9201.

Yu, W.W., Chang, E., Falkner, J.C., Zhang, J., Al-Somali, A.M., Sayes, C.M., et al., 2007. Forming biocompatible and nonaggregated nanocrystals in water using amphiphilic polymers. J. Am. Chem. Soc. 129 (10), 2871–2879.

Yuan, Q., Hein, S., Misra, R.D.K., 2010. New generation of chitosan-encapsulated ZnO quantum dots loaded with drug: synthesis, characterization and in vitro drug delivery response. Acta Biomater. 6 (7), 2732–2739.

Zhang, H.J., Xiong, H.M., Ren, Q.G., Xia, Y.Y., Kong, J.L., 2012. ZnO@ silica core–shell nanoparticles with remarkable luminescence and stability in cell imaging. J. Mater. Chem. 22 (26), 13159–13165.

Zhang, Y., Petibone, D., Xu, Y., Mahmood, M., Karmakar, A., Casciano, D., et al., 2014. Toxicity and efficacy of carbon nanotubes and graphene: the utility of carbon-based nanoparticles in nanomedicine. Drug Metab. Rev. 46 (2), 232–246.

Zheng, Z., Yin, W., Zara, J.N., Li, W., Kwak, J., Mamidi, R., et al., 2010. The use of BMP-2 coupled–nanosilver-PLGA composite grafts to induce bone repair in grossly infected segmental defects. Biomaterials 31 (35), 9293–9300.

Zhu, W., Xu, T., Wang, H., Zhang, C., Deotare, P.B., Agrawal, A., et al., 2017. Surface plasmon polariton laser based on a metallic trench Fabry-Perot resonator. Sci. Adv. 3 (10), e1700909.

Further reading

Benn, T.M., Westerhoff, P., 2008. Nanoparticle silver released into water from commercially available sock fabrics. Environ. Sci. Technol. 42 (11), 4133–4139.

Chen, J.S., Tan, Y.L., Li, C.M., Cheah, Y.L., Luan, D., Madhavi, S., et al., 2010a. Constructing hierarchical spheres from large ultrathin anatase TiO_2 nanosheets with nearly 100% exposed (0 0 1) facets for fast reversible lithium storage. J. Am. Chem. Soc. 132 (17), 6124–6130.

Kim, C.H., Thompson, L.T., 2005. Deactivation of Au/CeO_x water gas shift catalysts. J. Catal. 230 (1), 66–74.

Li, X.Q., Elliott, D.W., Zhang, W.X., 2006a. Zero-valent iron nanoparticles for abatement of environmental pollutants: materials and engineering aspects. Crit. Rev. Solid State Mater. Sci. 31 (4), 111–122.

Liu, J.-M., Hu, Y., Yang, Y.-K., Liu, H.L., Fang, G.-Z., Lu, X., et al., 2018. Emerging functional nanomaterials for the detection of food contaminants. Trends Food Sci. Technol. 71, 94–106.

Rodríguez, J.A., Ma, S., Liu, P., Hrbek, J., Evans, J., Perez, M., 2007. Activity of CeO_x and TiO_x nanoparticles grown on Au(1 1 1) in the water-gas shift reaction. Science 318 (5857), 1757–1760.

Wang, Y., Cao, J., Wang, S., Guo, X., Zhang, J., Xia, H., et al., 2008a. Facile synthesis of porous α-Fe_2O_3 nanorods and their application in ethanol sensors. J. Phys. Chem. C 112 (46), 17804–17808.

CHAPTER

Characterization of the physical and chemical properties of engineered nanomaterials

2

Davide Calestani
Institute of Materials for Electronics and Magnetism (IMEM)-CNR, Parma, Italy

2.1 Introduction

Characterization of engineered nanomaterials (ENMs) is fundamental to assess their effective impacts on the environment and human health, as well as their evolution in time and fate. The more complementary information we can obtain on the ENMs that are under investigation, combining the results from different techniques, the more complete can be the evaluation for a realistic forecast or risk assessment. Whenever it is possible, physical properties, chemical properties, and their related statistics have to be considered all together to build a single and consistent picture.

Particle size is the main key to define an ENM. Several organizations, such as the International Organization for Standardization, the International Union of Pure and Applied Chemistry, or the International Council for Science, agree to use the term "nanomaterials" to describe all the materials having dimensions (or at least one of their dimensions) on the nanoscale, that is, approximately in the range between 1 and 100 nm (ISO Technical Committee 229 Nanotechnologies, 2015; European Commission, 2011). Size makes ENMs different from macroscopic materials in terms of structure, properties, and interactions. Size allows ENMs to float in air and travel for thousands of kilometers transported by winds or penetrate through some kind of industrial filters, natural filters, or biological tissues.

But size is not the only important property to be characterized. Shape is another example of a remarkable property to be studied, as the sad and well-known example of asbestos-related materials can easily remind us: it is just because of their needle-like shape with nanoscale tip that they can stick and lodge in the lung tissue.

And then, material composition is the property that mainly defines its chemical interaction with the surroundings. Surface chemistry, above all, plays a fundamental role in ENMs, being the surface-to-volume ratio maximized in nanosized

Exposure to Engineered Nanomaterials in the Environment. DOI: https://doi.org/10.1016/B978-0-12-814835-8.00002-9

particles. A high surface energy can exceed more easily the activation energy barrier of a direct chemical reaction or make the nanomaterials play the role of catalyst for reactions in the surrounding environment.

Many chemical or natural substance or compound can be potentially dangerous for biological living things, mainly depending on the total amount they are exposed to. Some materials are dangerous even in very small quantities, others are barely dangerous only in ridiculously high doses. A rigorous toxicology study always requires an exact knowledge of the quantity of toxic agent that is involved in the experiment. And the same is true for ENMs (Srivastava et al., 2015; Bour et al., 2015).

It is important to stress once more that a complete ENMs characterization should cover the greatest possible number of physical and chemical properties. Not only those just mentioned above, but also many others such as their aggregation state in different environments, their possible evolution in time, their interaction with light or temperature, their magnetic or radioactive properties, and so on (ISO Technical Committee 229 Nanotechnologies, 2012; EFSA Scientific Committee, 2011; Chow et al., 2005).

Approaches and techniques to collect, separate, and determine their properties evolve continuously along with the unstoppable technological progress (Tiede et al., 2008; Hassellov et al., 2008; Mattarozzi et al., 2017; Calzolai et al., 2012; Domingos et al., 2009; Cornelis et al., 2014; Grillo et al., 2015; Laborda et al., 2016; Lopez-Serrano et al., 2014; Baalousha and Lead, 2015; Shukla and Iravani, 2018; Simeonidis et al., 2016; Sadik et al., 2014; Majestic et al., 2010; Fukuhara et al., 2008; Kowalczyk et al., 2011). In the following sections the main characterization techniques that are used today to characterize ENMs will be briefly described, trying to highlight the strengths and weaknesses of each of them, together with some useful suggestions about other important aspects such as sample collection, preparation, and quantitation.

2.2 Sample collection, preparation, separation, or fractionation

Characterization of ENMs generally starts with the collection of the nanosized particles. In most of the cases they are indeed not as easily available as, for example, at the end of their synthetic or production process. Over their lifetime, ENMs are often dispersed, mixed, incorporated, transformed by the interaction with the surrounding environment. So, if we want to study their fate, ENMs have to be collected, eventually separated from other matrices in which they have been incorporated, concentrated when a minimum amount is required by the characterization technique, or fractionated when the characterization technique is not specific.

It is also very important to verify whenever any of these operations can induce further modifications in the ENMs, because this may produce misleading investigation results.

A possible general approach to the sample preparation issue, although simplified, is outlined in Fig. 2.1. The scheme gives an indication of the steps that should be taken into consideration after sample collection and before its characterization by the proper technique. In practice some of the available preparation methods include more of these aspects at once.

As mentioned before, usually the first point to verify after sample collection is whether ENMs to be analyzed are mixed with other materials or embedded in complex matrices of different compounds. The presence of undesired materials/matrices in the collected sample, indeed, might interfere with the results of the characterization technique.

In this case, the main methods to separate, dissolve, or eliminate them are based on chemical reactions or thermal decomposition. Enzymatic reactions are sometimes used to remove organic biological matrices without affecting inorganic nanoparticles (Mattarozzi et al., 2017; Laborda et al., 2016; Beltrami et al., 2011).

Filtration and centrifugation are probably the most known low-cost techniques for the collection, preparation, and size fractionation of samples. Traditional centrifugation and filtration tools, however, generally are not suitable for nanosized materials. Commonly used filtration membranes, for example, have pores only down to 1 μm, or 200 nm in the best case, while sedimentation time can increase from seconds to hours when particle size decreases from 500 to just 50 nm. Ultrafiltration (UF) and ultracentrifugation (UC) are the technological evolution of these two low-cost tools for dealing with nanomaterials.

UC exploits very high spinning speed to obtain relative centrifugal force up to $10^6 g$ and, hence, increasing the sample collection speed. Some UC apparatus are also equipped with a real-time ultraviolet (UV) optical detection system to live monitor the sample concentration evolution. In addition, different fractions can be collected at set time intervals obtaining a partial size-based fractionation. However, for very fine particles (<10 nm), the fall is opposed by Brownian motion and counter-diffusion induced by concentration gradient. Even in UC the sedimentation rate becomes too slow for these small nanoparticles (Tsao et al., 2011; Lopez-Serrano et al., 2014).

A different approach, such as the one used in differential centrifugal sedimentation, also known as centrifugal liquid sedimentation, is exploited to separate particles down to 5 nm in size. In this technique a sample is placed at the center of a rotating disk with a fluid which contains a slight density gradient. Nanoparticles in the range of 5–1000 nm are separated during the spinning (Cascio et al., 2014).

UF, instead, exploits different and subsequent filters and membranes with pore size down to 1 nm, assisted by additional forces such as mechanical pressure, vacuum or ultrasound, especially in the last stage of the filtration process when membrane resistance increases. The main drawback of this technique is that

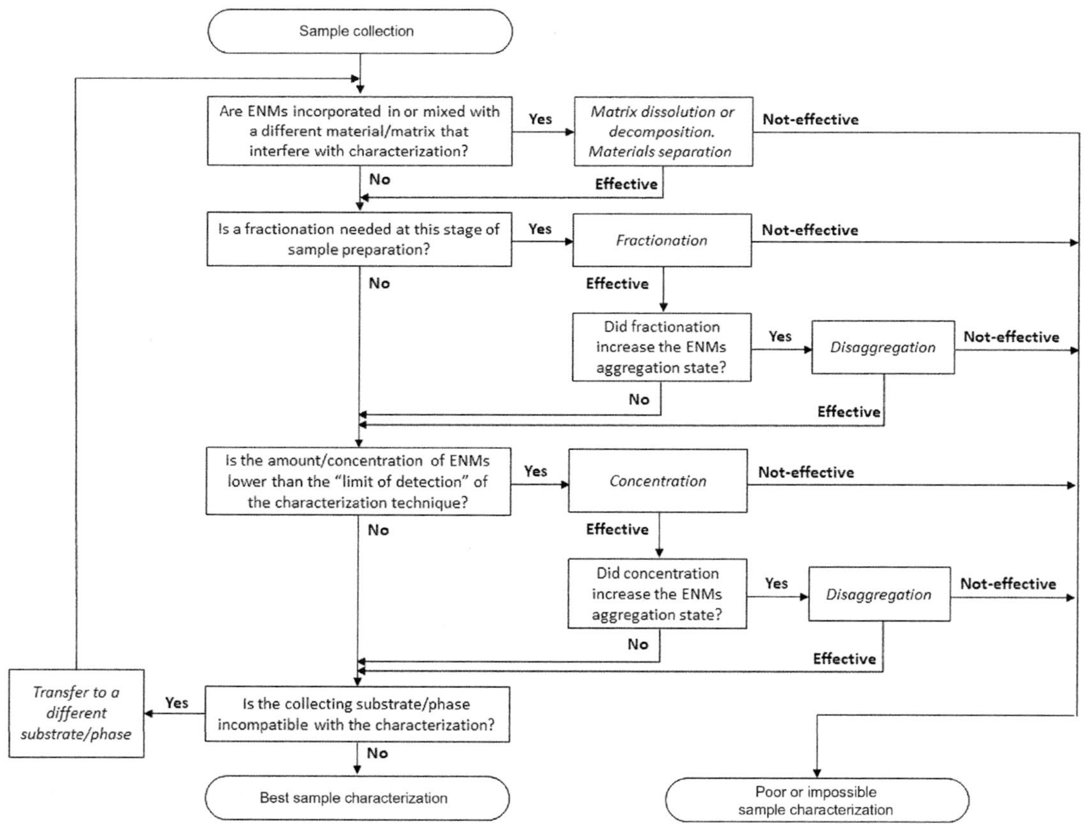

FIGURE 2.1

Outline of a possible general approach to the sample preparation issue.

progressive sedimentation can lead to membrane fouling, hence producing concentration polarization, "cake" formation or complete membrane blocking. Filtration parameters (membrane and pore size choice, flow velocity, pressure, temperature, etc.) are generally optimized in order to reduce this problem as much as possible (Hassellov et al., 2008; Tsao et al., 2011; Lopez-Serrano et al., 2014).

Cross-flow filtration reduces membrane fouling issues by recirculating or stirring the sample tangentially to a lateral filter. In each cycle only a fraction of the sample pass through the filter while the most of it is recirculated, washing away most of the sedimentation "cake." By reducing the fouling, also the formation of agglomerates, which were not present in the starting sample, is also reduced (Hassellov et al., 2008).

The most advanced family of techniques derived from these principles are those implementing field-flow fractionation (FFF). In an FFF technique, separation is produced by a field which is perpendicular to the main flow direction. This generic field, in practice, can be gravitational, centrifugal, based on thermal gradients, electric or magnetic fields, semipermeable membranes. The most known FFF technique today is asymmetric flow field fractionation (AF4), in which particles are separated on the basis of their diameter and diffusion coefficient only thanks to a semipermeable membrane at the bottom of the channel. The cross-flow is hence regulated only by fluid dynamics and gravity (and in some cases temperature), allowing a very mild and broad separation in an extremely wide range. AF4 is often combined, as primary and preliminary step, to sizing and quantification techniques (some of them will be discussed in the next sections) to obtain a mass- and diameter-based particle size distribution (Schimpf et al., 2000; Baalousha et al., 2011; Tiede et al., 2008; Calzolai et al., 2012).

Not so effective, by widely used, is also hydrodynamic chromatography (HDC). As in all chromatographic techniques, separation is obtained by the different retention times in a proper column with nonporous beads. Although HDC can separate a broad particle size range, it can be affected by unpredictable interaction with the beads (chemical interaction, aggregation, etc.) when the nature of the studied ENMs is unknown (Hassellov et al., 2008; Laborda et al., 2016; Tiede et al., 2009).

Separation, indeed, can be obtained also exploiting the chemical properties of the ENMs. For example, polar (hydrophilic) and nonpolar (hydrophobic) surface properties may be exploited to separate ENMs from other substances in a biphasic liquid. A more advanced way to exploit these properties is represented by techniques such as cloud point extraction (CPE). In CPE a neutrally charged surfactant is used in aqueous solution at concentration high enough to exceed the critical micellar concentration. This solution generally becomes turbid when an external condition (temperature, pH, pressure, etc.) is changed because it attains the so-called cloud point, leading to a separation of the original surfactant solution in a large aqueous diluted phase and a smaller one rich in surfactant micelles and the trapped analyte. CPE is often used to precipitate small nanoparticles, especially made of noble metals (such as Ag, Au, Pd, Pt) or metal oxides (such as TiO_2, ZnO, CuO, Fe, and its oxides) (Hagarova, 2017).

Electrically charged particles, instead, may be also collected by applying electrical fields, as well as magnetic particles by the use of proper magnetic fields.

Main advantages and limitation of some of the mentioned separation techniques are summarized in Table 2.1.

Table 2.1 A brief summary of main advantages and limitations of some of the mentioned separation techniques.

Technique	Acronym	Advantages	Limitations	Lower size limit (nm)
Ultracentrifugation	UC	Rather cheap apparatus; can process tens of mL per tube	Very low yield and slow process for smaller nanoparticles	10
Differential centrifugal sedimentation	DCS	High-resolution separation is achieved in a relatively short time	Residuals from the viscous media; best analytical results with integrated sensors (difficult sample collection)	5
Ultrafiltration	UF	Cheap and simple apparatus	Large sedimentation and "cake" formation issues; only wide size ranges are generally produced depending on the filter pore size	1
Cross-flow filtration	CFF	"Cake" sedimentation is prevented; can operate continuously for large amounts of feed slurry	Some fouling still occurs after long-time processing; separation with good resolution needs a large number of membranes	1–5
Asymmetric flow field fractionation	AF4	Very mild and broad separation in an extremely wide range; a single membrane is required	Diluted samples are needed; works mainly in-line with other coupled analytical techniques	3
Hydrodynamic chromatography	HDC	Good separation for a broad particle size range; can be applied in standard liquid chromatography configuration	Can be affected by unpredictable interaction with the stationary phase beads; less effective than competitive FFF techniques	20

In any case, once separated or fractionated, ENMs amount or concentration must fulfill working range requests of a specific characterization technique. If not, concentration or dilution procedures might be also necessary.

During any of these operations, large attention has to be devoted to two main aspects that should lead our choices. The first one is to avoid the formation packed agglomerates of ENMs, especially when the properties of single nanoparticles or the study of their original aggregation state are the main goal of the study. The second one is the choice of the collecting substrate or phase, whether it is a filtering membrane, a liquid or any solid support, to collect and transfer the ENMs. This substrate/phase must indeed be compatible with the following characterization technique. For example, a substrate for electron microscopy should be preferentially a good electrical conductor; liquid or wet phases often are not suitable when vacuum is required during the measurement; the substrate or precipitating agents should be soluble or degradable without leaving traces of elements/materials that could interfere with a chemical characterization. Whenever a "measure-friendly" substrate/phase is not available, a further transfer step to a better one has to be taken in consideration, paying attention to any possible further sample loss or transformation.

2.3 Size and shape definition, quantification

Sample preparation, together with fractionation/separation and the proper dilution/concentration procedures (Mattarozzi et al., 2017; Hassellov et al., 2008; Tsao et al., 2011), is usually considered a preliminary step to the following and most important characterization, which is the definition of ENMs size, quantity, and if possible shape. Some techniques can define very well all these characteristics at once, even if, in most of cases, they can process only small amounts of particles at a time. On the contrary, it is possible to characterize a large ensemble of ENMs to obtain an average size distribution by combining different techniques.

Some of the most commonly used techniques are discussed in the following subsections, and Table 2.2 reports a brief comparison of their main characteristics.

2.3.1 Electron microscopy

Microscopy is ideally the best technique to visualize ENMs and hence determine their shape and size (Goldstein et al., 2007; Fultz and Howe, 2013; Beltrami et al., 2011). Optical microscopy, however, is generally not enough to work at nanoscale.

Abbe's equation on diffraction limit, indeed, states that the maximum resolution limit for an ideal perfect microscope, because of the diffraction phenomena, is proportional to the wavelength of the used "illuminating" electromagnetic

Table 2.2 A brief comparison of characterization techniques described in this section. Since the working size range is often instrument-related, only up-to-date value are here reported. Details on limitations, pros, and cons for each of them are reported in the following subsections.

Technique	Acronym	Physical state	Size range	Shape definition
Scanning electron microscope	SEM	Dry solid	1 nm to 1 mm	Yes
Transmission electron microscope	TEM	Dry solid	0.1–100 nm	Yes
Atomic force microscopy	AFM	Dry solid	0.1 nm to 10 μm	Yes
Scanning mobility particle sizer	SMPS	Aerosol	3 nm to 1 μm	No
High-resolution time-of-flight aerosol mass spectrometer	HR-ToF-AMS	Aerosol	10/50 nm to 1 μm	No
Dynamic light scattering	DLS	Liquid suspension	1 nm to 100 μm	No
Nanoparticle tracking analysis	NTA	Liquid suspension	1 nm to 100 μm	No
Laser-induced breakdown detection	LIBD	Liquid suspension	15 nm to 1 μm	No
Single-particle inductively coupled plasma mass spectrometry	spICP-MS	Liquid suspension	10/100 nm to 1/10 μm	No
Multiangle light scattering	MALS	Liquid suspension	10 nm to 1 μm	No
X-ray diffraction (Scherrer analysis)	XRD	Dry solid	0.1–100 nm	No
Small-angle X-ray scattering	SAXS	Dry solid	1–100 nm	No

radiation. For visible light, in the 400–800 nm wavelength range this limit is about 200 nm. Electron microscopes (EM) are used instead of optical microscopes in order to reach the nanoscale. In this case an accelerated electron beam is used to "illuminate" the sample and the theoretical resolution limit of such a microscope can be as low as 10^{-3} nm, when accelerating voltage above 100 kV are used. Best available electron microscopes have today a lower practical resolution, because of the high complexity in making electromagnetic lenses to converge or diverge electron beams. This resolution is around 1 nm in a scanning electron microscope (SEM) and atomic resolution in a transmission electron microscope (TEM).

Thus if on one hand, their resolution is nominally high enough to get images of any ENM, on the other hand, several limitations have to be considered when electron microscopy is chosen. Sample preparation is a fundamental step for this kind of characterization, in order to reduce as much as possible the impact of these limitations.

First of all, EM samples must have a sufficient electrical conductivity to avoid accumulation of electrons from the beam on the analyzed surface, as they would give rise to interference with the beam itself up to a complete shielding effect. For this reason, samples that are not electrically conductive are often coated with a very thin conductive layer, generally made of gold or carbon.

Then, both SEM and TEM samples are placed in vacuum, where the microscope electron beam can be generated and driven on the sample without discharges and interferences. This means that samples with liquid inside have to be dried before the analysis. In the past the main alternative was to keep hydrated samples in a solid form by cryogenic conditions or to encapsulate them behind thin electron-transparent membranes, especially when the sample morphology or aggregation state could be strongly modified by a dehydration process. But nowadays electron microscopy evolved to provide "low vacuum" microscopes that allow the direct analysis of nonconductive and uncoated dry samples, and the so-called environmental or atmospheric SEM and TEM, through which it is possible to study humid or even liquid specimens by maintaining a saturated water pressure in a minimum volume surrounding the sample (Luo et al., 2013; Tiede et al., 2008).

Unfortunately, both SEM and TEM are almost surface-related analytical tools that work at very high magnification, which means that the volume that can be analyzed at once in electron microscopes is very small because it is limited to small areas and few micrometers or tens of nanometers in depth. Such a small volume may often have a detrimental impact on the number of ENM particles that can be visualized, making their detection more difficult. So, preconcentration or separation method should be taken into consideration whenever ENMs concentration is too low.

It is important to highlight that electron microscope can be equipped with other characterization tools that allow to study, almost contemporary, other properties. For example, energy-dispersive X-ray spectroscopy (EDS or EDX), wavelength-dispersive X-ray spectroscopy (WDS or WDXS), electron energy loss spectroscopy, or the novel combined Raman imaging + SEM are used to obtain compositional (elemental or chemical) information, electron diffraction (ED) or electron backscattered diffraction (EBSD) for crystallographic information, cathodoluminescence (CL) for optoelectronic properties, electron beam–induced current for electrical properties, SEM with polarization analysis for the study of spin and magnetic domains (Luo et al., 2013; Tiede et al., 2008; Thomas et al., 2017a).

2.3.2 Scanning probe and atomic force microscopy

Scanning probe microscopy (SPM) includes a wide family of techniques that share the characterization method, which is performed by means of a physical nanometric tip, which position is raster scanned over a limited surface area, exploiting different interaction with the sample. Some of the scanning probe microscopes can image more than one interaction simultaneously (mechanical and short-range forces, electric and magnetic properties, chemical and spectroscopic behaviors, thermal and piezoelectric response, etc.) (Meyer et al., 2000).

The most known is atomic force microscopy (AFM) that can image small objects, such as ENMs, on a solid substrate with very high resolution (Lu et al., 2015; Thomas et al., 2017b; Domingos et al., 2009). AFM images are formed by measuring the force or deflection to which the tip is subjected during the scan, as for example while it approaches and crosses a nanoparticle. AFM images can depict the contour and height (hence producing an almost 3D picture) of ENMs with extremely high precision (Fig. 2.2A and B), because its resolution is limited only by the size of the probe–sample interaction volume which can be as small as a fraction of nanometer, and without the need of vacuum. However, due to the tip shape, a few artifacts may be introduced during the sample rastering when sharp heights variations are present (Fig. 2.2C).

Unlike electron microscopy, high-resolution 3D images can be obtained with very few concerns about sample preparation, except for the almost indispensable flat substrate on which the nanomaterial should be deposited.

The flip side of this detailed images is a very slow scan rate, the small area that can be visualized at once, the impossibility to examine what is buried below the surface.

The main advantage of this technique is however represented by the easy coupling with other SPM measurements, in most of cases made available by modular system upgrades. A few examples, which represent however only a small and incomplete list of all the available techniques, are conductive AFM, magnetic force microscopy, chemical force microscopy, near-field scanning optical microscopy, nanoscale Fourier-transformed infrared (IR) (FTIR) (nano-FTIR) spectroscopy, piezoresponse force microscopy. In this way the obtained 3D image can be enriched with a wide range of complementary information (Meyer et al., 2000).

2.3.3 Particle counters and sizers for engineered nanomaterials in air

Traditional methods for counting and measuring particle size distribution in air, such as some of those used to have a simple PM10 measurement, are generally not suitable for ENMs in the nanoscale. Some of them, for example, are based on the light obscuration principle, that is, sensing the light decrease when particles exiting from a narrow nozzle pass between a light source and its detector, or on light adsorption phenomena. This cannot be applied to particles in the nanoscale, because their effect on the instrument signal is negligible. More sophisticate instruments are used to reach the size range of few nanometers.

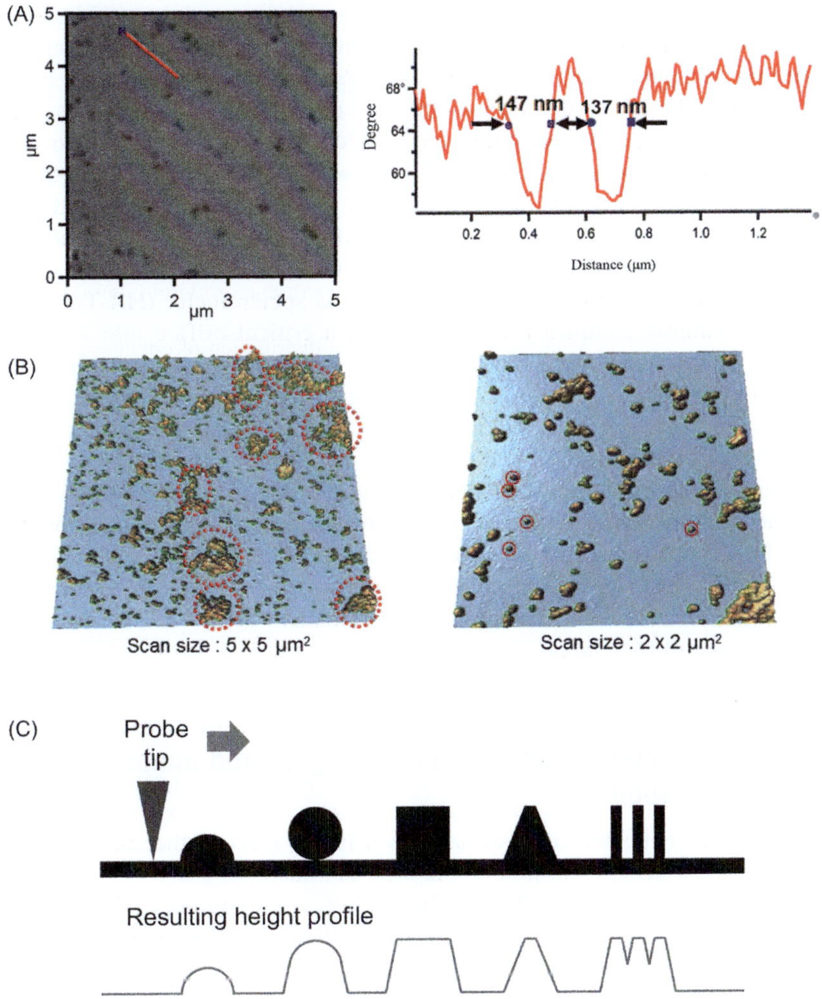

FIGURE 2.2

(A) AFM phase image; (B) three-dimensional topographical image of TiO_2 nanoparticle in commercial sunscreens (Lu et al., 2015); and (C) schematic representation of possible artifacts originating during AFM scan. *AFM*, Atomic force microscopy.

Mobility particle size spectrometers, for example, are instruments capable of determining submicrometer particle number and size exploiting the different mobilities of charged particles in an electric field. In a scanning mobility particle sizer (SMPS), nanoparticles entering the system are at first neutralized in order to obtain a Fuchs equilibrium charge distribution (Tiede et al., 2008;

Lee et al., 2013a). Then they enter a differential mobility analyzer (DMA) where the aerosol is classified according to electrical mobility, allowing only particles of a narrow range of mobility to exit through the output slit. The size of particle exiting through the slit is determined by the combination of their size, charge, central rod voltage, and flow within the DMA. This monodisperse distribution finally goes to a condensation particle counter which determines the particle concentration at that size. A full particle size distribution can be build up by exponentially scanning the voltage on the central rod. The newest instrumentation can measure nanoparticles with size down to 3 nm.

The high-resolution time-of-flight aerosol mass spectrometer (HR-ToF-AMS), instead, continuously samples particles through a critical orifice into an aerodynamic lens, which focuses and accelerates them into a narrow beam with a velocity inversely related to their vacuum aerodynamic diameter (Lee et al., 2013a,b). Particles in the beam are vaporized then ionized by electron impact. The produced ions are finally detected by a time-of-flight mass spectrometer. If the particle beam is modulated by a spinning chopper wheel, size-resolved mass spectra can be obtained because particle velocity is size-dependent. High-resolution mass spectra can be used to identify the chemical compositions related with nanoparticle of a specific size range. The lower size limit of this technique is generally a bit higher (30–50 nm) than the one of SMPS, but HR-ToF-AMS has the considerable advantage of the compositional information.

2.3.4 Particle counters and sizers for engineered nanomaterials in liquid suspension

When ENMs are collected or suspended in a liquid phase, other techniques are available to characterize them in terms of size and quantity.

In dynamic light scattering (DLS), also known as photon correlation spectroscopy or quasi-elastic light scattering, a laser is pointed at the sample with nanoparticles in liquid suspension (ASTM Subcommittee E56.02 on Physical and Chemical Characterization, 2015; Pecora, 2000; Calzolai et al., 2012). Scattered light intensity is subject to time-dependent fluctuations, generated by particles Brownian motion. Small particles move faster than big particles and hence generate faster intensity variations. An autocorrelator can measure these variations of speed and a diffusion coefficient can be calculated for the particles. Stokes–Einstein equation is then used to convert this coefficient in the corresponding hydrodynamic diameter, generally in the range from several microns to below 1 nm.

An evolution of this technique is represented by the nanoparticle tracking analysis (NTA) that, instead of collecting the integral intensity of the scattered light, live monitor the movement of the scattered light of each single particle (Fig. 2.3) (Jarzębski et al., 2017; Vasco et al., 2010). While particle concentration ranges are more limited and hence sample preparation is more complicated, better

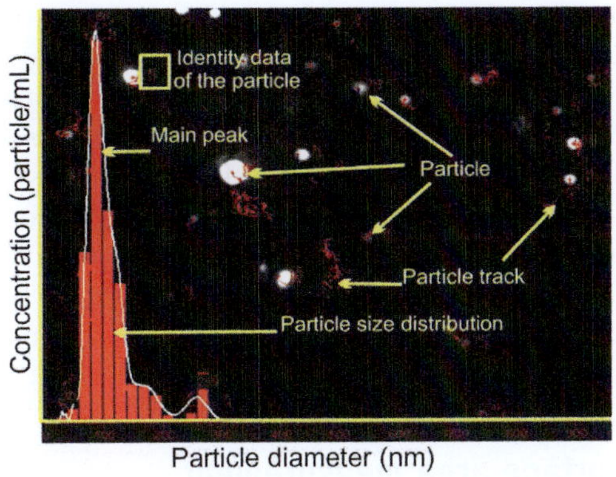

FIGURE 2.3

Example of an NTA software window during nanoparticle characterization in food (Jarzębski et al., 2017). *NTA*, Nanoparticle tracking analysis.

details can be obtained with NTA, for example, the differentiation of particles with similar size but made of different materials, a better resolution in the measurement of distribution, lower interferences from dirt or aggregates.

A different kind of interaction with a stronger laser beam can be also exploited. Laser-induced breakdown detection is based on the effect of a nanosecond pulsed laser when focused into a liquid-containing nanoparticles to be detected (Singh and Thakur, 2007; Kaegi et al., 2008). Basically, each time a particle crosses the laser beam in the focal region breakdown events can be initiated. Dielectric breakdown generates a plasma, in which emission and the succeeding pressure wave are detected with a computer-based image-detection system or a photoacoustic sensor, respectively. By means of a proper calibration with monodisperse reference spheres, the average size and concentration of particles can be derived. If acoustic detection is used, it is also possible to measure size distribution in the range 15–1000 nm.

Single-particle inductively coupled plasma mass spectrometry (spICP-MS), instead, is a recently developed ICP-MS operating mode that can provide particle counting and size distribution, together with the chemical information that is typical of this technique (Peters et al., 2015; Mattarozzi et al., 2017; Calzolai et al., 2012; Laborda et al., 2016). Although the analysis is performed on a nanoparticle aerosol, sample is generally requested in the form of a highly diluted suspension. In spICP-MS each particle that is ionized by the plasma produces proportional peaks that can be resolved in time for counting and attribute nominal size (in the hypothesis that the particle is spherical, and hence its peaks height proportional to the radius cubed). During the measurement the signal originating from the known solvent is neglected. The typical runtime is very fast, usually about 1 minute, and the lower size limit depends on many factors, among which the sensitivity of mass detector and the chemical nature (and hence the produced ions) of the particle, for example, 10–20 nm for gold or silver, 50 nm for titania, 200 nm for silica.

Less used, but often coupled with fractionation techniques such as AF4 or FFF, is multiangle light scattering (MALS) technique (Hassellov et al., 2008). MALS can be used to determine the average size of particles. Multiple detectors ensure light scattering data as a function of scattering angle, so the radius of gyration (Rg) or root mean squared radius can be calculated to give the size of molecules or particles. Typically, nanoparticles below particles 10 nm in diameter cannot be measured because their radius of gyration is so small that the angular dependence is negligible.

2.3.5 Specific surface area measurement

The measurement of the specific surface area of an ENM sample can also give a fast and easy access to the average size of the nanoparticles or of their aggregates. This evaluation has several limitations, but, in some cases, it can provide a reasonable result, especially when ENM sample is rather homogeneous.

BET method, which acronym originates from the names Brunauer, Emmett, and Teller who theorized it, is often used to calculate the specific surface area of solids (Tiede et al., 2008; Hassellov et al., 2008). The method is based on the adsorption of a specific gas, usually nitrogen, on the sample surface. The quantitation of the adsorbed gas is performed at its boiling point and, through the plot of adsorption/desorption isotherms and the knowledge of the gas molecular size, it is possible to calculate the surface area on which the gas molecule monolayer condensed.

BET method can be applied only on dry solid samples. For ENMs in liquid phase a different method has been proposed in the last years to evaluate the nanoparticles specific surface area. This method is based on nuclear magnetic resonance (NMR) and it exploits the property of liquid in close proximity to the particle surface, which has a relaxation time that is orders of magnitude shorter than the one of liquid far away from the particle surface.

When ENMs are small crystalline solids, the specific surface area might be correlated with the one that can be obtained with an X-ray powder diffraction (XRD) measurement. Beside the well-known chemical and crystallographic information, Scherrer equation can be used to calculate the average crystal size from the peak broadening at half the maximum intensity. This is generally possible only for nanocrystals smaller than 100 nm (Warren, 1990; Cao and Wang, 2011).

If Scherrer value is available and we associate it with the effective nanoparticle size, its comparison with the value calculated from specific surface area measurement can tell us if nanoparticles/nanocrystals are aggregated ($size_{[Scherrer]} < size_{[surface\text{-}area]}$) or not ($size_{[Scherrer]} \approx size_{[surface\text{-}area]}$).

Size might be also determined by means of small-angle X-ray scattering technique (Cao and Wang, 2011). This technique, although less easily accessible, can be performed in properly equipped X-ray diffractometers (or even synchrotron

X-ray facilities) and provides the tool for quantitating density differences at nanoscale, and hence also nanoparticle size distributions, generally in the range 1–100 nm.

2.4 Chemical composition and structure

On one hand, characterization of size, shape, aggregation state, and quantity of ENMs is fundamental to understand where nanoparticles can diffuse, through which physical or biological barrier can penetrate and how relevant can be their impact. On the other hand, characterization of chemical composition and structural properties is very important to understand the kind of interaction that ENMs can have with the surrounding environment and which possible modifications and fate they have undergone or we can predict. A wide range of characterization techniques are available for this kind of characterization. Some of them are standalone or integrated in other instrumentation for nanoparticle analysis, while other can be added as a final step of a characterization chain that starts with fractionation, follows with sizing and/or quantification, and ends with compositional analysis. The most common are discussed in the following subsections, and Table 2.3 reports a brief comparison of their main characteristics.

2.4.1 Composition and structure analysis in electron microscope

Since electron microscopy is often used to define at once several important properties of the analyzed ENMs, the correlated EDS or EDX and WDS or WDXS techniques should be described at first (Goldstein et al., 2007; Beltrami et al., 2011; Thomas et al., 2017a). This kind of elemental analysis is also generally referred to as X-ray microanalysis. Both techniques exploit X-rays generated when the microscope's electron beam hits the sample. Beam energy is indeed high enough to produce X-rays, energies of which are characteristic of the elements that are in the sample. WDS technique is more suitable to investigate with high-precision elements with low atomic number, such as those that are typical of organic materials, while EDS can provide a faster analysis on a wide range of elements (almost the entire periodic table). Both techniques usually cannot quantify the three lighter elements: hydrogen, helium, and lithium.

EDS and WDS can be also supported in SEM by the complementary EBSD technique, which can provide information about the crystal structure, orientation, or phase for each point of the observed sample through the resolution of the EBSD patterns that are collected on a charge-coupled device (CCD) array (Goldstein et al., 2007). In TEM, instead, transmitted ED is generally studied, because it produces diffraction patterns that carry a much higher level of information and detail. Patterns are collected in selected area ED using the simplest geometry, with a parallel beam of electrons incident on the specimen, while in

Table 2.3 A brief comparison of characterization techniques described in this section. Since the detection limits are often instrument-related, only up-to-date value are here reported. Details on limitations, pros, and cons for each of them are reported in the text.

Technique	Acronym	Physical state	Chemical information	Detection limit	Target[a]
Energy-dispersive X-ray spectroscopy on SEM/TEM	SEM/TEM-EDS	Dry solid	Quantitative elemental analysis	0.1%–1%	S
Wavelength-dispersive X-ray spectroscopy on SEM/TEM	SEM/TEM-WDS	Dry solid	Quantitative elemental analysis	0.1%–1%	S
Electron backscattered diffraction on SEM	SEM-EBSD	Dry solid	Crystallographic information	–	S
Selected area or convergent beam electron diffraction on TEM	TEM-SAED/CBED	Dry solid	Crystallographic information at atomic resolution	–	S
Cathodoluminescence on SEM/TEM	SEM/TEM-CL	Dry solid	Band structure of semiconductors and ceramics	–	S
Inductively coupled plasma mass spectrometry	ICP-MS	Liquid suspension	Quantitative compositional analysis	Below ppt	S/B
Inductively coupled plasma optical emission spectrometry	ICP-OES	Liquid suspension	Quantitative elemental analysis	Below ppb	B
Fourier-transformed infrared spectroscopy	FTIR	Dry solid	Chemical composition and bond/lattice information	5/100 μm wide area	B
Nano–Fourier-transformed infrared spectroscopy	Nano-FTIR	Dry solid	Chemical composition and bond/lattice information	10–20 nm wide area	S
Raman spectroscopy and micro-Raman spectroscopy	Raman, micro-Raman	Dry/liquid	Chemical composition and bond/lattice information	1 μm wide area	B
Tip-enhanced Raman spectroscopy	TERS	Dry solid	AFM/STM image + chemical composition and bond/lattice information	10 nm wide area	S
X-ray powder diffraction	XRD	Dry solid	Crystallographic information	1 μg and relative 1%–2% in mix	B
Nuclear magnetic resonance	NMR	Dry/liquid	Chemical composition information	1 μg	B
X-ray fluorescence	XRF-EDS/WDS	Dry/liquid	Quantitative elemental analysis	10–100 mg ppb	B
(Micro)Particle-induced X-ray emission	(Micro-)PIXE	Dry/liquid	Microscopic analysis of the distribution of trace elements	1 μm wide area	B
X-ray photoelectron spectroscopy	XPS	Dry solid	Chemical composition and chemical bond information	10 μm area 0.1%–1.0%	B
Rutherford backscattering spectrometry	RBS	Dry solid	Chemical composition and chemical bond information	100 μm wide area	B

AFM, *Atomic force microscopy;* SEM, *scanning electron microscope;* TEM, *transmission electron microscope.*
[a]*"S" = result from single or few particles; "B" = average result from bulk amount of particles.*

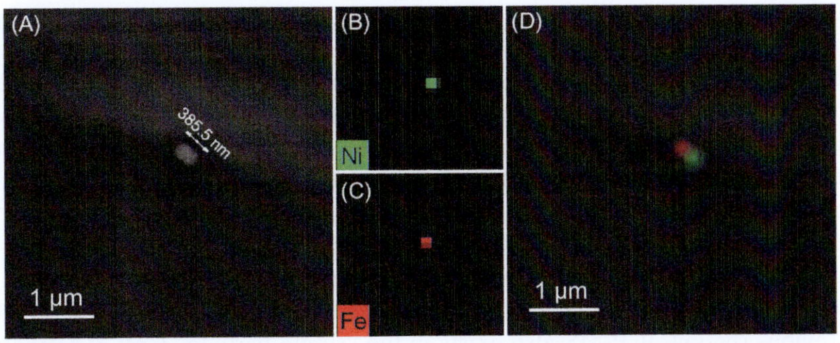

FIGURE 2.4

A sample with Fe and Ni nanoparticles within a carbonaceous matrix visualized by SEM: (A) backscattering image; (B) Ni elemental SEM-EDS map; (C) Fe elemental SEM-EDS map; and (D) combined image. *EDS*, Energy-dispersive X-ray spectroscopy; *SEM*, scanning electron microscope.

convergent beam ED they are collected by converging the electrons in a cone onto the specimen, in order to perform a diffraction experiment over several incident angles simultaneously (Fultz and Howe, 2013).

CL technique, instead, collects the light emitted by the sample while it is impinged by the electron beam (Kociak and Zagonel, 2017). Secondary electrons generated by the microscope beam can excite electron transitions from valence to conduction band of semiconductors, or even insulating materials, whenever a kinetic energy about three times higher than the band gap energy of the material is available. The wide range of accessible energies in an electron microscope allows CL to generate transitions even in oxides, ceramics, and other nonmetallic compounds. Whenever a radiative deexcitation process occurs, information about the band structure and hence about the material can be obtained.

All these analyses can be directly correlated, point-by-point, with the SEM/TEM sample image and, hence, to each visualized nanoparticle (Fig. 2.4). The only drawback is that some of these techniques may have a different lateral and in-depth resolution, so the obtained results can be uniquely attributed to single nanoparticles only if they are not too close each other on the substrate. Different useful color composition/phase-related images can be obtained by mixing the morphologic image and the chemical or structural information.

2.4.2 Inductively coupled plasma mass and emission spectrometries

As mentioned before, a chemical spectrometric method is often coupled with techniques for nanoparticles fractionation and/or sizing counting to provide additional information about quantification and, above all, about the chemical composition of the investigated ENMs (Montoro Bustos et al., 2013; Tiede et al., 2008).

ICP is often used together with MS (ICP-MS) to obtain compositional analysis on small amounts or even single nanoparticles, because this coupled technique can detect most of chemical elements (atomic mass ranging between 7 and 250, with exception of values related to Ar, used for the plasma, its related dimers/compounds and interferents from the liquid media) at extremely low concentrations (as low as one part in 10^{15}, i.e., ppq). In ICP-MS, excited ions produced by the plasma in which the sample is sprayed, are guided into a mass spectrometer, usually a quadrupole, where they are separated on the basis of their mass-to-charge ratio (Taylor, 2000; Thomas, 2013; Wagner et al., 2015). Ions reaching the detector generate a signal proportional to the concentration. As mentioned in the previous section, ICP-MS also exists in the "single-particle" (spICP-MS) operating mode (Peters et al., 2015; Mattarozzi et al., 2017; Calzolai et al., 2012; Laborda et al., 2016).

ICP is also used in combination with optical spectrometry. ICP optical emission spectrometry (ICP-OES), also known as ICP atomic emission spectrometry (ICP-AES), measures electromagnetic radiation emitted by the excited atoms and ions from the plasma (Laborda et al., 2016; Lopez-Serrano et al., 2014). Emitted radiation is separated in its components by a monochromator or a diffraction grid and then measured with a phototube or proper CCD arrays. Wavelengths, typically measured in the UV–visible range between 180 and 800 nm, are characteristic of each element. Intensity is indicative of the concentration of the element within the sample. ICP-OES/AES has good limit of detection (as low as one part in 10^{12}, i.e., ppt, for some elements) and a linear dynamic range, although spectral interferences may occur when many emission lines from different elements overlap.

Both ICP-MS and ICP-OES/AES generally require diluted analyte in liquid solution/suspension. So, they are a particularly appropriate choice when it is difficult to have a dry sample.

2.4.3 Infrared and Raman spectroscopies

IR spectroscopy is a very common tool for chemical analysis of organic compounds through the study of the vibrational modes of their molecules, but it can be also applied to many inorganic materials (Larkin, 2017; Thomas et al., 2017b). So, it can be employed in the characterization of several ENMs.

Vibrational modes are generally studied by means of two techniques: IR absorption/transmission measurement, especially in the high-resolution configuration of FTIR spectroscopy, and Raman spectroscopy. They can provide both information about the chemical composition and type of bond or crystal lattice. The first one measures the IR radiation that is adsorbed by the material when specific vibrational modes are activated. The second one measures the energy that incident photons (usually from a laser beam in the visible range) loose or gain by inelastic scattering with the sample; this energy is characteristic of specific vibrational modes of molecules or crystal lattice in the sample. So many materials generate a characteristic footprint in the absorption or Raman spectrum (Larkin, 2017; Nyquist, 2001; Thomas et al., 2017b).

Both techniques are suitable to measure large ensemble of nanoparticles, but they are also available in combination with microscopic analysis. A micro-FTIR apparatus allows samples to be observed and spectra measured from regions about 5 μm wide. However, the spatial resolution of FTIR can be further improved below this scale by integration of a scanning near-field optical microscopy platform. The so-called nano-FTIR allows to perform broadband spectroscopy with 10–20 nm spatial resolution. The resolution of a micro-Raman apparatus, instead, is mainly related to the wavelength of the incident radiation (that can range from near-UV to near-IR) and used optics. In the best case it can reach a value around 100 nm, but it is often around 1 μm. A nano-Raman measurement, or tip-enhanced Raman spectroscopy, can be instead performed in an AFM/SPM apparatus thanks to the use of proper tips and tools, providing images that both have the chemical specificity of Raman spectroscopy and a spatial resolution typically down to 10 nm (Nicklaus, 2014; Thomas et al., 2017b).

A Raman apparatus is generally more expensive than a FTIR one, but Raman measurements can be performed both on solid/dry samples or samples suspended in water, while IR absorption measurements are strongly affected when water or other solvents with intense IR adsorption are used.

2.4.4 Other characterization techniques for engineered nanomaterial ensembles

Other techniques can be used only for the chemical analysis of large amounts of nanoparticles, form hundreds of micrograms to a few grams. We can mention here the most used ones.

XRD can provide very important information about the crystalline phase of the investigated materials (Cao and Wang, 2011; Warren, 1990). XRD can be applied only to dry nanocrystalline materials. Below 100 nm, it is also possible to inversely correlate the diffraction peak width nanocrystals size. This technique usually requires a few milligrams of sample, even if the best modern diffractometers and long-time acquisitions can get information from samples weighing less than 1 μg. In most of the cases, if the spectrum refinement mathematical method converges to a satisfactory result, it is possible to distinguish different materials mixed together, although a relative 1%–2% detection limit has to be considered in this case. In order to increase the measurement resolution by orders of magnitude, it is necessary to use a different X-ray source, such as synchrotron high-intensity and highly monochromatic beam (Cao and Wang, 2011; Warren, 1990). Although these facilities are less easily accessible and experiments are more complex, synchrotron XRD can provide information even from highly diluted phases.

NMR spectroscopy is also widely employed to get chemical information about organic compounds, although several papers report also about characterization of metallic and inorganic nanoparticles (Nyquist, 2001; Thomas et al., 2017b). As for XRD, a few milligrams sample is generally preferred but with the proper equipment and long acquisition time, even few micrograms can be characterized. NMR is generally applicable both to solid dry samples and liquid samples. It is noteworthy to mention also that a diffusion-ordered spectroscopy NMR has been developed to separate the NMR signals of different species according to their diffusion coefficient and, hence, different sizes.

X-ray fluorescence (XRF) can be measure with the EDS and WDS technique also without an electron microscope by using an X-ray excitation source (Haschke, 2014; Tiede et al., 2008). The main advantage is that while electrons usually can penetrate the investigated sample only for a few micrometers, X-rays can penetrate much more in depth and produce an XRF signal also from nanoparticles buried deeper below the surface. Light elements are more difficult to be detected. Several milligrams (or even grams) are generally required, but trace elements down to ppm levels can be detected if compositionally similar, well-characterized standards are available. It is worthy of mention also the existence of particle-induced X-ray emission or proton-induced X-ray emission (PIXE) technique, in which characteristic secondary X-ray emissions are produced by an ion beam (Lozano et al, 2013; Gontier et al., 2008). Although it is not so widespread, a micro-PIXE technique is also available, making use of tightly focused beams (down to 1 μm) to get a microscopic analysis of the distribution of trace elements in a wide range of samples.

Less used in ENMs characterization, X-ray photoelectron spectroscopy (XPS) can however be exploited to get information both on chemical composition and chemical bonds (Sarma et al., 2013). It operates on areas that range from 10 μm to some millimeters wide, so it cannot be focused on single nanoparticles. An XPS spectrum is obtained by irradiating the sample with an X-ray beam and measuring the kinetic energy and number of electrons that escape from the surface (usually 0–10 nm in depth). XPS has a 0.1%–1.0% detection limit for most of elements and, working in high-vacuum conditions, can be applied only to dry samples.

Finally, Rutherford backscattering spectrometry (RBS), which is also rarely used for ENMs but can also provide chemical information (Gontier et al., 2008). The working principle is based on the analysis of backscattered ions (usually He^{2+}) after hitting the sample with a 1–3 MeV ion beam. Also, RBS is a surface measurement (not more than 10 nm) that generally operates on areas that are at least 100 μm wide (because of the minimum beam size) and high vacuum.

2.5 Surface-related properties in nanomaterials and other worth investigating properties

Atoms on a surface have dangling bonds that generate a high local energy. This is particularly effective in nanoparticles, where the surface-to-volume ratio is extremely high. This energy can be high enough to have a strong influence on some of the physical properties of the whole nanoparticle. Phase transition temperature or solubility, for example, could easily promote a change of state or higher dissolution rates at temperatures much lower than the known ones for their bulk counterpart. Similarly, nanoparticles often agglomerate to reduce this energy or easily react with the surrounding environment. Surface charge can also modify the interaction between ENMs and biological tissues/membranes.

A surface can easily charge because of protonation/deprotonation processes, because of the adsorption of ions or molecules, or because an external electrical field is applied. Surface charge is the main responsible for repulsion or attraction of nanoparticles. Moreover, when a nanoparticle is immersed in a solution containing electrolytes, it develops a net surface charge because of ionic adsorption (Verma and Stellacci, 2010; Jiang et al., 2009; Hellack et al., 2017; Cornelis et al., 2014).

In particular, aqueous solutions contain cations and anions that can interact with partial charges on the surface. This net surface charge results in a surface potential that is generally counteracted by a cloud of ions of the opposite charge and in repulsion between particles. This ion/counterion layer, also present in colloids systems, is known as the electric double layer. The so-called zeta potential is defined as the electric potential in the interfacial double layer. Zeta potential is widely used for quantification of the magnitude of the charge and can be generally calculated from electrophoretic mobility value. This can be obtained, for example, also in a DLS measurement when combined with an applied electric field (this technique is also known as electrophoretic light scattering) (Xu, 2002; Tiede et al., 2008; Cornelis et al., 2014). Zeta potential is a key value to predict the aggregation states of nanoparticles/colloids in different conditions. For particles that are small enough, a high zeta potential means that the solution or dispersion will resist aggregation. On the contrary, when the potential is small, attractive forces may exceed this repulsion and aggregation/flocculation may occur.

A solution's pH can also greatly affect surface charge because functional groups on the surface of particles often contain atoms such as oxygen or nitrogen, which can be protonated or deprotonated to become charged. So the surface charge and zeta potential of the particles often proportionally change with pH value. When the average surface charge is equal to zero for a certain pH value, the "point of zero charge" (PZC) is defined. When the pH is lower than this value, water donates more protons than hydroxide groups and so the surface is positively charged, while above PZC the surface is negatively charged.

The overall surface state can then determine the hydrophilic or hydrophobic behavior of nanoparticles. This property can be exploited to separate them from other undesired materials or to predict their possible fate in the environment.

The high surface energy of nanoparticles, moreover, can strongly promote the formation of temporary intermediate adsorbates and hence accelerate specific reactions. This means that often nanoparticles can act as catalyst for different reaction. The study and quantification of the possible catalytic effects of ENMs is very important to determine their possible effect on the surrounding environment. Photocatalysis extend even more the range of possible induced catalytic processes. All nanoparticles that can absorb incoming photons and transform them into active electron–hole pairs with sufficiently long lifetime might also produce reactive free radicals, such as hydroxyl radicals in presence of water, or reaction intermediates. Catalytic and photocatalytic activity measurements have been known for years and are often tested with simple discoloration experiments of reference dyes solution or more sophisticated experiments, for example, involving mass spectroscopy quantification, which can be applied also to gases and vapors (Astruc, 2008).

Rather rare in ENMs, but of strong impact on the surrounding environment, is also radioactivity (see, for example, Streil et al., 2007). Special purpose ENMs may contain radioactive isotopes or may become radioactive after exposure to specific conditions/experiments. Even low-activity elements, in form of nanoparticles, can potentially reach and stick for long time in specific parts of living being, where, from inside, even the weakest alpha-decay can produce obvious damages to the surrounding tissues. All radioactivity data are known in detail, while the combined effect of radioactive nanoparticle accumulation and fate in living tissues and the possible mutagen effects are a today widely studied and debated, as for example in the case of depleted uranium aerosol produced in the impact of armor-piercing rounds and special bullets.

Nanoscale may induce significant modification also in optical or magnetic properties.

Charge carriers in semiconductor nanoparticles whose size is smaller than twice the size of its exciton Bohr radius (i.e., usually few nanometers) are subject to quantum confinement. For this reason, these nanoparticles are generally referred to as "quantum dots" (Manoj et al., 2018). This phenomenon leads to light emission with wavelength proportional to the nanoparticle size. Beside the material engineering potential, the color of these nanoparticles can be also reversely used to have a quick information about their size, usually within the range 2–10 nm. A similar effect is observed in gold nanoparticles in the range 2–200 nm, although the physical reason behind this is different because it is related with surface plasmons in metallic materials. Colloidal suspension of gold nanoparticles absorbs and scatters light, resulting in colors ranging from red to blue, to black or even colorless, depending on particle size, shape, local refractive index, and aggregation state.

Other nanoparticles show size-dependent magnetic properties, such as in case of some common iron oxides (magnetite Fe_3O_4 or maghemite γ-Fe_2O_3). Nanoparticles of these oxides have superparamagnetic properties. This means that they have paramagnetic properties if they are smaller than a critical size (around 10 nm) or ferromagnetic/ferrimagnetic properties if they are larger or in closely packed aggregate form (Kowalczyk et al., 2011).

Both optical and magnetic properties can be characterized with the classical characterization techniques if a minimum amount (usually a few milligrams) is available.

Key points

- Many physical and chemical properties of engineered nanomaterials (ENMs), all together, determine the behavior they have with the surrounding environment and, hence, their fate. So, in order to make realistic forecasts or risk assessments, it is useful to know as many of these properties as possible.
- Some of these are fundamental, such as their size, chemical composition, and quantity (weight or concentration). But, if possible, it is also very important to know their shape, crystallinity, aggregation state, or their tendency to form aggregates in different conditions. Then, all the other physical or chemical properties that can be determined, are useful to have a more precise and complete picture of the studied nanoparticles.
- Several characterization methods are available today, but none of them is sufficient to have all this information at once. Characterization techniques have to be chosen, beside their availability, depending on the available nanoparticles quantity, their physical state (dry, in liquid suspension, dispersed inside a different matrix, etc.) and the size or concentration range that fits our needs. Moreover, the cross-check between multiple characterizations of the same property could sometimes help to reduce the analysis uncertainty.
- Quite often a pretreatment is required to make the ENMs "ready" for the characterization. The pretreatment may involve the dissolution of complex matrices in which the nanoparticles are embedded, the prevention of undesired aggregation, the separation of materials with different composition or size, as well as drying, dilution, concentration, or fractionation processes.
- Pretreatment should be designed to not affect the investigated ENMs properties.
- Usually techniques based on electron microscopy, atomic force microscopy, and mass spectrometry are the most widely used, mainly because of the large versatility provided by their different configurations and available tools. However, in this chapter, also other techniques, which could be more suitable for specific ENM samples, were described and compared.

References

ASTM Subcommittee E56.02 on Physical and Chemical Characterization, 2015. Standard guide for measurement of particle size distribution of nanomaterials in suspension by photon correlation spectroscopy (PCS). ASTM E2490–09(2015). ASTM International, West Conshohocken, PA.

Astruc, D. (Ed.), 2008. Nanoparticles and Catalysis. Wiley, Weinheim, Germany.

Baalousha, M., Lead, J.R., 2015. Characterization of Nanomaterials in Complex Environmental and Biological Media, Frontiers of Nanoscience, vol. 8. Elsevier, Amsterdam, The Netherlands.

Baalousha, M., Stolpe, B., Lead, J.R., 2011. Flow field-flow fractionation for the analysis and characterization of natural colloids and manufactured nanoparticles in environmental systems: a critical review. J. Chromatogr. A 1218 (27), 4078–4103.

Beltrami, D., Calestani, D., Maffini, M., Suman, M., Melegari, B., Zappettini, A., et al., 2011. Development of a combined SEM and ICP-MS approach for the qualitative and quantitative analyses of metal microparticles and sub-microparticles in food products. Anal. Bioanal. Chem. 401 (4), 1401–1409.

Bour, A., Mouchet, F., Silvestre, J., Gauthier, L., Pinelli, E., 2015. Environmentally relevant approaches to assess nanoparticles ecotoxicity: a review. J. Hazard. Mater. 283, 764–777.

Calzolai, L., Gilliland, D., Rossi, F., 2012. Measuring nanoparticles size distribution in food and consumer products: a review. Food Addit. Contam. A 29 (8), 1183–1193.

Cao, G.Z., Wang, Y., 2011. Nanostructures and Nanomaterials: Synthesis, Properties, and Applications, second ed. World Scientific, Singapore.

Cascio, C., Gilliland, D., Rossi, F., Calzolai, L., Contado, C., 2014. Experimental evaluation of key methods to detect, size and quantify nanoparticulate silver. Anal. Chem. 86, 12143–12151.

Chow, J.C., Savage, N., Salomon, C.J., Cheng, Y.S., McMurry, P.H., Corey, L.M., et al., 2005. Nanoparticles and the environment. J. Air Waste Manage. Assoc. 55, 1411–1417.

Cornelis, G., Hund-Rinke, K., Kuhlbush, T., van den Brink, N., Nickel, C., 2014. Fate and bioavailability of engineered nanoparticles in soils: a review. Crit. Rev. Environ. Sci. Technol. 44, 2720–2764.

Domingos, R.F., Baalousha, M.A., Ju-Nam, Y., Reid, M.M., Tufenkji, N., Lead, J.R., et al., 2009. Characterizing manufactured nanoparticles in the environment: multimethod determination of particle sizes. Environ. Sci. Technol. 43 (19), 7277–7284.

EFSA Scientific Committee, 2011. Scientific opinion—guidance on the risk assessment of the application of nanoscience and nanotechnologies in food and the feed chain. EFSA J. 9 (5), 2140.

European Commission, 2011. Commission recommendation of 18 October 2011 on the definition of nanomaterial. Off. J. Eur. Union 2011, L275/38–L275/40.

Fukuhara, N., Suzuki, K., Takeda, K., Nihei, Y., 2008. Characterization of environmental nanoparticles. Appl. Surf. Sci. 255 (4), 1538–1540.

Fultz, B., Howe, J., 2013. Transmission Electron Microscopy and Diffractometry of Materials, fourth ed. Springer, Berlin, Germany.

Goldstein, J., Newbury, D., Joy, D., Lyman, C., Echlin, P., Lifshin, E., et al., 2007. Scanning Electron Microscopy and X-Ray Microanalysis, third ed. Springer, New York.

Gontier, E., Ynsa, M.D., Biro, T., Hunyadi, J., Kiss, B., Gaspar, K., et al., 2008. Is there penetration of titania nanoparticles in sunscreens through skin? A comparative electron and ion microscopy study. Nanotoxicology 2 (4), 218–231.

Grillo, R., Rosa, A.H., Fraceto, L.F., 2015. Engineered nanoparticles and organic matter: a review of the state-of-the-art. Chemosphere 119, 608–619.

Hagarova, I., 2017. Separation and quantification of metallic nanoparticles using cloud point extraction and spectrometric methods: a brief review of latest applications. Anal. Methods 9 (24), 3594–3601.

Haschke, M., 2014. Laboratory Micro-X-Ray Fluorescence Spectroscopy: Instrumentation and Applications. Springer, Berlin, Germany.

Hassellov, M., Readman, J.W., Ranville, J.F., Tiede, K., 2008. Nanoparticle analysis and characterization methodologies in environmental risk assessment of engineered nanoparticles. Ecotoxicology 17 (5), 344–361.

Hellack, B., Nickel, C., Albrecht, C., Kuhlbusch, T.A.J., Boland, S., Baeza-Squiban, A., et al., 2017. Analytical methods to assess the oxidative potential of nanoparticles: a review. Environ. Sci. Nano 4 (10), 1920−1934.

ISO Technical Committee 229 Nanotechnologies, 2012. Guidance on physico-chemical characterization of engineered nanoscale materials for toxicologic assessment. ISO/TR 13014:2012, ICS: 07.120 Nanotechnologies. International Organization for Standardization, Geneva, Switzerland.

ISO Technical Committee 229 Nanotechnologies, 2015. Lists terms and definitions related to particles in the field of nanotechnologies. ISO/TS 80004-2:2015, ICS: 01.040.07 Natural and applied sciences (Vocabularies)—07.120 Nanotechnologies. International Organization for Standardization, Geneva, Switzerland.

Jarzębski, M., Bellich, B., Białopiotrowicz, T., Śliwa, T., Kościński, J., Cesàr, A., 2017. Particle tracking analysis in food and hydrocolloids investigations. Food Hydrocolloids 68, 90−101.

Jiang, J., Oberdörster, G., Biswas, P., 2009. Characterization of size, surface charge, and agglomeration state of nanoparticle dispersions for toxicological studies. J. Nanopart. Res. 11 (1), 77−89.

Kaegi, R., Wagner, T., Hetzer, B., Sinnet, B., Tzetkov, G., Böller, M., 2008. Size, number and chemical composition of nanosized particle in drinking water determined by analytical microscopy and LIBD. Water Res. 42 (10−11), 2778−2786.

Kociak, M., Zagonel, L.F., 2017. Cathodoluminescence in the scanning transmission electron microscope. Ultramicroscopy 176, 112−131.

Kowalczyk, B., Lagzi, I., Grzybowski, B.A., 2011. Nanoseparations: strategies for size and/or shape-selective purification of NPs. Curr. Opin. Colloid Interface Sci. 16 (2), 135−148.

Laborda, F., Bolea, E., Cepriá, G., Gómez, M.T., Jiménez, M.S., Perez-Arantegui, J., et al., 2016. Detection, characterization and quantification of inorganic engineered nanomaterials: a review of techniques and methodological approaches for the analysis of complex samples. Anal. Chim. Acta 904, 10−32.

Larkin, P., 2017. Infrared and Raman Spectroscopy: Principles and Spectral Interpretation, second ed. Elsevier, Amsterdam, The Netherlands.

Lee, B.P., Li, Y.J., Flagan, R.C., Lo, C., Chan, C.K., 2013a. Sizing characterization of the fast-mobility particle sizer (FMPS) against SMPS and HR-ToF-AMS. Aerosol Sci. Technol. 47 (9), 1030 1037.

Lee, B.P., Li, Y.J., Yu, J.Z., Louie, P.K.K., Chan, C.K., 2013b. Physical and chemical characterization of ambient aerosol by HR-ToF-AMS at a suburban site in Hong Kong during springtime 2011. J. Geophys. Res. Atmos. 118 (15), 8625−8639.

Lopez-Serrano, A., Olivas, R.M., Landaluze, J.S., Camara, C., 2014. Nanoparticles: a global vision. Characterization, separation, and quantification methods. Potential environmental and health impact. Anal. Methods 6 (1), 38−56.

Lozano, O., Oliver, T., Dogné, J.M., Lucas, S., 2013. The use of PIXE for engineered nanomaterials quantification in complex matrices. J. Phys. Conf. Ser. 429, 012010.

Lu, P.J., Huang, S.C., Chen, Y.P., Chiueh, L.C., Shih, D.Y.C., 2015. Analysis of titanium dioxide and zinc oxide nanoparticles in cosmetics. J. Food Drug Anal. 23, 587−594.

Luo, P., Morrison, I., Dudkiewicz, A., Tiede, K., Boyes, E., O'Toole, P., et al., 2013. Visualization and characterization of engineered nanoparticles in complex environmental and food matrices using atmospheric scanning electron microscopy. J. Microsc. 250 (1), 32−41.

Majestic, B.J., Erdakos, G.B., Lewandowski, M., Oliver, K.D., Willis, R.D., Kleindienst, T. E., et al., 2010. A review of selected engineered nanoparticles in the atmosphere sources, transformations, and techniques for sampling and analysis. Int. J. Occup. Environ. Health 16 (4), 488–507.

Manoj, K., Arzan, A., Megha, M., Prajin, J., 2018. Toxicological impact of quantum dots. In: Sridharan, K. (Ed.), Emerging Trends of Nanotechnology in Environment and Sustainability: A Review-Based Approach. Springer, Cham, Switzerland.

Mattarozzi, M., Suman, M., Cascio, C., Calestani, D., Weigel, S., Undas, A., et al., 2017. Analytical approaches for the characterization and quantification of nanoparticles in food and beverages. Anal. Bioanal. Chem. 409 (1), 63–80.

Meyer, E., Hug, H.J., Bennewitz, R., 2000. Scanning Probe Microscopy: The Lab on a Tip. Springer, New York.

Montoro Bustos, A.R., Ruiz Encinar, J., Sanz-Medel, A., 2013. Mass spectrometry for the characterization of nanoparticles. Anal. Bioanal. Chem. 405 (17), 5637–5643.

Nicklaus, M., 2014. Tip-Enhanced Raman Spectroscopy for Nanoelectronics. BoD, Norderstedt, Germany.

Nyquist, R., 2001. Interpreting Infrared, Raman, and Nuclear Magnetic Resonance Spectra. Academic Press, San Diego, CA.

Pecora, R., 2000. Dynamic light scattering measurement of nanometer particles in liquids. J. Nanopart. Res. 2 (2), 123–131.

Peters, R., Herrera-Rivera, Z., Undas, A., van der Lee, M., Marvin, H., Bouwmeestera, H., et al., 2015. Single particle ICP-MS combined with a data evaluation tool as a routine technique for the analysis of nanoparticles in complex matrices. J. Anal. At. Spectrom. 30 (6), 1274–1285.

Sadik, O.A., Du, N., Kariuki, V., Okello, V., Bushlyar, V., 2014. Current and emerging technologies for the characterization of nanomaterials. ACS Sustain. Chem. Eng. 2 (7), 1707–1716.

Sarma, D.D., Santra, P.K., Mukherjee, S., Nag, A., 2013. X-ray photoelectron spectroscopy: a unique tool to determine the internal heterostructure of nanoparticles. Chem. Mater. 25 (8), 1222–1232.

Schimpf, M.E., Caldwell, K., Giddings, J.C., 2000. Field-Flow Fractionation Handbook. John Wiley & Sons, New York.

Shukla, A., Iravani, S., 2018. Green Synthesis, Characterization and Applications of Nanoparticles, first ed. Elsevier, Amsterdam, The Netherlands.

Simeonidis, K., Mourdikoudis, S., Kaprara, E., Mitrakas, M., Polavarapu, L., 2016. Inorganic engineered nanoparticles in drinking water treatment: a critical review. Environ. Sci. Water Res. 2 (1), 43–70.

Singh, J.P., Thakur, S.N., 2007. Laser-Induced Breakdown Spectroscopy. Elsevier, Amsterdam, The Netherlands.

Srivastava, V., Gusain, D., Sharma, Y.C., 2015. Critical review on the toxicity of some widely used engineered nanoparticles. Ind. Eng. Chem. Res. 54, 6209–6233.

Streil, T., Oeser, V., Rambousky, R., Buchholz, F.W., 2007. Alarm inhalation dosemeter for long living radioactive dust due to an uncontrolled release. In: Paschoa, A.S., Steinhausler, F. (Eds.), Natural Radiation Environment, 8th International Symposium on Natural Radiation Environment, Buzios, Brasil, October 07–12, 2007. AIP Conf. Proc. 1034, 445.

Taylor, H., 2000. Inductively Coupled Plasma-Mass Spectrometry: Practices and Techniques, first ed. Academic Press, San Diego, CA.

Thomas, R., 2013. Practical Guide to ICP-MS: A Tutorial for Beginners, third ed. CRC Press, Boca Raton, FL.

Thomas, S., Thomas, R., Zachariah, A., Mishra, R.K., 2017a. first ed. Microscopy Methods in Nanomaterials Characterization, vol. 1. Elsevier, Amsterdam, The Netherlands.

Thomas, S., Thomas, R., Zachariah, A., Mishra, R.K., 2017b. Spectroscopic Methods for Nanomaterials Characterization, vol. 2. Elsevier, Amsterdam, The Netherlands.

Tiede, K., Boxall, A.B.A., Tear, S.P., Lewis, J., David, H., Hassellov, M., 2008. Detection and characterization of engineered nanoparticles in food and the environment. Food Addit. Contam. A 25 (7), 795–821.

Tiede, K., Tear, S.P., David, H., Boxall, A.B.A., 2009. Imaging of engineered nanoparticles and their aggregates under fully liquid conditions in environmental matrices. Water Res. 43, 3335–3343.

Tsao, T.M., Chen, Y.M., Wang, M.K., 2011. Origin, separation and identification of environmental nanoparticles: a review. J. Environ. Monit. 13 (5), 1156–1163.

Vasco, F., Hawe, A., Jiskoot, W., 2010. Critical evaluation of nanoparticle tracking analysis (NTA) by NanoSight for the measurement of nanoparticles and protein aggregates. Pharm. Res. 27 (5), 796–810.

Verma, A., Stellacci, F., 2010. Effect of surface properties on nanoparticle–cell interactions. Small 6 (1), 12–21.

Wagner, S., Legros, S., Loeschner, K., Liu, J., Navratilova, J., Grombe, R., et al., 2015. First steps towards a generic sample preparation scheme for inorganic engineered nanoparticles in a complex matrix for detection, characterization, and quantification by asymmetric flow-field flow fractionation coupled to multi-angle light scattering and ICP-MS. J. Anal. At. Spectrom. 30 (6), 1286–1296.

Warren, B.E., 1990. X-Ray Diffraction. Dover Publications, New York.

Xu, R.L., 2002. Electrophoretic light scattering. Particle Characterization: Light Scattering Methods. Kluwer, New York (Chapter 6).

SECTION

II

ENMs in the environment: fate, transfer and interactions with organisms

CHAPTER

3 Fate of engineered nanomaterials in natural environments and impacts on ecosystems

Monika Mortimer and Patricia A. Holden

Bren School of Environmental Science & Management, Earth Research Institute, University of California Center for Environmental Implications of Nanotechnology (UC CEIN), University of California, Santa Barbara, Santa Barbara, California, United States

3.1 Introduction

Natural environments are essential to human well-being in that they provide valuable ecosystem services, including clean air, water, climate regulation, biodiversity, nutrient cycling, waste degradation, food, and esthetic benefits (Costanza et al., 1997). Yet, worldwide, even protected lands and waters are experiencing pressures of human activity in ways that can diminish the very ecosystem services upon which humans depend (Jones et al., 2018). Managed ecosystems, such as agriculture and fisheries, balance needs to conserve soil and other natural resources—often while intentionally employing potentially polluting chemicals and materials—with meeting increasing demands for human food and other products. Management practices can also unintentionally introduce novel pollutants, termed contaminants of emerging concern (CECs), which are potent toxicants at low concentrations, are widely distributed, and have uncertain environmental consequences (Wilkinson et al., 2017). Such is the case for phthalates, which are manufacturing additives that become CECs as they leach, for example, from plastic mulches in agriculture (Steinmetz et al., 2016). Plastics in agriculture also physically fragment into micro- (10^{-6} m) to nano (10^{-9} m)-sized particles, which are also considered broadly as CECs and as carriers of other CECs (Wilkinson et al., 2017).

While manmade plastics can incidentally fragment into nanoparticles (NPs) in the environment, engineered nanomaterials (ENMs) are intentionally produced at the nanoscale for use in widely varying applications. Making and using ENM-containing products allow ENMs to enter the environment and become CECs. ENMs are in waste streams, such as sewage, and thus enter the environment with fates that are similar to specific plastics (Hüffer et al., 2017) such as synthetic microfibers (Hartline et al., 2016) that can become nanoscale. The point is that

Exposure to Engineered Nanomaterials in the Environment. DOI: https://doi.org/10.1016/B978-0-12-814835-8.00003-0

ENMs are CECs, just as plastics and plasticizers, although the latter are more well recognized and more common to the broad discourse in environmental pollution (Machado et al., 2018). While plastics have been manufactured since the 1950s (Geyer et al., 2017), ENMs are newer and rapidly evolving (Roco, 2011). As the development and disposal of ENMs increase, it is timely to consider their potential effects on humans and the environment such that their continued evolution can be sustainable and maximize benefit, while minimizing harm (National Research Council, 2012).

In this chapter the focus is on ENMs that enter natural environments with potential impacts to ecosystems. This chapter provides an overview of the major types of ENMs in use, and how they enter the environment based on manufacture, application, and disposal. This sets the stage for reviewing the evidence for ENM exposure in the environment, based on sample recovery and quantitative analysis. Issues of discerning ENMs from natural nanomaterials are discussed, with attention to method development. Modeling the entry and fates of ENMs in the environment are reviewed, reinforcing the importance of material flow analysis and mechanistic fate models to improve estimations of environmental exposures. Still, ENMs have multiple fates in the environment, which can alter their integrity and bioavailability to biological receptors. The major fate interactions are discussed, including agglomeration, dissolution, degradation, and organismal uptake. Under these conditions of exposure, we discuss the potential for ENMs to affect organisms and ecosystems. We end with key concepts and points, including an overview of uncertainties and opportunities to advance understanding in the future. We find that while there exists an impressive body of literature on ENM fate and effects in natural environments, there is a need for more research in certain areas such as the sensitivity and selectivity of ENM measurement methods, uncertainties in modeling approaches, effects of ENM aging on ENM fate, transport and biological effects, understanding long-term effects of ENM exposures to agricultural ecosystems, and linking ENM-induced molecular-level toxic effects to cell, population, community, and ecosystem level effects.

3.2 Deposition and transport: how do engineered nanomaterials enter and move within the natural environment?

ENMs enter natural environments via different pathways, with some more prevalent for certain types of ENMs than for others, depending on ENM production volumes and applications. For example, ZnO and TiO_2 are the most commonly used ENMs in personal care products (PCPs), representing ~94% of ENM use in PCPs, such as sunscreens, facial moisturizers, and foundation, and are estimated to pass through wastewater treatment plants (WWTPs) and then discharge into water bodies and soil via effluent and biosolids, respectively (Keller et al., 2014;

Liu et al., 2015). While carbon black (CB) is a massively produced and used nanomaterial (Holden et al., 2014a), it is not engineered for use in PCPs and thus on a mass basis, TiO_2, SiO_2, and ZnO may be the most produced such ENMs worldwide (Keller et al., 2013). Considering both global production volumes and number of applications, TiO_2 has been estimated to have the highest concentration in all environmental compartments. Yet, according to the Nanotechnology Consumer Products Inventory, nanosized Ag is by far the most often used ENM in consumer goods, present in over 400 different products, despite its global annual production being only 2% of that of TiO_2 (Vance et al., 2015). Due to its favorable physicochemical and antimicrobial properties, nanosized Ag is used in textiles, coatings and paints, packaging, and medical applications (Vance et al., 2015). Considering the nature of its applications, the majority of nanosized Ag is expected to enter WWTPs (Sun et al., 2014). On the other hand, as the main application areas of Cu-based ENMs are marine antifouling paints and pesticides, nanosized Cu is expected to be released to the environment via applications in agriculture and directly in water bodies, while the amounts of Cu ENM entering WWTPs and subsequent application in soils via biosolids are expected to be low (Keller et al., 2017). A significantly different material flow is expected for carbon nanotubes (CNTs) and nanosized SiO_2. Due to applications in electronics and optics, composite materials, energy and environmental applications, and automotive industry, only a small fraction of CNTs is expected to pass through WWTPs (6%) or waste incinerator plants (5.5%) with most of the disposed material going directly to landfills (Keller et al., 2013). A similar material flow scenario is expected for nanosized SiO_2, used in energy and environmental applications, catalysts, and electronics and optics automotive industry (Keller et al., 2013).

Environmental release assessments of ENMs must consider a variety of emission sources that cover the whole life cycle of ENMs and ENM-containing products: manufacturing, product use, and recycling or disposal (Fig. 3.1). The synthesis and handling of ENM powders in the manufacturing process poses the highest likelihood for ENM release because a large amounts of pure material are handled. ENM total release during manufacturing (Fig. 3.1A) has been estimated to be 0.1%–2% of which 20%–80% is released directly into the environment (10%–40% to air and 10%–40% to water prior to WWTP; Keller et al., 2013). The rest of ENMs released in the manufacturing process are expected to end up in landfills (Fig. 3.1B). Overall, considering all life-cycle stages from manufacturing, to application to disposal, the majority (63%–91%) of ENMs are estimated to be disposed in landfills. ENM release into the environment from well-managed and controlled landfills is expected to be low; however, in less controlled landfills, such as these in developing countries, ENM release via leachates and gas emission could be significant (Nowack et al., 2013). In secondary processing or fabrication of ENM-enabled products (Fig. 3.1C), release might occur during mechanical processes such as drilling, cutting and sanding, or thermal and high-energy processes that could destabilize composites and release ENMs (Nowack et al., 2013). During product use (Fig. 3.1D), ENM release depends on the form

FIGURE 3.1

Conceptual environmental release scenarios for ENMs across their life cycles. Potential release sites (with highest expected relative NM concentration indicated in the parentheses) include (A) primary ENM manufacturing (high NM concentration); (B) landfill with solid waste including nanoenabled electronics, consumer goods, and permitted industrial waste (low NM concentration); (C) secondary processing or goods manufacturing sites using ENMs (medium NM concentration); (D) consumer (household) use of ENM-enabled products (low NM concentration); (E) agricultural ENM-enabled product use (medium NM concentration); (F) agricultural runoff and marine or freshwater ENM-enabled product use, including coatings and bathing (low NM concentration); (G) waste treatment with aqueous effluent and solids residuals that may contain ENMs or transformation products thereof (medium NM concentration); (H) WIP. According to the legend, *colored circles (in the online version)* adjacent to each location indicate the highest expected relative NM concentration; the NM forms are as-produced (°1) or in products (°2) or in mixed waste streams (°3); release destinations include waste infrastructure and major environmental compartments (soil, water, air). *ENM*, Engineered nanomaterial; *NM*, nanomaterial; *WIP*, waste incineration plant.

Modified from Holden, P.A., Gardea-Torresdey, J.L., Klaessig, F., Turco, R.F., Mortimer, M., Hund-Rinke, K., et al., 2016. Considerations of environmentally relevant test conditions for improved evaluation of ecological hazards of engineered nanomaterials. Environ. Sci. Technol. 50, 6124–6145.

and type of application: significant release is expected from ENM-containing liquids, pastes, creams, powders, and aerosol sprays, with lower release from products where ENMs are embedded in a solid material (Clark et al., 2012). For most ENMs, the estimates of emissions to soils are based on the assumption that a fraction of WWTP sludge is used as a fertilizer in agriculture (Fig. 3.1E). However, only 50% of biosolids produced in the United States are land-applied, and these biosolids are used on less than 1% of agricultural land in the United States (http://www.epa.gov/biosolids) and even less in the European Union (Evans, 2012). A smaller fraction of ENMs reaches soils via direct applications, such as pesticides, fertilizers, or environmental remediation (Fig. 3.1E). ENMs may reach water bodies from soils through agricultural runoff. Other routes of ENMs to water bodies include direct release from paints and coatings or cosmetics (Fig. 3.1F) and indirect discharge through WWTP effluent (Fig. 3.1G). Likely, many insoluble ENMs aggregate in natural waters and will settle out into sediments (Wang et al., 2015). Global flow models across all ENM life-cycle stages have shown that, among environmental compartments, the largest fraction of ENMs is deposited in soils (8%–28%), in water (0.4%–7%), and in the atmosphere (0.2%–1.5%) (Fig. 3.1; Keller et al., 2013). The rest is disposed in landfills.

3.3 Distribution: where are the engineered nanomaterials and what is the evidence?

ENM release and resulting concentrations in environmental compartments can be determined by (1) experimentally measuring the release of ENMs from products, characterizing ENM physicochemical properties that determine ENM distribution in environmental compartments and measuring ENM concentrations in environmental matrices or (2) developing models that account for total life-cycle release of ENMs from different applications.

3.3.1 Experimental quantitation of engineered nanomaterials in natural environments

While the methods for analysis and characterization of ENMs in suspensions and even in complex media under laboratory conditions are relatively well developed, experimental measurements of ENMs in natural samples are challenged by the presence of natural NPs, formed by geological, atmospheric, and biological processes, and by low concentrations of ENMs in the environment. Particulate materials within environmental samples, including nanosized particles, are often composed of the same elements as the ENMs. Consequently, measurement techniques that focus on a certain size fraction fail to differentiate between ENMs and natural NPs and do not detect ENMs adsorbed to larger particles (Nowack et al., 2015). In addition to enabling discrimination between ENMs and naturally occurring particles, appropriate analytical methods have to be sensitive enough to quantify low, parts per billion (ppb, i.e., μg/L or μg/kg) range ENM concentrations expected to occur in the environment (Sun et al., 2014). In general the detection limits of most methods are the lowest in pristine water samples and higher in soils, sediments, and biological tissues (Petersen et al., 2016). The choice of method for quantifying, identifying, and characterizing ENMs thus depends on several factors, including ENM chemistry, concentrations, and environmental compartments under study.

Based on ENM chemistry, ENM detection and quantitation can be based on the thermal or spectroscopic properties of ENMs, quantity of metallic elements in ENMs (metal-based ENMs and catalytic impurities in CNTs), or the presence of stable or radioactive isotopes in ENMs. In addition, various microscopic techniques, either alone or coupled with spectroscopic analysis, can be used for determining ENM biological uptake and compartmentalization and obtaining ENM particle number concentrations in environmental samples. Among spectroscopy-based techniques, ultraviolet-visible (UV–vis) spectroscopy is most commonly used for ENM quantification in laboratory samples. However, with detection limits in the tens of μg/L to mg/L (Goodwin et al., 2018), UV–vis spectroscopy may not be suitable for directly quantifying ENMs in natural waters where the modeled ENM concentrations are in the low ng/L range. Other thermal and spectroscopic methods that are specifically suitable for quantitation of organic ENMs, such as CNTs and graphene, include thermogravimetric, total organic carbon, and programed thermal analysis, microwave-induced heating (Doudrick et al., 2015; He et al., 2017) and fluorimetry, Raman and near-infrared fluorescence spectroscopy (Goodwin et al., 2018; Petersen et al., 2016; Schierz et al., 2012). Most of these methods have the advantages of being readily available in laboratories and their ease of use. However, because of low ENM concentrations, heteroagglomeration, and the presence of other carbonaceous materials in natural environments, none of these methods are directly applicable for ENM quantification in complex matrices. Examples of interferences include the presence of thermally or spectroscopically similar carbonaceous materials (e.g., soot) in the environment, biomolecules [e.g., natural organic matter (NOM) and nutrients] that absorb light in the UV region or are fluorescent, and natural colloids and biopolymers that form heteroagglomerates with ENMs. Thus efficient and selective extraction methods are often needed for separation of ENMs from the matrix prior to ENM quantification in environmental compartments, especially in soils and sediments.

Recent advances with extraction protocols for selective isolation and concentration of ENMs prior to analysis allow improved measurements of ENMs in complex environmental matrices. For example, sodium pyrophosphate was recently demonstrated as a potent agent for the extraction of TiO_2 ENMs from TiO_2-clay heteroagglomerates (Loosli et al., 2018). Density gradient centrifugation, a method often used in biology for separating subcellular components with differing buoyant densities, has also proven efficient for separation and isolation of ENMs from complex matrices (Faust et al., 2016). The same approach also works for the isolation and concentration of CNT-exposed cells prior to quantification of CNT accumulation in or on live cells (Mortimer et al., 2016a). Another method often used for ENM extraction and concentration in natural waters (Majedi et al., 2014), wastewater (Li et al., 2013), and soil (Hadri and Hackley, 2017) is cloud point extraction (CPE). CPE involves the addition of a surfactant at a concentration above its critical micelle concentration, subsequent temperature increases to initiate micelle formation via dehydration and the formation of surfactant-rich and surfactant-depleted phase (Duester et al., 2016). ENMs in the surfactant-rich phase are separated from the supernatant by centrifugation. The extraction method has been optimized for Ag, CuO, ZnO, and Au ENMs (Duester et al., 2016; Hadri and Hackley, 2017). Other methods include chemical matrix digestion (Doudrick et al., 2013) and thermal or chemical oxidation of organic compounds that are more labile than ENMs (Plata et al., 2012).

Methods that can be used for the analysis of ENMs in aqueous samples, under certain conditions without prior ENM extraction and separation steps, include ICP-MS-based techniques. To date, these are the most selective and sensitive methods for analysis of ENMs in natural samples (Gondikas et al., 2014; Mitrano et al., 2014a). spICP-MS can be used to detect ENMs at very low number concentrations that are often found in environmental samples (in the range 10^3-10^5 cm^{-3}; Montoro Bustos and Winchester, 2016). The shortcomings of spICP-MS include the interference from high dissolved metal backgrounds in the sample and that the technique only measures the total mass of only one isotope in a single analysis (Nowack et al., 2015). For example, nanosized Ag particles cannot be distinguished from Ag_2S particles or TiO_2 ENMs from natural particles containing Ti. Coupling spICP-MS with field flow fractionation allows determining the number of particles per aggregate, which enables distinguishing heteroagglomerated TiO_2 ENM from natural Ti-containing particles (Nowack et al., 2015; von der Kammer et al., 2012). Some of the latest developments in spICP-MS such as single-particle time-of-flight ICP-MS (Borovinskaya et al., 2013) or using reduced data collection times by employing 100 μs dwell times (Montaño et al., 2014) allow simultaneous multielement detection and introduce new possibilities for differentiating ENMs from their naturally occurring analogs. Recently, the use of 100 μs dwell times in spICP-MS was successfully employed for the detection of single-walled CNTs (SWCNTs), based on CNT yttrium content (Wang et al., 2016a). In the latter study the capacity of spICP-MS as a powerful method for environmental exposure assessment of SWCNTs was demonstrated by measuring SWCNT association with water flea *Daphnia magna* at μg/L SWCNT concentrations (Wang et al., 2016a).

Microscopic techniques, when coupled with spectroscopic analysis, can serve as indispensable tools for identifying ENMs in complex matrices. For example, electron microscopy (EM) with electron energy loss spectroscopy can be used for identifying CNTs (Edgington et al., 2014) and EM with energy-dispersive X-ray spectroscopy for identifying heavier elements (Mielke et al., 2013;

Werlin et al., 2011). Quantitative microscopic analysis of ENMs, however, requires ENM collection from a known volume to EM grid so the ENM number per area can be converted into particle number concentration. It has been reported that atomic force microscopy and transmission EM (TEM) can both be applied successfully to measure particle number concentration from a suspension at environmentally realistic ENM concentrations (c. 0.2–100 μg/L Au NPs; Prasad et al., 2015). Still, due to low predicted ENM concentrations in the environment, obtaining a statistically significant count of ENMs in microscopy samples remains challenging and the results from EM analysis are of limited utility for ENM quantification.

However, in biota, where ENMs accumulate in the digestive system, agglomerated ENMs have successfully been quantified using image analysis of TEM micrographs. For example, TiO_2 ENM accumulation in protozoan food vacuoles was quantified by analysis of scanning TEM images (Mielke et al., 2013). Although conventional optical microscopy cannot resolve single ENMs, it is suitable for visualizing ENM agglomerates that are often larger than the resolution limit of light microscopes with a conventional lens, that is, ~200 nm. In addition, ENMs are packed into agglomerates in the digestive system (i.e., food vacuoles) of particle feeding (phagocytosing) single-celled species (Mortimer et al., 2016b). This phenomenon provides a good opportunity for using quantitative optical microscopy for ENM uptake and depuration kinetics measurements. Coupling dark field or fluorescence microscopy with hyperspectral analysis further provides spectral information on ENM localization, allowing both identification and quantification of agglomerated ENMs in cells (Mortimer et al., 2014a). Fluorescent ENMs, such as quantum dots (QDs), can be traced inside the cells using confocal fluorescence microscopy and quantified using flow cytometry (Holbrook et al., 2008; Mortimer et al., 2014b; Salvati et al., 2018; Wang et al., 2013). Since microscopy is not an analytical technique that provides mass-based ENM concentrations, assumptions and calculations have to be used to convert ENM area measurements in 2D images into mass-based concentrations (Mielke et al., 2013). The validation of the calculated results with an analytical method is recommended, when feasible, to confirm the reliability of microscopy-based quantification. For instance, the validity of image analysis-based quantification of multiwall CNTs (MWCNTs) that had accumulated in protozoan food vacuoles was confirmed by analytical measurements using radioactively labeled (^{14}C) MWCNTs and liquid scintillation counting (Mortimer et al., 2016b).

Although not applicable to industrial materials, isotopic labeling of ENMs provides definitive quantification of ENMs in complex matrices and can be used as an orthogonal technique to develop other analytical techniques. In addition, radioactive or stable isotope labeling allows detection of ENMs at relatively low concentrations and facilitates selective tracing of ENMs along the food chain and in mesocosm studies. Radioactive labeling is especially helpful in studying carbon-based ENMs in complex matrices such as soils (Zhang et al., 2012) or quantification in biota at low-exposure concentrations (Dong et al., 2018; Mortimer et al., 2016b).

While liquid scintillation counting enables quantification of ^{14}C-labeled carbonaceous NMs with the detection limit around 0.1 mg/L, techniques such as accelerator mass spectrometry can measure CNTs at concentrations below ppb level (Mortimer et al., 2016b). Radioactive labeling has been used to quantify and trace also metal ENMs (Yin et al., 2017). The advantages of using radioactively labeled metal ENMs include the ease of detection by quantitation directly in whole organisms (Jung et al., 2014) and in situ imaging at organism, tissue, cell, and subcellular levels (Yin et al., 2017). Both radioactive and stable isotope-labeled metal ENMs allow sensitive detection of ENMs composed of naturally occurring metals (Croteau et al., 2014). Stable isotopes are favored over radioactive isotopes as tracers due to possible radiocontamination and special authorizations needed to work with radionuclides. Also, using multiple stable isotopes enables simultaneous monitoring of different ENMs (Yu et al., 2016).

The use of the methods described in this section has enabled researchers to characterize the fate and transport of ENMs in different environmental matrices, such as natural water, sediments, soils, and biota. However, given the lack of sufficient sensitivity and selectivity of the existing analytical methods as described earlier, it is difficult or in some cases impossible to directly measure ENM environmental concentrations or trace ENMs found in the environment back to original emission sources. Thus ENM flows to the environment and resulting ENM environmental concentrations are often estimated using modeling.

3.3.2 Estimation and modeling of engineered nanomaterial concentrations in natural environments

Environmental exposure and distribution analysis can be conducted at various levels of complexity, with screening level assessments carried out with multimedia models to identify major exposure pathways and higher tier analyses with detailed single-medium models (Liu et al., 2015). Since the intermedia transport of ENMs is governed by physical transport processes of particulate matter, particle size distribution and its dependence on the transport processes in technical (e.g., WWTP) and environmental (e.g., air, water, sediment, and soil) compartments need to be accounted for in modeling. Another important factor in assessing the environmental multimedia distribution of ENMs is their release rates. In order to estimate ENM release rates, life-cycle assessment (LCA) approaches have been developed to track ENM mass from production and use to final disposal and release into the environment (Keller et al., 2013; Sun et al., 2014). LCA approaches are based on ENM production rates and empirical transfer coefficients that quantify the fraction of ENM mass transferred between compartments. Some examples of other parameters needed for ENM life-cycle material flow modeling, depending on the aim of the specific model (Keller and Lazareva, 2014; Keller et al., 2013, 2014; Lazareva and Keller, 2014; Sun et al., 2014), include, but are not limited to

- ENM release during product manufacturing, use, and end of life;
- ENM-enabled product use rate in the population;
- the degree of wastewater treatment (primary, secondary, or higher) by region or country;
- ENM transformation in the wastewater treatment process;
- biosolids management (application to agricultural soils, incineration, and landfilling); and
- ENM release from waste incineration plants to the air.

The sources for ENM release variables and transfer coefficients include scientific literature, market information, surveys, and assumptions based on quantitative analyses using publicly available statistics (Clark et al., 2012; Keller et al., 2014; Lazareva and Keller, 2014). However, accurate data, for example, on ENM production volumes are not readily available due to antitrust law requirements (Holden et al., 2014a) or ambiguities in the definition of a nanomaterial (Wigger et al., 2018). Consequently, using market research estimates of production volumes, that can vary by two to three orders of magnitude, as starting points for modeling environmental ENM concentrations can lead to large uncertainties in the predicted values (Holden et al., 2014a). This issue is expected to be addressed to some extent by the European Commission's (EC) recommendation for a definition of "nanomaterials" (EC, 2011). Namely, based on the EC recommended definition, regulatory agencies in several European countries have implemented mandatory registration requirements for ENMs and ENM-containing products. Recently, Wigger et al. (2018) compared the data on registered production volumes of ENMs, published in a French nanoregistry report, to known production volumes of total or "conventional" materials. They found that for some materials, such as SiO_2, the reported nanoform production volumes were equal to total production volumes, whereas for other materials, such as TiO_2 and ZnO, the reported ENM production volumes were only a fraction of the total production volumes. By applying the knowledge that the global production of SiO_2 is in fact all nanosilica, the authors refined the material flow model for nanosilica and showed that estimated input mass flow of SiO_2 to environmental compartments was four to five times larger than previously modeled (Wigger et al., 2018). Considering that the US Environmental Protection Agency(EPA) has also recently established rules that require the producers of ENM-enabled products and ENMs to report certain information including production volumes (EPA, 2017), it is expected that more accurate data will be available for improved ENM exposure modeling in the near future.

Other sources of uncertainties in most ENM life-cycle material flow models derive from the failure to include variables, such as ENM release from non-nanosources, changes in ENM surface chemistry, and chemical transformations that are possible at each environmentally relevant stage of ENM life cycle. Also, so far, mainly static models have been used that do not consider the rapid development of ENM production nor the fact that many ENM are entering an in-use stock and are released with a lag phase. Examples of the latter applications include paints and coatings where ENMs remain mostly as in-use stock for years and may gradually be released over time (Kaegi et al., 2008) or disposed of at the end of product life. Only recently, in 2016–17, dynamic ENM life-cycle material flow models have been reported (Song et al., 2017; Sun et al., 2016). These models are applied to predict ENM flows to the environment and to quantify ENM amounts in sinks such as the in-use stock and environmental sinks such as soil and sediment. Song et al. (2017) predicted using a dynamic model that the fraction of the end-of-life ENM release among total ENM release flows would increase from 11% in 2002 to 43% in 2020. Consideration of time lag and stock between ENM production and release have a significant effect on the release estimates of ENMs: compared to static models, dynamic modeling predicts about an order of magnitude lower values for the amount of ENM released from paints and coatings (Song et al., 2017). However, ENM stocks are expected to continue to accumulate.

Although there are uncertainties in the ENM LCA approaches, such as uncertainties in the estimated ENM production rates and transfer coefficients as mentioned earlier, these methods are currently considered reasonably suitable for assessing ENM release rates. Based on models and measurements, the reported ENM environmental concentrations range from a low of ≤ 0.001 parts per million (ppm) to a high of >1000 ppm (Gottschalk et al., 2013; Holden et al., 2014a; Keller and Lazareva, 2014; Liu and Cohen, 2014). Across environmental compartments of water (surface water or WWTP effluent) and solid media (soil, sediments, and biosolids), the lowest ENM concentration estimates are well below 0.001 ppm, that is, many of the estimated surface water concentrations are below parts per trillion. The highest for water are in WWTP effluent (0.11–1 ppm), and the highest overall are for biosolids (>1000 ppm; Holden et al., 2014a).

The data provided by mass flow models can be used as input values for more sophisticated environmental fate models that incorporate a mechanistic description of fate processes, for example, agglomeration and sedimentation. Unlike flow models, fate models consider the dynamics of physical and chemical reactions of ENMs in environmental compartments (Liu and Cohen, 2014). Increasingly sophisticated modeling approaches—improved by incorporating fundamental understanding of ENM environmental behaviors and fates—simultaneous to more sensitive and specific means of quantifying ENMs in complex matrices, suggest that ENM exposure assessments based on theory and measurement can converge. A case is nano-TiO_2 (Box 3.1) for which, with increasingly advanced research approaches, modeled and measured estimates of far-field environmental exposures have better reconciled.

Box 3.1 Converging modeling and measurement in exposure assessment: case of TiO_2 engineered nanomaterials (ENMs)

Risk assessment is an important practice that can guide ENM safe use while allowing society to benefit from ENM-based technologies. An ongoing challenge in assessing risks of ENMs in the environment is knowing actual exposures to environmental receptors, that is, "near field" exposures (Holden et al., 2016). More progress has been made with exposure assessments that are "far field," that is, dealing with manufacturing amounts, use in—and release from—products, environmental transport and transformation, and environmental measurements that support modeling estimates (Holden et al., 2016). Challenges remain, particularly for ENMs such as titania that are physicochemically similar to mineral phases in nature (von der Kammer et al., 2012). Still, there is progress stemming from improved technologies for separating and measuring ENMs in complex matrices, coupled with refined modeling that uses new understanding of ENM behaviors as related to environmental fates.

Early reports of TiO_2 ENMs in the environment were pivotal in raising awareness, for example, showing that building façade paint with embedded TiO_2 ENMs could shed up to 600 μg/L Ti into flowing water (Kaegi et al., 2008), and that urban estuarine sediments could harbor up to 6 g/kg Ti, with NPs observed (Luo et al., 2011). In both cases, EM with energy-dispersive spectroscopy was an important analytical technique. Yet these concentrations were at least one order of magnitude higher than those predicted across several original modeling efforts (Holden et al., 2014a), possibly owing to limitations of measurement technologies. Modeling ENM environmental concentrations has evolved to incorporate particle-associated phenomena, such as heteroagglomeration (Therezien et al., 2014) and media-specific settling (Keller et al., 2010), into mechanistically based predictions of how estimated releases (Lazareva and Keller, 2014) move and deposit across multiple environmental media (Liu and Cohen, 2014). Such multimedia modeling predicts TiO_2 ENM concentrations at sub-ppb levels in surface waters, with the regional release of these ENMs mainly being to water, and not to soil (Liu and Cohen, 2014). While this magnitude of surface water concentration is predicted across many recent and original modeling studies (Peters et al., 2018), advances in ENM separations from complex matrices, coupled with state-of-the-art analysis by spICP-MS, can now reveal surface water concentrations that are aligned with model predictions (Peters et al., 2018). Thus far-field exposure assessment by measurement and modeling is converging owing to advances in each endeavor, for the widely used (Weir et al., 2012) ENM, nano-TiO_2.

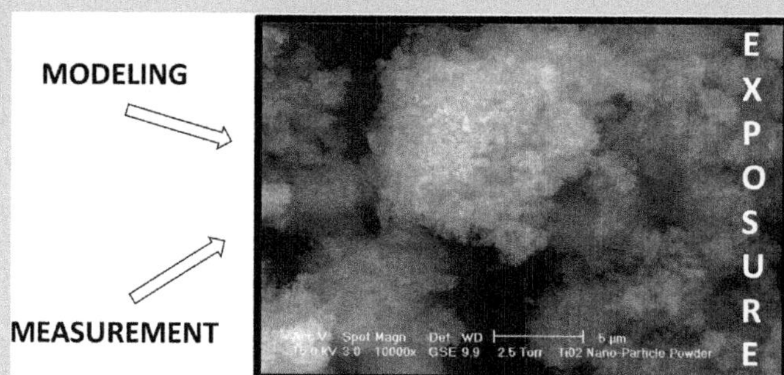

Concept of how ENM release and transport modeling has converged with ENM measurement for resolving exposure assessments, using nano-TiO_2 as an example. P25 (Evonik) powder was imaged with environmental scanning electron microscope (ESEM). Image credit: Sage Davis, MEIAF, UCSB (https://www.bren.ucsb.edu/facilities/MEIAF/). *ENM*, Engineered nanomaterial.

3.4 Fates: what happens to engineered nanomaterials in the natural environment?

One of the major aspects that influences the success of defining "environmental relevance" of ENM hazard assessment regards ENM chemical transformations and changes in ENM surface chemistry (Holden et al., 2014a). Studies of ENMs released from textiles, paints, and nanocomposites suggest that released particles significantly transform and age, and exhibit different environmental behavior and effects compared to pristine ENMs (Mitrano et al., 2014b, 2015; Nowack et al., 2013). The physicochemical transformations vary with ENM types and their properties, and ambient conditions. Since changes in ENM properties may change bioavailability (Priester et al., 2013) and toxicity (Juganson et al., 2013; Stevenson et al., 2013), it is imperative to understand the transformations of ENMs.

ENM environmental transformations are greatly affected by media chemistry, physical characteristics, and additives, for example, soil pH (Schlich and Hund-Rinke, 2015), temperature (Adeleye and Keller, 2014) or organic matter content (Farkas et al., 2015), and dispersing agents in aqueous media (Sauer et al., 2015). Ionic strength and organic compound content of media influence the agglomeration pattern of ENMs, through modifying the surface chemistry of ENMs that often results in reduction of NP number and concentration and, consequently, decreased NP–cell contact and NP-specific toxicity (Wang et al., 2015). In the case of soluble metal-based ENMs (e.g., CuO, ZnO, Ag, Fe_3O_4, and CdSe), medium composition, including the concentration of organic compounds, govern the extent of ENM dissolution, speciation, and bioavailability of dissolved metal ions (Ivask et al., 2014; Priester et al., 2009). In the following paragraphs the major ENM environmental transformation processes are briefly discussed.

3.4.1 Agglomeration

The term "agglomeration" is used here to describe a reversible process of contact and adhesion whereby dispersed particles are held together by weak physical interactions ultimately leading to phase separation by the formation of precipitates of larger than colloidal size (McNaught and Wilkinson, 1997). The terminology is justified by the tendency of ENMs to agglomerate in environmental matrices, instead of forming aggregates that by definition are composed of strongly and irreversibly bonded colloidal particles (McNaught and Wilkinson, 1997). ENMs can either form homoagglomerates, that is, agglomerates composed of ENMs of the same type, or heteroagglomerates, that is, bind to other types of ENMs, natural colloids, or microorganisms (Wang et al., 2015). In environmental media which usually contain high concentrations of natural colloids, colloidal organic matter, viruses, and bacteria, ENMs typically heteroagglomerate (Boyes et al., 2017). The attractive forces that drive heteroagglomerate formation include electrostatic interactions, van der Waals forces, hydrogen and chemical bonding, and bridging via high-molecular-weight polymers (Wang et al., 2015). In complex environmental systems, heteroagglomerate diameters, stability, and formation rates are often governed by multiple mechanisms. For example, in Venice Lagoon water that was sampled along a salinity and dissolved organic carbon (DOC) gradient, the trend of increasing agglomeration of nanosized TiO_2 with increasing salinity was counteracted in samples containing high levels of DOC, resulting in TiO_2 agglomerates with reduced hydrodynamic diameters (Perstrimaux et al., 2015). Similarly, Horst et al. (2010) showed that preferential biosorption of nanosized TiO_2 onto bacterial cell surfaces contributed to dispersion of TiO_2 homoagglomerates. Agglomeration of ENMs, and specifically heteroagglomeration, changes the physicochemical properties and sedimentation of ENMs. These changes include modification of soluble metal-based ENM dissolution, for example, as a result of metal ions binding to NOM sorbed to ENMs. Also, agglomeration may play a major role in the bioavailability of carbonaceous ENMs in soils: low concentrations of MWCNTs, graphene, and CB were shown to negatively impact symbiotic dinitrogen fixation in plants more severely than high concentrations due to extensive agglomeration of ENMs in soil water at high ENM concentrations (Wang et al., 2017).

3.4.2 Dissolution

Certain types of metal-based ENMs, for example, Ag, oxides of Zn and Cu, and Cd-based QDs, have been shown to exert toxicity via dissolved metal ions, making it imperative for environmental hazard assessment purposes to understand the mechanisms of ENM dissolution in various environmental matrices with differing physicochemical properties. For example, ZnO ENMs have been shown to dissolve almost completely in artificial aqueous media, while Ag dissolved only marginally (Odzak et al., 2014; Priester et al., 2014) and dissolution of nanosized CuO was strongly pH dependent (Odzak et al., 2014). Theoretically, it may be assumed that ENM solubility increases as particle size decreases; however, this does not always hold true experimentally because of the influence of various capping agents used to coat ENMs and the tendency of ENMs to agglomerate in aqueous matrices (Misra et al., 2012). In addition to intrinsic ENM properties, characteristics of the surrounding media (e.g., pH, ionic strength, and water hardness) and the presence of organic components (e.g., NOM, polysaccharides, and proteins) can affect suspension stability, affecting the exposed surface area and dissolution of the particles. For example, dispersion of nanosized CuO in the medium containing high concentrations of organic compounds resulted in enhanced dissolution of copper, due to increased dispersion of CuO particles and complexation of copper ions with the organic compounds (Adeleye et al., 2014; Kaekinen et al., 2011). These simultaneous processes influenced the level of bioavailable copper and consequently nanotoxicity, compared to that in media with little to no organic compound content (Kaekinen et al., 2011). CuO dissolution in saline waters has been shown to be very slow ($\leq 1\%$ in months), even though the formation of insoluble complexes increases dissolution (Keller et al., 2017). Also free Ag ion concentrations are expected to be very low in freshwater or estuary environments that are likely to contain chloride, phosphate, and carbonate ions (Boyes et al., 2017). However, contact with organisms and cells may increase the release of silver ions at close proximity to biological membranes, increasing the exposure to toxic Ag ions and thus hazard of nanosized Ag (Bondarenko et al., 2013). Hence both abiotic and biotic factors influence the environmental risk level of soluble metal-based ENMs.

3.4.3 Chemical transformations

In complex media environmental transformations of ENMs have important effects on their transport and toxicity. Transformations are typically dependent on the chemical and biological conditions, so reaction rates and products vary throughout ENM life cycles. Due to extensive use of ZnO and Ag ENMs in consumer products, such as PCPs and textiles, the fate and chemical transformations of these ENMs in sewerage networks and WWTPs have been extensively studied (Brunetti et al., 2015; Kaegi et al., 2013). The studies have shown that both ZnO and Ag are mostly reduced to the respective sulfides while Ag also binds to cysteine and histidine (Brunetti et al., 2015). The generation of sparingly soluble Ag_2S has often been considered a natural antidote to the toxicity of Ag ENMs (Levard et al., 2013). However, recent reports indicate that nanosized Ag_2S are toxic to soil microorganisms and were bioavailable for plants as indicated by the presence of Ag in plant roots (Schlich et al., 2018). Similarly, Li et al. (2015) reported that CuO sulfidation yielded CuS NPs that were more toxic to Japanese medaka than CuO NPs. However, in the latter study CuS was formed via a dissolution–precipitation mechanism instead of direct solid-state–shell mechanism, which is common in low redox conditions and at high S^{2-} concentrations (Keller et al., 2017). This outlines the importance of chemical transformation mechanisms in the environmental hazards of ENMs.

Chemical transformations of carbonaceous ENMs in the environment can be initiated by sunlight and increase the toxicity of these ENMs, for instance, by photoproduction of singlet oxygen from CNTs. Environmental phototransformation of ENMs occurs by both direct and indirect processes (Boyes et al., 2017). Direct processes include absorption of UV and visible light which, for example, can transform graphene oxide (GO) into significantly less photoreactive reduced GO. Indirect photolysis facilitated by NOM then contributes to the secondary phototransformation of reduced GO to polycyclic aromatic hydrocarbons and CO_2 (Hou et al., 2015).

3.4.4 Nanoparticle formation

Many ENMs (e.g., Ag, iron-oxide, CuO, and ZnO) are highly dynamic and may undergo redox reactions, dissolve, and interact with other ions and reprecipitate as new NMs (Wang et al., 2016b). As a result, solutions of ionic or bulk forms of a given metal may eventually produce particulate matter similar to ENMs over time. In environmental matrices, NOM has been reported as a contributing factor to ENM reformation from metal ions, due to various functional groups such as carbonyl, sulfhydryl, and phenolic and enolic hydroxyl groups, which all can act as reduction sites of metal ions (Wang et al., 2016b). Humic substances were found to promote the formation of nanosized iron oxide by inhibiting the development of hydroxide nuclei and reducing the aggregation of iron NPs (Pédrot et al., 2011). Also extra- and intracellular biomolecules synthesized by microorganisms, plants, and fungi can reduce metal ions and promote formation of nanosized metallic particles. These "green synthesis" reactions of ENMs have been reported for various different organism and metal combinations and the reduction mechanisms may vary with biological species and starting metal (Hebbalalu et al., 2013; Vaseghi et al., 2018). For Ag NP formation from silver ions in a solution of protozoan extracellular substances, exposure to light was essential (Juganson et al., 2013). Nam et al. (2008) reported the same for Ag NP synthesis in genetically engineered yeast cultures: spontaneous reduction of silver was facilitated by carboxylic acid groups of yeast surface-displayed glutamic and aspartic acid peptides in the presence of ambient light. In addition to natural systems, Ag NP formation also occurs as a result of anthropogenic activities, such as household laundering of Ag-containing textiles. Ag ions released from textiles were reduced to nanoparticulate metallic Ag, AgCl, and Ag_2S in the washing liquid (Mitrano et al., 2014b). In addition, the study showed that textiles treated with "conventional" silver had an equal or greater propensity to form nanosized Ag particles during washing than those treated with "nano"-silver. These findings and the fact that metal ENMs can be formed in the environment or generated during product use should be considered when assessing exposure and hazards of ENMs.

3.4.5 Sorption of biomolecules

Upon release into natural environments, ENMs will be exposed to complex mixtures of organic compounds originating from the decomposition of plant and animal residues, soluble microbial products, and extracellular polymeric substances (EPS). This mixture of organic compounds is often referred to as NOM. NOM is composed of nonhumic (polysaccharides, lignin, proteins, and polypeptides) and humic substances (high-molecular-weight microbial degradation products of plant material; Lin et al., 2017). Prevalent acidic functional groups in NOM, that is, carboxyl and phenolic groups, render NOM negatively charged in soils and natural waters. Due to the high adsorption affinity of NOM for mineral surfaces and $\pi-\pi$ stacking on hydrophobic surfaces, NOM readily sorbs to ENMs, acting as a dispersing and stabilizing agent of ENMs in the aqueous phase (Chen et al., 2010). NOM coating plays a crucial role in ENM environmental fate and toxicity because of increased dispersibility and alternation in the polarity of ENMs.

NOM, surfactants, and polymers have been shown to stabilize Cu-based ENM dispersions via electrostatic and steric influences (Adeleye et al., 2014). EPS secreted by marine phytoplankton decreased the aggregation and sedimentation of copper-based ENMs (Adeleye et al., 2014); however, bovine serum albumin (BSA) provided the highest steric stabilization of nanosized CuO compared to alginate and activated sludge EPS (Miao et al., 2015). The efficiency of BSA in stabilizing ENMs can be attributed to the relatively low molecular weight and globular shape of the protein that enables it to adsorb as a thin layer on the surface of CuO (Miao et al., 2015). Sorption of NOM to the surface of CNTs, graphene, and other hydrophobic ENMs is driven by hydrophobic interactions. MWCNTs were readily dispersed in the presence of NOM and were shown to be stable as an aqueous dispersion for over 1 month (Hyung et al., 2007). The favorable properties of NOM and its components have been employed for preparing homogeneous aqueous dispersions of intrinsically hydrophobic ENMs to uniformly maximize ENM bioavailability during mechanistic nanotoxicity studies (Chang et al., 2015). For example, Wang et al. (2018b) used alginic acid for aqueous dispersion of three types of CNTs, two types of graphene nanoplatelets, boron nitride nanotubes, and hexagonal boron nitride flakes. The developed dispersion protocol was effective for several different ENMs and could, if applied across studies, improve the reproducibility of the toxicity test results and facilitate data comparisons across studies.

3.4.6 Interactions with other contaminants

In addition to NOM and natural minerals, environmental matrices contain metals and synthetic chemicals of anthropogenic origin. Toxic metals and synthetic chemicals, regarded as environmental contaminants, may also interact or sorb to ENMs released into the environment. Interaction with ENMs may either exacerbate or mitigate harmful effects of other contaminants (Deng et al., 2017) or induce no changes in cocontaminant toxicity or bioaccumulation (Baun et al., 2008). Owing to their large specific surface area and potential to act as vectors for other contaminants, the role of ENMs in modulating the toxicity of cocontaminants and vice versa is an important aspect in assessing ENM hazards.

The adsorption of other contaminants by ENMs profoundly affects their toxicity and fate. The concentration of the contaminant on ENM surface may be orders of magnitude higher than in the environment and, if ENMs are taken up by biota, the ENMs act as carriers of contaminants into the organisms. If the ENM-adsorbed contaminant is released from the ENM inside the organism, ENMs contribute to increased bioavailability and toxicity of the contaminant. For example, nanosized TiO_2 was shown to enhance arsenic (As) accumulation and methylation in freshwater algae by acting as a carrier for As (Luo et al., 2018). On the other hand, if ENMs are not taken up by biota, or the adsorbed contaminant is not desorbed from the ENMs once inside the organism, ENMs may mitigate the toxicity of cocontaminants. TiO_2 ENMs have been shown to prevent the phytotoxicity of tetracycline, a bacteriostatic wide-spectrum antibiotic, by sorption of the antibiotic on ENMs (Ma et al., 2017). Other mechanisms by which ENMs may affect cocontaminant fate and toxicity include facilitating metabolism to more reactive metabolites or, on the contrary, promoting degradation of contaminants, impacts to the structure and function of cell membranes, making cells more susceptible to contaminants or competition with contaminants for membrane receptor binding sites, and decreasing the uptake of contaminants (Deng et al., 2017). ENM-adsorbed contaminants can affect the fate of ENMs by changing the surface charge or polarity of the ENMs and thus either increase or decrease ENM uptake and interactions with biota (Ma et al., 2017).

In aqueous solutions, carbonaceous ENMs can readily adsorb organic compounds by hydrophobic or electrostatic interactions. Sanchís et al. (2016) investigated the effects of binary mixtures of fullerene soot and organic cocontaminants malathion, glyphosate, diuron, triclosan, and nonylphenol using a standard toxicity assay—*D. magna* immobilization. Their results indicated the following trends: (1) the most hydrophilic organic compounds were less likely immobilized by carbonaceous ENMs and thus ENMs did not contribute to their transfer to the organisms, (2) the most nonpolar compounds with aromatic rings (e.g., diuron) strongly bound to carbonaceous ENMs and consequently the bioavailability of the organic compound was reduced, (3) the organic compounds with intermediate polarity sorbed to carbonaceous ENMs, which then entered the organisms and released organic compounds inside the organism—like a Trojan horse—resulting in harmful effects and bioaccumulation (Sanchís et al., 2016). Similarly, the effects of GO on the activity of antibiotics were shown to be dependent on the adsorption and desorption capacities of antibiotics to GO (Gao et al., 2017). The higher the number of aromatic rings and functional groups in the antibiotic structure, the less likely they were to desorb from the GO surface and the lower their antibacterial activity. On the other hand, if the attachment of antibiotic molecules to GO was strong enough to be carried with GO to the vicinity of the bacterial membrane and then released, GO increased the antibacterial activity of the compound (Gao et al., 2017). The latter approaches of using ENMs as delivery vehicles for conventional antibiotics are actively studied for combating antibiotic-resistant bacteria (Allahverdiyev et al., 2011). Another area of application of ENM adsorptive properties is contaminant removal from water and wastewater, including antibiotic removal from sewage effluent and contaminated waters (Ahmed et al., 2015). MWCNTs especially have been frequently studied as adsorbents of heavy metals (e.g., Zn^{2+}, Cu^{2+}, Co^{2+}, Cd^{2+}, Pb^{2+}, and Fe^{3+}) and organic compounds (e.g., polycyclic aromatic hydrocarbons or PAHs and dyes; Boncel et al., 2015). In general a higher sorption capacity of metals is achieved with negatively charged (oxidized) MWCNTs, whereas organic compounds bind to CNT surfaces via electrostatic interactions, van der Waals interactions, and $\pi-\pi$ stacking, and usually have orders of magnitudes higher sorption capacity on CNTs than metals (Boncel et al., 2015).

3.4.7 Transformations at the biological receptors and uptake by biota

Multiple studies have established the role of organisms and their secreted biomolecules in modifying the fate and effects of ENMs. Depending on the physicochemical properties of the ENMs and the test organisms, ENM interactions with cells have been shown to either enhance the dissolution, and hence, toxicity of the ENMs (e.g., Ag NP adsorption to bacterial membranes; Bondarenko et al., 2013), or reduce the dissolution of ENMs (e.g., ZnO NP adsorption to unicellular algae; Chen et al., 2012). Further, bacteria have been reported to promote dispersion of TiO_2 NP agglomerates by ENM adsorption to cells, while similar effects were not achieved using bacterial EPS (Horst et al., 2010), suggesting the importance of cell-ENM contact.

With the possible exception of very small QDs, ENMs in particulate form cannot be internalized in cells and organisms with intact cell walls (e.g., some unicellular algae) due to the rigidity of the structure and the diameter of the cell wall pore of 50 nm (Kettler et al., 2014). However, soluble metal ENMs can shed metal ions in the close vicinity of the cell, resulting in the uptake of the soluble fraction of the ENMs. For example, silver accumulation in Ag NP-exposed algae was demonstrated even though Ag NPs did not pass the cell wall (Leclerc and Wilkinson, 2014). Intracellularly, Ag^+ may be reduced to metallic Ag which may aggregate into the particulate form (Leclerc and Wilkinson, 2014), indicating that biological reduction reactions may play important roles in ENM environmental fates.

In more complex systems such as planted soil mesocosms, complicated relationships between plants, soil microbes, and soil-amended ENMs were revealed (Ge et al., 2014). Compared to the effects observed in unplanted soils, CeO_2 NPs had more pronounced effects on soil bacterial communities in soybean-planted soils, whereas soybean plants seemed to suppress the effects of ZnO NPs to soil bacterial communities. Thus plants—likely through their effects on belowground biogeochemistry—could either promote or limit ENM effects on soil microorganisms, depending on the physicochemical properties of the ENMs. From soil, both ZnO and CeO_2 NPs were shown to be taken up by soybeans and transferred into pods (Hernandez-Viezcas et al., 2013). However, chemical forms of each ENM found inside the plant were different: Zn was detected as complexed with biomolecules, likely citrate, while Ce was stored in the pods in the form of CeO_2 NPs. A small percentage of cerium as Ce^{4+}, the oxidation state of Ce in CeO_2 NPs, was biotransformed to Ce^{3+} (Hernandez-Viezcas et al., 2013). A recent study of CeO_2 NP transformation mechanisms and uptake in barley roots showed that the reduction of CeO_2 NPs to Ce^{3+} in the root−soil interface was necessary for plant uptake of Ce. However, Ce^{3+} was likely rapidly reoxidized to Ce^{4+} inside the roots because only limited amounts of Ce^{3+} were detected in the root tissue (Rico et al., 2018). The study provided new insights on the role of plant root exudates and rhizosphere in modulating the uptake and fate of ENMs.

In filter feeders (e.g., crustaceans, bivalves, and protozoa) and other aquatic organisms (e.g., fish) that can ingest particulate matter, including ENMs, uptake and depuration rates of ENMs have been shown to depend on the availability of the food source. However, there appears to be no clear trend. For example, higher amounts of Au NPs accumulated in mussels upon direct aqueous exposure compared to feeding on algae that were incubated with Au NPs (Larguinho et al., 2014), and TiO_2 NPs in zebrafish compared to feeding on TiO_2 NP-exposed daphnids (Zhu et al., 2010). However, protozoa accumulated similar masses of TiO_2 NPs by direct exposure in the medium and via feeding on TiO_2 NP-encrusted bacteria (Mielke et al., 2013). Also, marine mussels accumulated CeO_2 NPs in equal amounts, regardless of whether the ENMs were associated with phytoplankton or as free particles in the water column (Conway et al., 2014). On the other hand, freshwater snails accumulated higher amounts of CuO NPs via dietary intake compared to waterborne exposure (Croteau et al., 2014).

Once ingested, the translocation of ENMs across the intestinal epithelium or food vacuole membranes into organisms' tissues or cell cytoplasm, respectively, is generally limited (Heinlaan et al., 2011; Khan et al., 2014; Mortimer et al., 2016b), but ENMs may become trapped in the digestive tract and not depurated even after feeding by the organisms (Chen et al., 2014; Petersen et al., 2009). In these cases, ENMs could still be considered as being accumulated in biota (Bour et al., 2015). It has been reported, however, that ENM excretion from the digestive tract is facilitated by food intake. ENM depuration from the gut of daphnids was faster when the organisms had access to food, whereas limited or no access to food decreased particle excretion (Khan et al., 2014; Petersen et al., 2011). A few studies have reported ENM translocation to tissues and circulatory system. For example, a 2-week exposure of zebrafish to MWCNTs resulted in uptake and retention of MWCNTs and translocation in the fish blood and muscles, indicating the potential for ENM transfer in the food chain (Maes et al., 2014).

3.4.8 Trophic transfer

ENM interactions with biota can include either adsorption on, or uptake in, cells and organisms as discussed earlier. Both types of interaction encompass an environmentally relevant pathway for ENM trophic transfer. For example, it was shown that regardless of whether ENMs were attached to bacterial cell envelopes, in case of TiO_2 (Mielke et al., 2013) or MWCNTs (Mortimer et al., 2016b), or internalized, in case of Cd QDs (Werlin et al., 2011), ENMs were taken up by bacteria-grazing protozoa and thus were transferred to the next trophic level. ENM transfer in the food chain has been widely reported in both terrestrial (De la Torre Roche et al., 2015; Gardea-Torresdey et al., 2014; Servin et al., 2017) and aquatic environments (Mielke et al., 2013; Mortimer et al., 2016b; Tangaa et al., 2016; Werlin et al., 2011). During the transfer, ENMs may be subject to transformations such as agglomeration or dissolution in the organism's digestive tract, that is, gut or food vacuoles (Tangaa et al., 2016). Such physicochemical changes alter the environmental fate of ENMs, if depurated by the organism. In case of uptake by higher trophic level organisms the agglomerated ENMs may disagglomerate or be subject to enzymatic degradation, depending on the physiology of the organism.

Other scenarios for trophic transfer of ENMs include ENM transport in terrestrial systems upon addition of biosolids to agricultural soils or via the application of nanoenabled agricultural products (Gardea-Torresdey et al., 2014; Kah et al., 2018). Contamination of agricultural soils with ENMs creates challenges for crop production and compromises food safety. As discussed earlier, crop plants have been shown to take up ENMs from soils and transport ENMs or their transformation products into aboveground plant tissues, including fruits. To date, a few studies have reported ENM transfer to secondary consumers that feed on primary consumers exposed to ENMs via plants. Hawthorne et al. (2014) demonstrated CeO_2 NP uptake from soil into zucchini leaves, accumulation of Ce in crickets that consumed the leaves, and transfer of Ce into wolf spiders (*Lycosidae*) from the crickets that served as prey. Another study which also employed a three-step terrestrial food chain showed transfer of La_2O_3 NPs from lettuce leaves to crickets and then to mantises (De la Torre Roche et al., 2015). These results demonstrate that ENMs can be transferred in the terrestrial food chain, and thus there is a likelihood of human exposure through dietary uptake. Such warrants careful risk assessment of any potential ENM-enabled agricultural product and of field application of biosolids that may contain ENMs.

3.4.9 Degradation

ENMs released into the environment are subject to various chemical (e.g., photodegradation) and biological processes (biodegradation) that may result in either partial or complete degradation of ENMs. The degradation rate and completeness depend on the composition and structure of ENMs: metal-based ENMs can, at most, completely dissolve, however, the ionic metal will persist and be subject to complexation and chemical or biological oxidation–reduction reactions, determining the fate and toxicity of the contaminant. In oxygenated waters, metallic ENMs are likely to be oxidized. For example, oxidation of Ag^0 releases Ag^+ which is considered antibacterial and toxic to aquatic organisms (Priester et al., 2014). Ag^+ can react with Cl^- to form AgCl resulting in precipitation of silver. Oxidation of Cu ENMs releases Cu^+ ions that in turn are oxidized into Cu^{2+} and usually rapidly complexed as carbonates or hydroxides: $CuCO_3$, $Cu(CO_3)_2^{2-}$, $Cu(OH)_2$, $Cu(OH)^+$, and $Cu_2(OH)_2^{2+}$ (Keller et al., 2017). These secondary compounds and ions are not the end-products of metal ENM transformation. The process can be reverted, or the formed secondary ions can be further oxidized or complexed with NOM (Das et al., 2017). The oxidation–reduction reaction rates in environmental waters are highly dependent on the oxygen content, but also other factors such as the influence of light which can cause biochemical (e.g., by photosynthesis) or photochemical production of metal species independent of the water oxygen levels (Luoma and Rainbow, 2008).

Complete degradation is plausible in case of organic ENMs, that is, carbonaceous ENMs (Chen et al., 2017b) and ENMs synthesized from biodegradable block copolymers (Zweers et al., 2004), and organic coatings of metal ENMs (Jacobs et al., 2016). While the degradation of organic ENMs in certain applications (e.g., drug carriers) is desired and designed to be a controlled process (Morachis et al., 2012), ENMs released in the environment may undergo uncontrolled degradation, resulting in production of possibly more toxic byproducts (Boyes et al., 2017). Examples include environmental photodegradation of GO (Hou et al., 2015) and cometabolism of MWCNTs by bacterial communities (Chen et al., 2017b; Zhang et al., 2013) which have been shown to produce polycyclic aromatic hydrocarbons as intermediate products in addition to the final degradation product CO_2. Almost complete degradation of certain types of MWCNTs and SWCNTs has been reported using horseradish peroxidase and hydrogen peroxide (H_2O_2) over 80 days (Zhao et al., 2011). However, others have reported much lower degradation levels (below 10% of original ENM mass) with other types of MWCNTs (Zhang et al., 2013). This suggests that CNT degradation rates and efficiencies are strongly dependent on the number of defects and functional groups in CNT structures. Photofragmentation of GO was shown to be rapid, producing reduced photoproducts similar to reduced GO nanomaterials that were significantly smaller in size as compared to the parent GO (Hou et al., 2015). However, once formed, these reduced photoproducts were much more resistant to further phototransformation than GO (Hou et al., 2015). Low efficacy of complete biodegradation of carbonaceous ENMs suggests that accumulation of potentially persistent and toxic intermediate products in the environmental compartments should be considered as a realistic scenario in ENM risk assessment.

3.4.10 Aging

ENM aging, that is, ENM physicochemical transformation through the ENM life cycle, is an important factor contributing to ENM fate and toxicity. However, the concept of aging has only recently gained more attention in ENM risk assessment and toxicity studies (Mitrano et al., 2015). Understanding the impact of aging-related transformations over time may help forecast the benefits and risks associated with the use of products containing ENMs. ENM aging through a product life cycle and after it is released includes several interconnected steps of all the earlier described transformations such as degradation, chemical modifications, biotransformation, sorption of natural and synthetic compounds, dissolution, agglomeration, and others (Mitrano and Nowack, 2017).

The types and extents of transformations that an ENM or ENM-enabled product may undergo are ultimately determined by the specific application and the ENM physicochemical properties. In hazard assessment, it is important to consider how fast aging occurs, relative to the study periods. For example, nanosized

ZnO can dissolve quickly in environmental waters, including the soil solution (Reddy Pullagurala et al., 2018). If plants grow slowly in soils relative to ENMs dissolving in the soil solution then administering as-produced ENMs may be environmentally relevant. For ENMs that are embedded in composites with polymers, conducting hazard assessments using pristine ENMs may still be environmentally relevant if ENMs are released from composites (Holden et al., 2014a). On the other hand, ENMs that are relatively stable may exert similar toxicities over prolonged exposure, while other ENMs may acquire a mitigating cap, as in Ag ENMs becoming sulfidated (Levard et al., 2013).

However, if aging happens at a longer time scale, the process may result in ENMs that have significantly different reactivity and physicochemical properties than as-produced ENMs of the same material. For instance, aged nanosized Cu exhibited unique chemistry including oxide phases that formed and different surface adsorption properties compared to new Cu and CuO ENMs (Mudunkotuwa et al., 2012). The aging of $Al(OH)_3$-coated nano-TiO_2, a component of sunscreens, in simulated swimming pool water was shown to redistribute the coating and reduce its protective properties, thereby increasing reactivity and potential phototoxicity of ENMs (Al-Abed et al., 2016).

Since major applications of ENMs are as functional fillers embedded in a solid matrix such as plastics, tires, and coatings, it is important to understand the extent of degradation of the solid matrix and release of ENMs and fragments during the use, so that ENM fates and hazards can be assessed more realistically. ENM release from composite materials is considered a rather uncontrollable source of emission into the environment, especially by aging, and degradation, in the long run (Nowack et al., 2013). Recently, ENM release rates from 27 materials of different combinations of ENMs and matrices were assessed after weathering that was conducted using a prevalidated method (Wohlleben and Neubauer, 2016). The results indicated that the release rates of ENMs were primarily determined by the matrix: polyethylene resulted in lowest release rates and epoxy in highest release rates, with polyamide and polyurethane at intermediate rates. The release rate from a given matrix was modulated by less than an order of magnitude by embedded ENMs (Wohlleben and Neubauer, 2016), indicating that fate and transport assessment of ENM-enabled products can in some cases be extrapolated from the knowledge on similar products that do not contain ENMs.

The environmental hazards of aged ENMs can be assessed either by performing long-term studies with ENMs, possibly including repeated ENM applications, so that time-dependent transformation under realistic conditions could occur (Holden et al., 2016). Alternatively, preaged ENMs could be used. However, only a few aging protocols have been developed and the first standardized method for aged ENMs was published only recently (Lankone et al., 2017). In addition, the appropriate aging protocol would depend on the envisioned exposure scenario (Holden et al., 2016) as recently reviewed and proposed (Nowack and Mitrano, 2018).

3.5 Effects: how do engineered nanomaterials affect biota and ecosystems?

An ecosystem is defined as "all living organisms together with the physical and chemical components of their environment" (Madigan et al., 2015). The processes that occur within ecosystems deliver "services" upon which humans depend, including pollination, food production, climate regulation, and nutrient cycling (Costanza et al., 1997). When ecosystem services are disrupted, for example if water supplies dwindle due to political strife or drought, human suffering can result (Butler and Oluoch-Kosura, 2006). The concept of "Planetary Health" recognizes that there are complex interconnections within the environment that affect human survival (Seltenrich, 2018). An example of how Planetary Health can be disrupted lies at the nexus of climate change, fisheries, and human health: where changed ocean conditions no longer support local fishery stocks, indigenous people are forced to use other, perhaps less nutritious, food sources—with possible long-term human health effects (Seltenrich, 2018).

Pollution is a major cause of noncommunicable diseases that reduces human lifespans (Landrigan et al., 2018), but pollution is also a threat to Planetary Health (Seltenrich, 2018). As such, how and to what extent contaminants disrupt ecosystem services' provisions is important to understand (Rohr et al., 2017). The relationships between pollutants, toxicity, ecosystem services, and human health are complex. Yet, modeling frameworks are emerging that relate specific CEC effects and their mechanisms to exposed biota and the ecosystem services that they deliver (Forbes et al., 2017). Such modeling frameworks are envisioned to assist in future environmental risk assessments for ENMs (Holden et al., 2013).

ENM toxicity to environmental biota is an important and well-studied concern that may drive broader ecological effects of ENMs. Regardless, ecological implications of ENMs merit study because of potential relationships to ongoing delivery of ecosystem services (Holden et al., 2013). In this section, potential effects of ENMs on biota and ecosystems are discussed for the aquatic environment with food webs, and the terrestrial environment with agricultural crops. Both environments have been studied for potential ecological effects of ENMs. The discussions apply to managed ecosystems as well as to protected areas that are also essential resources for ecosystem services provisioning (Terraube et al., 2017).

3.5.1 Aquatic environments and food webs

Aquatic environments provide ecosystem services including supporting biodiversity, food production, nutrient cycling, waste treatment, climate regulation, and recreation (Costanza et al., 1997). Yet, in marine environments, human impacts, including pollution, seriously impinge on the delivery of such ecosystem services (Crain et al., 2009). A question in ENM risk assessment has regarded if ENMs, in that they are CECs, threaten aquatic populations, communities, and ecosystems, possibly through food web interactions, such that ecosystem services are negatively impacted.

In general, aquatic invertebrates—which are typically at the base of food webs—are susceptible to Ag^+ ions released from Ag ENMs; thus toxic effects relate to Ag ENM solubility under the conditions of aquatic exposure (Schaumann et al., 2015). For photocatalytic TiO_2 ENMs, toxicity in the aquatic environment is highly related to illumination, that is, UV light that can excite the ENMs, leading to free-radical formation and direct toxicity. Toxicity is highly affected by heteroagglomeration and thus ENM bioavailability, but also coatings that are acquired in the environment that may mitigate or enhance photocatalytic activity (Schaumann et al., 2015). Interactions with biofilm EPS and species differences, in addition to ENM composition and physicochemical characteristics, can profoundly affect—besides ENM dose—measured toxicities of ENMs to aquatic microorganisms (Schaumann et al., 2015). Ions of metallic ENMs, including many types of Cu-based ENMs (Keller et al., 2017), or the NPs themselves, can react with oxygen to form highly destructive ROS: hydrogen peroxide (H_2O_2), or the superoxide ($\cdot O_2^-$) or hydroxyl ($\cdot OH$) radicals. ROS-mediated cell damage is observed under laboratory conditions for aquatic invertebrates (Schaumann et al., 2015) and microorganisms (Ge et al., 2016a) exposed to metallic and metal oxides ENMs.

In an overview of aquatic biotic hazards associated with carbonaceous ENMs such as MWCNTs, SWCNTs, fullerenes, and graphene, it was observed that exposure concentrations were typically high in laboratory assessments relative to actual field predictions and that toxicity across freshwater organisms (algae, bacteria, crustaceans, and fish) occurred with mg/L, as opposed to field predictions of ng/L, exposure levels (Freixa et al., 2018). The use of high-exposure concentrations in hazard assessment has generally been true for most ENM ecotoxicity studies (Holden et al., 2014a), although increasingly it is recognized that studies should and can be performed using exposure concentrations that more closely mimic those predicted to occur under specific scenarios in the environment (Holden et al., 2016). At the low carbonaceous ENM concentrations expected to occur in the aquatic environment, longer term and sublethal effects may be indicated, as well as interactive effects with other pollutants (Freixa et al., 2018). The latter was demonstrated when coadministering a metal oxide ENM, nano-TiO_2, with 2,3,7,8-tetrachlorodibenzo-*p*-dioxins (TCDD) to the marine bivalve (mussel, *Mytilus galloprovincialis*): nano-TiO_2 became an efficient carrier of sorbed TCDD and allowed for enhanced TCDD toxicity as a result (Canesi et al., 2014). Thus ENMs may be carriers of other pollutants and exert toxicity as such, rather than acting as direct toxicants.

Many ENM toxicity studies have been performed in laboratory batch (i.e., bottle) systems, without the complexity of realistic aquatic environmental conditions including complex water chemistry and the influences of other organisms. Yet, as suggested earlier, factors such as sunlight and NOM can profoundly affect ENM integrity and released solute speciation, which in turn affect bioavailability with differing "triggers" of toxic effects (Schaumann et al., 2015). Mesocosms, that is, larger experimental systems that recreate many environmental complexities—including multiple trophic levels of biota—allow for observing endpoints that occur from the integration of all cooccurring processes (Holden et al., 2016). A seminal example of mesocosm experimentation of ENM aquatic distribution concerned a model estuarine system with added nano-Au which was found to become highly concentrated within benthic biofilms, yet also distributed throughout food webs (Ferry et al., 2009). The value of examining the full spectrum of biotic and abiotic processes and outcomes via realistic mesocosm studies was also demonstrated by Colman et al. (2014) in a model aquatic wetland experiment to which various forms of Ag, including nano-Ag, were administered. As indicated by laboratory studies, the dissolution of Ag salts or nano-Ag released toxic Ag^+ ions. However, in demonstration of how ENMs can affect specific ecosystem services (here, climate and gas regulation), it was observed that Ag toxicity to wetland plants caused hyperrelease of DOC that fueled excessive methanogenesis (Colman et al., 2014). As such, the potential for ENMs in aquatic systems to alter important ecosystem services was shown.

The transformation of ENMs in complex aquatic food webs may affect hazard levels and can also be assessed in mesocosm conditions. Nano-CeO_2 is an established toxicant in laboratory studies with microorganisms and aquatic species, but at doses often exceeding those predicted to occur in the environment (Collin et al., 2014). In one study, nano-CeO_2 was administered at a realistically low concentration (1 mg/L), as either bare ENMs or ENMs with a citrate cap, to freshwater mesocosms containing phytoplankton (algae) and benthic snails (Tella et al., 2014). The ceria NPs agglomerated and settled, becoming more bioavailable to snails via sediments. Uptake occurred into snails with both types of CeO_2 ENMs, and transformation was followed using X-ray synchrotron approaches to trace Ce (IV) into the Ce(III) oxidation state. Known as a "prooxidant," the evidence for oxidative stress in snails (as per lipid peroxidation data) was more pronounced for uncoated versus coated ENMs (Tella et al., 2014). This demonstrates that mesocosm experiments can reveal ENM design characteristics that may be more ecologically protective.

As suggested earlier, ENM bioavailability is a strong determinant, as is exposure dose and time, for ecologically relevant effects to manifest in aquatic systems. CuO ENMs that were isotopically labeled were effective tools to trace particulate ENMs into freshwater snails, showing that NP uptake was as efficient a Cu delivery system into the organisms as was Cu-based salts as ionic controls (Croteau et al., 2014). With TiO_2 ENMs associated with various freshwater algal biofilms, the degree of association affected ENM translocation into herbivorous predatory snails, but with no apparent impact to the snails (Kulacki et al., 2012). In a simplified microbial food web using bacteria and protozoa, accumulation of nano-TiO_2 inside protozoan food vacuoles slowed population growth ever so slightly, owing to digestive impairments (Mielke et al., 2013). There is a need for more study of such effects that, rather than being indicative of overt toxicity, would logically affect population dynamics in predator–prey interactions plus poise higher trophic levels in food webs for dietary exposure to bioaccumulated ENMs in prey.

3.5.2 Terrestrial environments with agricultural crops

In agriculture, based on many experimental observations, there is the potential for uptake of various types of ENMs into food crops, with subsequent ENM translocation and modification in plant tissues, damage to plants, and alterations to food quality (Ma et al., 2018). However, at low concentrations of ENMs composed of either Ag, Zn, or Ti, the addition of ENMs to soil via biosolids seemed to mitigate effects of ENMs to either the plant (*Medicago truncatula*) or soil microbial communities (Chen et al., 2017a). In the case of nano-Ag, its conversion in biosolids to sulfidated Ag (AgS_2), which is very stable, likely prevents its uptake into crops, even over long time frames (Wang et al., 2018a). Yet for other ENMs, such as ZnO, sewage sludge contents may be harmful to terrestrial receptors such as earthworms, and more so relative to ionic forms of Zn that are typically regulated in land-applied sludges (Lahive et al., 2017).

For ENMs that do enter plants, there are well-documented effects that depend on plant species, ENM type and dose, and exposure conditions including whether via hydroponic growth or soil cultivation (Du et al., 2017; Gardea-Torresdey et al., 2014; Mukherjee et al., 2016; Reddy et al., 2016; Zuverza-Mena et al., 2017). For example, nano-CeO_2 can enter soybean (*Glycine max*) from soil via the roots and translocate and transform throughout the plant (Hernandez-Viezcas et al., 2013). Although the exact implications to the crop as a food source are not fully understood, such uptake and transformation of nano-CeO_2 changes the elemental composition and thus the nutrition of soybean plants (Peralta-Videa et al., 2014). Yet ENMs applied to soil, which is a most ecologically relevant exposure scenario, have the potential to impact not only plants, but also microbial communities in bulk and rhizosphere soils, microbial symbionts of plants, and plant–microbe interactions that occur across the above- and belowground compartments in agriculture (Dimkpa, 2014).

When applied to unplanted soils, there are differential effects of various ENMs to soil microbial communities, which are known catalysts of the high-value ecosystem service of nutrient cycling (Costanza et al., 1997). For example, a seminal study of fullerenes showed virtually no effects to microbial communities in agricultural soils (Tong et al., 2007). A similar result was found for MWCNT, graphene, and CB over a long-term exposure in relatively dry grassland soils (Ge et al., 2016b). In both cases the bioavailability of carbonaceous ENMs could be low, owing possibly to interactions of the ENMs with soil organic matter, as suggested before (Tong et al., 2007). For metal oxides of ZnO and TiO_2, soil bacterial community diversity shifted and biomass declined with ENM dose and time (Ge et al., 2011), and the abundances of taxa known to contribute to N and C cycling were altered (Ge et al., 2012), likely owing—at least in the case of nano-TiO_2—to direct toxicity effects (Ge et al., 2013). Yet CeO_2 NPs added to agricultural soils in the absence of plants had insignificant effects on soil microbial communities; it was only under planted (with soybean) conditions that microbial communities shifted (Ge et al., 2014). The opposite was observed for nano-ZnO: effects to agricultural microbial communities were relatively great in unplanted soils but appeared mitigated in the presence of soybean (Ge et al., 2014). Such results point to the importance of plant–microbe interactions in possibly governing ENM effects to agricultural ecosystems, including emissions of plant root exudates that affect ENM bioavailability in the rhizosphere.

As recently critically reviewed (Holden et al., 2018), several studies, including with carbonaceous (Wang et al., 2017), metal (Judy et al., 2015), or metal oxide (Priester et al., 2012) ENMs applied to soil, point to root nodules in leguminous plants as particularly sensitive to ENMs. The reasons for such sensitivity may vary with ENM type yet are poorly understood. One explanation specific to nano-CeO_2 was that soybean plants that had translocated NPs into leaves (that relatedly suffered oxidative stress) were hoarding photosynthate aboveground to mitigate toxicity; the consequence was that root nodules were relatively starved (Priester et al., 2017). A mechanistic model of the interplay between host and symbiont with regards to damage and resource allocation supported this interpretation (Klanjšček et al., 2017). Still, observations of ENMs inside root nodules could also indicate that direct toxicity to bacteroids from ENM exposure can occur (Priester et al., 2012; Wang et al., 2017). If it were determined that N_2-fixing bacteria in soil were susceptible to ENM toxicity outside of symbioses then such inocula could be meaningful targets for screening toxicity of ENMs to ecologically relevant receptors (Holden et al., 2014b).

Taken together, many scientific reports suggest that ENMs in agriculture can be bioavailable to plants and soil microbes. In nanoagricultural applications such as fertilizers or pesticides, this may be advantageous. In the context of waste disposal or incidental soil exposure to ENMs, uncontrolled long-term addition could negatively alter agricultural ecosystem performance. Specific reports of potential impacts to plant root symbioses that fix dinitrogen from the atmosphere suggest broader implications to changed synthetic fertilizer needs and thus greenhouse gas emissions. More research would be needed to vet such possibilities in a more thorough manner needed for risk assessment.

3.6 Summary: key concepts and points

Recent literature discussed in this chapter indicates that the scientific knowledge on fate of ENMs in natural environments and impacts on ecosystems has grown steadily over the past decade. Scientists have a reasonable understanding of ENM sources, transport between environmental compartments, and fate and transformations that can be predicted based on current knowledge on ENM physicochemical properties and environmental conditions. Models and experimental data indicate that the major sinks of ENMs are expected to be soils and sediments due to the tendency of ENMs to agglomerate in environmental matrices. Smaller fractions of released ENMs are suspended in water bodies and dispersed in atmosphere. Regardless of environmental compartment, ENMs are subject to physicochemical transformations induced by photoreactions, chemical or biological redox reactions and physicochemical interactions. As a result, ENM mobility, bioavailability, and toxicity may be altered, for example, via acquiring protective coating such as Ag ENMs becoming sulfidated, or via formation of degradation products (e.g., dissolved Ag, Zn, and Cu ions, or PAHs), which may exert increased bioavailability and toxicity than the as-synthesized ENMs. ENMs may be transferred between environmental compartments also via uptake in biota and subsequent trophic transfer in the food chain. For instance, ENMs deposited in sediments may be taken up by filter feeders and transferred in aquatic systems to higher trophic levels by consumers who feed on filter feeders.

The focus of current research in the field of ENM fate and transport in natural environments is on addressing the challenges with the sensitivity and selectivity of ENM measurement methods and uncertainties in modeling approaches. Recent major advances in measurement techniques of ENMs in environmental matrices include improved extraction methods to separate ENMs from natural and incidental NPs and other matrix components, and development of improved spICP-MS methods for detection and quantification of ENMs at low concentrations as predicted to occur in environmental samples. Modeling efforts have recently been directed toward including the dynamic aspect of ENM life cycle that would consider the rapid development of ENM production and some ENMs entering an in use stock from where they are released with a lag phase. New regulations and rules, established by the regulatory agencies in several European countries and in the United States, that require the producers to report ENM production volumes, allow to make more accurate predictions using already developed models. Another trend in modeling is the development of environmental fate models that, unlike mass flow models, would consider the dynamics of physical and chemical reactions of ENMs in environmental compartments. Experimental studies have established how ENM transformations are affected by ENM types and ambient conditions. Recent efforts have been focused on assessing these physicochemical transformations over time, to understand how ENMs and ENM-enabled products age and how aging affects ENM fate, transport, and effects to biota. Advances in this field are evidenced by the recent emergence of the first standardized method for aged ENMs.

Many studies have regarded the potential toxicity of ENMs to various receptors, using exposure conditions that range from idealized to as environmentally realistic as possible. Studying high exposure concentrations of ENMs has revealed the potential for ROS-mediated toxicity across a broad range of ENM-receptor combinations. While not the focus of this chapter, a broad range of "omics" methods have been brought to bear on understanding mechanisms of ENM effects.

Results from such studies are invaluable when designing endpoints for measurement in complex laboratory studies including mesocosms. Ecological effects, as pertain to impairing ecosystem services delivery, are indicated from results of various aquatic and terrestrial mesocosm studies. Biota, including primary producers in aquatic environments and food webs they fuel, are susceptible to ENM toxicity, although bioavailability is modulated by aquatic chemistry and ENM propensities to agglomerate and settle. Where exposure occurs, ROS-mediated toxicity is frequently found, as is specific ion toxicity from dissolving ENMs. Similar mechanisms may apply in terrestrial agricultural ecosystems. Yet a current focus is also on food crop production and, given the observation that some ENMs interfere with symbiotic dinitrogen fixation, the broad ecological consequences of ENMs affecting fertilizer requirements and greenhouse gas emissions.

Highlights in brief

- ENM pathways for entering the environment depend on ENM applications and the life cycle of the products.
- The majority of ENMs are estimated to be disposed in landfills; among environmental compartments, the largest fraction of ENMs is deposited in soils (8%–28%), in water (0.4%–7%), and in the atmosphere (0.2%–1.5%).
- Major advances in ENM measurements in the environment include improved methods for extracting ENMs from complex environmental matrices and advancement of single-particle inductively coupled plasma mass spectrometry (spICP-MS).
- Major limitations of experimental methods are the lack of sufficient sensitivity and selectivity.
- Recent advances in ENM life-cycle material flow modeling include the development of dynamic models and improved estimations due to the availability of more accurate data on ENM production volumes because of mandatory registration requirements for ENMs.
- The sources of uncertainties in ENM life-cycle material flow models derive from insufficient data on ENM production rates and transfer coefficients, and the failure to include variables such as ENM release from non-nanosources and ENM transformations.
- The inclusion of more detail in fate and transport models, plus improved approaches for discerning low concentrations of ENMs in environmental matrices, forecasts a more routinely refined understanding of ENM environmental exposures in the future.
- ENM physicochemical transformations (e.g., agglomeration, dissolution, reduction and oxidation, sorption of biological and synthetic compounds, and degradation), induced by biotic and abiotic environmental factors, influence ENM fates in the natural environment (including uptake in biota and transfer in the food chain).

- Understanding ENM transformations over time (i.e., ENM aging) will assist forecasting the benefits and risks associated with the use of products containing ENMs.
- ENMs are emitted into aquatic environments where various biota including primary producers, predators, and higher trophic levels are exposed to, share, and affected by nanomaterials and their degradants. Reactive oxygen species (ROS) and shed ions from dissolving ENMs may explain toxicity, but compartmentalization (sediments vs water column) is crucial to understand such that realistic biotic receptors are evaluated.
- ENMs in agricultural ecosystems can affect plants, microbes, and their interactions, but studies of effects in isolation of each compartment risks misunderstanding how ENMs realistically move through and impact terrestrial biota. When plants and microbes are studied simultaneously in the context of ENM-contaminated soils, important lessons have been learned including the potential for ENMs to impair biotic dinitrogen fixation, which is a beneficial way to improve soil fertility and reduce the use of greenhouse gas-intensive synthetic fertilizers.

Acknowledgments

The authors acknowledge financial support from the US National Science Foundation and the US Environmental Protection Agency under Cooperative Agreement Number DBI-0830117. Any opinions, findings, and conclusions or recommendations expressed in this material are those of the author(s) and do not necessarily reflect the views of either the National Science Foundation or the Environmental Protection Agency. This work has not been subjected to Environmental Protection Agency review, and no official endorsement should be inferred. The authors thank Sage Davis for performing environmental scanning electron microscopy in the Micro-Environmental Imaging and Analysis Facility at University of California Santa Barbara (www.bren.ucsb.edu/facilities/MEIAF/).

References

Adeleye, A.S., Keller, A.A., 2014. Long-term colloidal stability and metal leaching of single wall carbon nanotubes: effect of temperature and extracellular polymeric substances. Water Res. 49, 236–250.

Adeleye, A.S., Conway, J.R., Perez, T., Rutten, P., Keller, A.A., 2014. Influence of extracellular polymeric substances on the long-term fate, dissolution, and speciation of copper-based nanoparticles. Environ. Sci. Technol. 48, 12561–12568.

Ahmed, M.B., Zhou, J.L., Ngo, H.H., Guo, W., 2015. Adsorptive removal of antibiotics from water and wastewater: progress and challenges. Sci. Total Environ. 532, 112–126.

Al-Abed, S.R., Virkutyte, J., Ortenzio, J.N.R., McCarrick, R.M., Degn, L.L., Zucker, R., et al., 2016. Environmental aging alters $Al(OH)_3$ coating of TiO_2 nanoparticles enhancing their photocatalytic and phototoxic activities. Environ. Sci. Nano 3, 593–601.

Allahverdiyev, A.M., Kon, K.V., Abamor, E.S., Bagirova, M., Rafailovich, M., 2011. Coping with antibiotic resistance: combining nanoparticles with antibiotics and other antimicrobial agents. Expert Rev. Anti Infect. Ther. 9, 1035–1052.

Baun, A., Sørensen, S.N., Rasmussen, R.F., Hartmann, N.B., Koch, C.B., 2008. Toxicity and bioaccumulation of xenobiotic organic compounds in the presence of aqueous suspensions of aggregates of nano-C60. Aquat. Toxicol. 86, 379–387.

Boncel, S., Kyzioł-Komosińska, J., Krzyżewska, I., Czupioł, J., 2015. Interactions of carbon nanotubes with aqueous/aquatic media containing organic/inorganic contaminants and selected organisms of aquatic ecosystems – a review. Chemosphere 136, 211–221.

Bondarenko, O., Ivask, A., Kaekinen, A., Kurvet, I., Kahru, A., 2013. Particle-cell contact enhances antibacterial activity of silver nanoparticles. PLoS One 8, 1–12.

Borovinskaya, O., Hattendorf, B., Tanner, M., Gschwind, S., Günther, D., 2013. A prototype of a new inductively coupled plasma time-of-flight mass spectrometer providing temporally resolved, multi-element detection of short signals generated by single particles and droplets. J. Anal. At. Spectrom. 28, 226–233.

Bour, A., Mouchet, F., Silvestre, J., Gauthier, L., Pinelli, E., 2015. Environmentally relevant approaches to assess nanoparticles ecotoxicity: a review. J. Hazard. Mater. 283, 764–777.

Boyes, W.K., Thornton, B.L.M., Al-Abed, S.R., Andersen, C.P., Bouchard, D.C., Burgess, R.M., et al., 2017. A comprehensive framework for evaluating the environmental health and safety implications of engineered nanomaterials. Crit. Rev. Toxicol. 47, 771–814.

Brunetti, G., Donner, E., Laera, G., Sekine, R., Scheckel, K.G., Khaksar, M., et al., 2015. Fate of zinc and silver engineered nanoparticles in sewerage networks. Water Res. 77, 72–84.

Butler, C.D., Oluoch-Kosura, W., 2006. Linking future ecosystem services and future human well-being. Ecol. Soc. 11, 30.

Canesi, L., Frenzilli, G., Balbi, T., Bernardeschi, M., Ciacci, C., Corsolini, S., et al., 2014. Interactive effects of n-TiO_2 and 2,3,7,8-TCDD on the marine bivalve *Mytilus galloprovincialis*. Aquat. Toxicol. 153, 53–65.

Chang, X., Henderson, W.M., Bouchard, D.C., 2015. Multiwalled carbon nanotube dispersion methods affect their aggregation, deposition, and biomarker response. Environ. Sci. Technol. 49, 6645–6653.

Chen, R., Ratnikova, T.A., Stone, M.B., Lin, S., Lard, M., Huang, G., et al., 2010. Differential uptake of carbon nanoparticles by plant and mammalian cells. Small 6, 612–617.

Chen, P.Y., Powell, B.A., Mortimer, M., Ke, P.C., 2012. Adaptive interactions between zinc oxide nanoparticles and *Chlorella* sp. Environ. Sci. Technol. 46, 12178–12185.

Chen, Q.Q., Yin, D.Q., Li, J., Hu, X.L., 2014. The effects of humic acid on the uptake and depuration of fullerene aqueous suspensions in two aquatic organisms. Environ. Toxicol. Chem. 33, 1090–1097.

Chen, C., Tsyusko, O.V., McNear, D.H., Judy, J., Lewis, R.W., Unrine, J.M., 2017a. Effects of biosolids from a wastewater treatment plant receiving manufactured nanomaterials on *Medicago truncatula* and associated soil microbial communities at low nanomaterial concentrations. Sci. Total Environ. 609, 799–806.

Chen, M., Qin, X., Zeng, G., 2017b. Biodegradation of carbon nanotubes, graphene, and their derivatives. Trends Biotechnol. 35, 836–846.

Clark, K., van Tongeren, M., Christensen, F.M., Brouwer, D., Nowack, B., Gottschalk, F., et al., 2012. Limitations and information needs for engineered nanomaterial-specific exposure estimation and scenarios: recommendations for improved reporting practices. J. Nanopart. Res. 14, 1–14.

Collin, B., Auffan, M., Johnson, A.C., Kaur, I., Keller, A.A., Lazareva, A., et al., 2014. Environmental release, fate and ecotoxicological effects of manufactured ceria nanomaterials. Environ. Sci. Nano 1, 533–548.

Colman, B.P., Espinasse, B., Richardson, C.J., Matson, C.W., Lowry, G.V., Hunt, D.E., et al., 2014. Emerging contaminant or an old toxin in disguise? Silver nanoparticle impacts on ecosystems. Environ. Sci. Technol. 48, 5229–5236.

Conway, J.R., Hanna, S.K., Lenihan, H.S., Keller, A.A., 2014. Effects and implications of trophic transfer and accumulation of CeO_2 nanoparticles in a marine mussel. Environ. Sci. Technol. 48, 1517–1524.

Costanza, R., d'Arge, R., de Groot, R., Farber, S., Grasso, M., Hannon, B., et al., 1997. The value of the world's ecosystem services and natural capital. Nature 387, 253–260.

Crain, C.M., Halpern, B.S., Beck, M.W., Kappel, C.V., 2009. Understanding and managing human threats to the coastal marine environment. Ann. N.Y. Acad. Sci. 1162, 39–62.

Croteau, M.-N., Misra, S.K., Luoma, S.N., Valsami-Jones, E., 2014. Bioaccumulation and toxicity of CuO nanoparticles by a freshwater invertebrate after waterborne and dietborne exposures. Environ. Sci. Technol. 48, 10929–10937.

Das, R., Vecitis, C.D., Schulze, A., Cao, B., Ismail, A.F., Lu, X., et al., 2017. Recent advances in nanomaterials for water protection and monitoring. Chem. Soc. Rev. 46, 6946–7020.

De la Torre Roche, R., Servin, A., Hawthorne, J., Xing, B., Newman, L.A., Ma, X., et al., 2015. Terrestrial trophic transfer of bulk and nanoparticle La_2O_3 does not depend on particle size. Environ. Sci. Technol. 49, 11866–11874.

Deng, R., Lin, D., Zhu, L., Majumdar, S., White, J.C., Gardea-Torresdey, J.L., et al., 2017. Nanoparticle interactions with co-existing contaminants: joint toxicity, bioaccumulation and risk. Nanotoxicology 11, 591–612.

Dimkpa, C.O., 2014. Can nanotechnology deliver the promised benefits without negatively impacting soil microbial life? J. Basic Microbiol. 54, 889–904.

Dong, S., Xia, T., Yang, Y., Lin, S., Mao, L., 2018. Bioaccumulation of ^{14}C-labeled graphene in an aquatic food chain through direct uptake or trophic transfer. Environ. Sci. Technol. 52, 541–549.

Doudrick, K., Corson, N., Oberdörster, G., Eder, A.C., Herckes, P., Halden, R.U., et al., 2013. Extraction and quantification of carbon nanotubes in biological matrices with application to rat lung tissue. ACS Nano 7, 8849–8856.

Doudrick, K., Nosaka, T., Herckes, P., Westerhoff, P., 2015. Quantification of graphene and graphene oxide in complex organic matrices. Environ. Sci. Nano 2, 60–67.

Du, W., Tan, W., Peralta-Videa, J.R., Gardea-Torresdey, J.L., Ji, R., Yin, Y., et al., 2017. Interaction of metal oxide nanoparticles with higher terrestrial plants: physiological and biochemical aspects. Plant Physiol. Biochem. 110, 210–225.

Duester, L., Fabricius, A.L., Jakobtorweihen, S., Philippe, A., Weigl, F., Wimmer, A., et al., 2016. Can cloud point-based enrichment, preservation, and detection methods help to bridge gaps in aquatic nanometrology? Anal. Bioanal. Chem. 408, 7551–7557.

EC, 2011. Commission recommendation of 18 October 2011 on the definition of nanomaterial. Off. J. Eur. Union 275, 38–40. 2011/696/EU.

Edgington, A.J., Petersen, E.J., Herzing, A.A., Podila, R., Rao, A., Klaine, S.J., 2014. Microscopic investigation of single-wall carbon nanotube uptake by *Daphnia magna*. Nanotoxicology 8, 2–10.

EPA, 2017. Chemical substances when manufactured or processed as nanoscale materials; TSCA reporting and recordkeeping requirements. Fed. Regist. 82, 22088–22098. FR 22088.

Evans, T.D., 2012. Biosolids in Europe. In: Proceedings of the Water Environment Federation 2012. pp. 108–117.

Farkas, J., Peter, H., Ciesielski, T.M., Thomas, K.V., Sommaruga, R., Salvenmoser, W., et al., 2015. Impact of TiO_2 nanoparticles on freshwater bacteria from three Swedish lakes. Sci. Total Environ. 535, 85–93.

Faust, J.J., Doudrick, K., Yang, Y., Capco, D.G., Westerhoff, P., 2016. A facile method for separating and enriching nano and submicron particles from titanium dioxide found in food and pharmaceutical products. PLoS One 11, e0164712.

Ferry, J.L., Craig, P., Hexel, C., Sisco, P., Frey, R., Pennington, P.L., et al., 2009. Transfer of gold nanoparticles from the water column to the estuarine food web. Nat. Nanotechnol. 4, 441–444.

Forbes, V.E., Salice, C.J., Birnir, B., Bruins, R.J.F., Calow, P., Ducrot, V., et al., 2017. A framework for predicting impacts on ecosystem services from (sub)oganismal responses to chemicals. Environ. Toxicol. Chem. 36, 845–859.

Freixa, A., Acuna, V., Sanchis, J., Farre, M., Barcelo, D., Sabater, S., 2018. Ecotoxicological effects of carbon based nanomaterials in aquatic organisms. Sci. Total Environ. 619, 328–337.

Gao, Y., Wu, J., Ren, X., Tan, X., Hayat, T., Alsaedi, A., et al., 2017. Impact of graphene oxide on the antibacterial activity of antibiotics against bacteria. Environ. Sci. Nano 4, 1016–1024.

Gardea-Torresdey, J.L., Rico, C.M., White, J.C., 2014. Trophic transfer, transformation, and impact of engineered nanomaterials in terrestrial environments. Environ. Sci. Technol. 48, 2526–2540.

Ge, Y., Schimel, J.P., Holden, P.A., 2011. Evidence for negative effects of TiO_2 and ZnO nanoparticles on soil bacterial communities. Environ. Sci. Technol. 45, 1659–1664.

Ge, Y., Schimel, J.P., Holden, P.A., 2012. Identification of soil bacteria susceptible to TiO_2 and ZnO nanoparticles. Appl. Environ. Microbiol. 78, 6749–6758.

Ge, Y., Priester, J.H., De Werfhorst, L.C.V., Schimel, J.P., Holden, P.A., 2013. Potential mechanisms and environmental controls of TiO_2 nanoparticle effects on soil bacterial communities. Environ. Sci. Technol. 47, 14411–14417.

Ge, Y., Priester, J.H., Van De Werfhorst, L.C., Walker, S.L., Nisbet, R.M., An, Y.-J., et al., 2014. Soybean plants modify metal oxide nanoparticle effects on soil bacterial communities. Environ. Sci. Technol. 48, 13489–13496.

Ge, Y., Horst, A.M., Kim, J., Priester, J.H., Welch, Z.S., Holden, P.A., et al., 2016a. Toxicity of manufactured nanomaterials to microorganisms. In: Xing, B., Vecitis, C.D., Senesi, N. (Eds.), Engineered Nanoparticles and the Environment: Biophysicochemical Processes and Toxicity. John Wiley & Sons, Inc, pp. 320–346.

Ge, Y., Priester, J.H., Mortimer, M., Chang, C.H., Ji, Z., Schimel, J.P., et al., 2016b. Long-term effects of multiwalled carbon nanotubes and graphene on microbial communities in dry soil. Environ. Sci. Technol. 50, 3965–3974.

Geyer, R., Jambeck, J.R., Law, K.L., 2017. Production, use, and fate of all plastics ever made. Sci. Adv. 3, e1700782.

Gondikas, A.P., von der Kammer, F., Reed, R.B., Wagner, S., Ranville, J.F., Hofmann, T., 2014. Release of TiO_2 nanoparticles from sunscreens into surface waters: a one-year survey at the Old Danube recreational lake. Environ. Sci. Technol. 48, 5415–5422.

Goodwin, D.G., Adeleye, A.S., Sung, L., Ho, K.T., Burgess, R.M., Petersen, E.J., 2018. Detection and quantification of graphene-family nanomaterials in the environment. Environ. Sci. Technol. 52, 4491–4513.

Gottschalk, F., Sun, T.Y., Nowack, B., 2013. Environmental concentrations of engineered nanomaterials: review of modeling and analytical studies. Environ. Pollut. 181, 287–300.

Hadri, H.E., Hackley, V.A., 2017. Investigation of cloud point extraction for the analysis of metallic nanoparticles in a soil matrix. Environ. Sci. Nano 4, 105–116.

Hartline, N.L., Bruce, N.J., Karba, S.N., Ruff, E.O., Sonar, S.U., Holden, P.A., 2016. Microfiber masses recovered from conventional machine washing of new or aged garments. Environ. Sci. Technol. 50, 11532–11538.

Hawthorne, J., De la Torre Roche, R., Xing, B., Newman, L.A., Ma, X., Majumdar, S., et al., 2014. Particle-size dependent accumulation and trophic transfer of cerium oxide through a terrestrial food chain. Environ. Sci. Technol. 48, 13102–13109.

He, Y., Al-Abed, S.R., Dionysiou, D.D., 2017. Quantification of carbon nanotubes in different environmental matrices by a microwave induced heating method. Sci. Total Environ. 580, 509–517.

Hebbalalu, D., Lalley, J., Nadagouda, M.N., Varma, R.S., 2013. Greener techniques for the synthesis of silver nanoparticles using plant extracts, enzymes, bacteria, biodegradable polymers, and microwaves. ACS Sustain. Chem. Eng. 1, 703–712.

Heinlaan, M., Kahru, A., Kasemets, K., Arbeille, B., Prensier, G., Dubourguier, H.-C., 2011. Changes in the *Daphnia magna* midgut upon ingestion of copper oxide nanoparticles: a transmission electron microscopy study. Water Res. 45, 179–190.

Hernandez-Viezcas, J.A., Castillo-Michel, H., Andrews, J.C., Cotte, M., Rico, C., Peralta-Videa, J.R., et al., 2013. In situ synchrotron X-ray fluorescence mapping and speciation of CeO_2 and ZnO nanoparticles in soil cultivated soybean (*Glycine max*). ACS Nano 7, 1415–1423.

Holbrook, R.D., Murphy, K.E., Morrow, J.B., Cole, K.D., 2008. Trophic transfer of nanoparticles in a simplified invertebrate food web. Nat. Nanotechnol. 3, 352–355.

Holden, P.A., Nisbet, R.M., Lenihan, H.S., Miller, R.J., Cherr, G.N., Schimel, J.P., et al., 2013. Ecological nanotoxicology: integrating nanomaterial hazard considerations across the subcellular, population, community, and ecosystems levels. Acc. Chem. Res. 46, 813–822.

Holden, P.A., Klaessig, F., Turco, R.F., Priester, J.H., Rico, C.M., Avila-Arias, H., et al., 2014a. Evaluation of exposure concentrations used in assessing manufactured nanomaterial environmental hazards: are they relevant? Environ. Sci. Technol. 48, 10541–10551.

Holden, P.A., Schimel, J.P., Godwin, H.A., 2014b. Five reasons to use bacteria when assessing manufactured nanomaterial environmental hazards and fates. Curr. Opin. Biotechnol. 27, 73–78.

Holden, P.A., Gardea-Torresdey, J.L., Klaessig, F., Turco, R.F., Mortimer, M., Hund-Rinke, K., et al., 2016. Considerations of environmentally relevant test conditions for improved evaluation of ecological hazards of engineered nanomaterials. Environ. Sci. Technol. 50, 6124–6145.

Holden, P.A., Mortimer, M., Wang, Y., 2018. Engineered nanomaterials and symbiotic dinitrogen fixation in legumes. Curr. Opin. Environ. Sci. Health 6, 54–59.

Horst, A.M., Neal, A.C., Mielke, R.E., Sislian, P.R., Suh, W.H., Madler, L., et al., 2010. Dispersion of TiO_2 nanoparticle agglomerates by *Pseudomonas aeruginosa*. Appl. Environ. Microb. 76, 7292–7298.

Hou, W.-C., Chowdhury, I., Goodwin, D.G., Henderson, W.M., Fairbrother, D.H., Bouchard, D., et al., 2015. Photochemical transformation of graphene oxide in sunlight. Environ. Sci. Technol. 49, 3435–3443.

Hüffer, T., Praetorius, A., Wagner, S., Von Der Kammer, F., Hofmann, T., 2017. Microplastic exposure assessment in aquatic environments: learning from similarities and differences to engineered nanoparticles. Environ. Sci. Technol. 51, 2499–2507.

Hyung, H., Fortner, J.D., Hughes, J.B., Kim, J.-H., 2007. Natural organic matter stabilizes carbon nanotubes in the aqueous phase. Environ. Sci. Technol. 41, 179–184.

Ivask, A., Elbadawy, A., Kaweeteerawat, C., Boren, D., Fischer, H., Ji, Z., et al., 2014. Toxicity mechanisms in *Escherichia coli* vary for silver nanoparticles and differ from ionic silver. ACS Nano 8, 374–386.

Jacobs, D.S., Huang, S.R., Cheng, Y.L., Rabb, S.A., Gorham, J.M., Krommenhoek, P.J., et al., 2016. Surface degradation and nanoparticle release of a commercial nanosilica/polyurethane coating under UV exposure. J. Coat. Technol. Res. 13, 735–751.

Jones, K.R., Venter, O., Fuller, R.A., Allan, J.R., Maxwell, S.L., Negret, P.J., et al., 2018. One-third of global protected land is under intense human pressure. Science 360, 788–791.

Judy, J.D., McNear, D.H., Chen, C., Lewis, R.W., Tsyusko, O.V., Bertsch, P.M., et al., 2015. Nanomaterials in biosolids inhibit nodulation, shift microbial community composition, and result in increased metal uptake relative to bulk/dissolved metals. Environ. Sci. Technol. 49, 8751–8758.

Juganson, K., Mortimer, M., Ivask, A., Kasemets, K., Kahru, A., 2013. Extracellular conversion of silver ions into silver nanoparticles by protozoan *Tetrahymena thermophila*. Environ. Sci. Process. Impacts 15, 244–250.

Jung, Y.-J., Kim, K.-T., Kim, J.Y., Yang, S.-Y., Lee, B.-G., Kim, S.D., 2014. Bioconcentration and distribution of silver nanoparticles in Japanese medaka (*Oryzias latipes*). J. Hazard. Mater. 267, 206–213.

Kaegi, R., Ulrich, A., Sinnet, B., Vonbank, R., Wichser, A., Zuleeg, S., et al., 2008. Synthetic TiO_2 nanoparticle emission from exterior facades into the aquatic environment. Environ. Pollut. 156, 233–239.

Kaegi, R., Voegelin, A., Ort, C., Sinnet, B., Thalmann, B., Krismer, J., et al., 2013. Fate and transformation of silver nanoparticles in urban wastewater systems. Water Res. 47, 3866–3877.

Kaekinen, A., Bondarenko, O., Ivask, A., Kahru, A., 2011. The effect of composition of different ecotoxicological test media on free and bioavailable copper from $CuSO_4$ and CuO nanoparticles: comparative evidence from a Cu-selective electrode and a Cu-biosensor. Sensors 11, 10502–10521.

Kah, M., Kookana, R.S., Gogos, A., Bucheli, T.D., 2018. A critical evaluation of nanopesticides and nanofertilizers against their conventional analogues. Nat. Nanotechnol. 13, 677–684.

Keller, A.A., Lazareva, A., 2014. Predicted releases of engineered nanomaterials: from global to regional to local. Environ. Sci. Technol. Lett. 1, 65–70.

Keller, A.A., Wang, H.T., Zhou, D.X., Lenihan, H.S., Cherr, G., Cardinale, B.J., et al., 2010. Stability and aggregation of metal oxide nanoparticles in natural aqueous matrices. Environ. Sci. Technol. 44, 1962–1967.

Keller, A.A., McFerran, S., Lazareva, A., Suh, S., 2013. Global life cycle releases of engineered nanomaterials. J. Nanopart. Res. 15, 1–17.

Keller, A.A., Vosti, W., Wang, H., Lazareva, A., 2014. Release of engineered nanomaterials from personal care products throughout their life cycle. J. Nanopart. Res. 16, 2489.

Keller, A.A., Adeleye, A.S., Conway, J.R., Garner, K.L., Zhao, L., Cherr, G.N., et al., 2017. Comparative environmental fate and toxicity of copper nanomaterials. NanoImpact 7, 28–40.

Kettler, K., Veltman, K., van de Meent, D., van Wezel, A., Hendriks, A.J., 2014. Cellular uptake of nanoparticles as determined by particle properties, experimental conditions, and cell type. Environ. Toxicol. Chem. 33, 481–492.

Khan, F.R., Kennaway, G.M., Croteau, M.-N., Dybowska, A., Smith, B.D., Nogueira, A.J. A., et al., 2014. In vivo retention of ingested Au NPs by *Daphnia magna*: no evidence for trans-epithelial alimentary uptake. Chemosphere 100, 97–104.

Klanjšček, T., Muller, E.B., Holden, P.A., Nisbet, R.M., 2017. Host-symbiont interaction model explains non-monotonic response of soybean growth and seed production to nano-CeO_2 exposure. Environ. Sci. Technol. 51, 4944–4950.

Kulacki, K.J., Cardinale, B.J., Keller, A.A., Bier, R., Dickson, H., 2012. How do stream organisms respond to, and influence, the concentration of titanium dioxide nanoparticles? A mesocosm study with algae and herbivores. Environ. Toxicol. Chem. 31, 2414–2422.

Lahive, E., Matzke, M., Durenkamp, M., Lawlor, A.J., Thacker, S.A., Pereira, M.G., et al., 2017. Sewage sludge treated with metal nanomaterials inhibits earthworm reproduction more strongly than sludge treated with metal metals in bulk/salt forms. Environ. Sci. Nano 4, 78–88.

Landrigan, P.J., Fuller, R., Hu, H., Caravanos, J., Cropper, M.L., Hanrahan, D., et al., 2018. Pollution and global health – an agenda for prevention. Env. Health Perspect. 126, 084501.

Lankone, R.S., Challis, K.E., Bi, Y., Hanigan, D., Reed, R.B., Zaikova, T., et al., 2017. Methodology for quantifying engineered nanomaterial release from diverse product matrices under outdoor weathering conditions and implications for life cycle assessment. Environ. Sci. Nano 4, 1784–1797.

Larguinho, M., Correia, D., Diniz, M.S., Baptista, P.V., 2014. Evidence of one-way flow bioaccumulation of gold nanoparticles across two trophic levels. J. Nanopart. Res. 16, 2549.

Lazareva, A., Keller, A.A., 2014. Estimating potential life cycle releases of engineered nanomaterials from wastewater treatment plants. ACS Sustain. Chem. Eng. 2, 1656–1665.

Leclerc, S., Wilkinson, K.J., 2014. Bioaccumulation of nanosilver by *Chlamydomonas reinhardtii*-nanoparticle or the free ion? Environ. Sci. Technol. 48, 358–364.

Levard, C., Hotze, E.M., Colman, B.P., Dale, A.L., Truong, L., Yang, X.Y., 2013. Sulfidation of silver nanoparticles: natural antidote to their toxicity. Environ. Sci. Technol. 47, 13440–13448.

Li, L., Hartmann, G., Doblinger, M., Schuster, M., 2013. Quantification of nanoscale silver particles removal and release from municipal wastewater treatment plants in Germany. Environ. Sci. Technol. 47, 7317–7323.

Li, L., Hu, L., Zhou, Q., Huang, C., Wang, Y., Sun, C., et al., 2015. Sulfidation as a natural antidote to metallic nanoparticles is overestimated: CuO sulfidation yields CuS nanoparticles with increased toxicity in medaka (*Oryzias latipes*) embryos. Environ. Sci. Technol. 49, 2486–2495.

Lin, S., Mortimer, M., Chen, R., Kakinen, A., Riviere, J.E., Davis, T.P., et al., 2017. NanoEHS beyond toxicity – focusing on biocorona. Environ. Sci. Nano 4, 1433–1454.

Liu, H.H., Cohen, Y., 2014. Multimedia environmental distribution of engineered nanomaterials. Environ. Sci. Technol. 48, 3281–3292.

Liu, H.H., Bilal, M., Lazareva, A., Keller, A., Cohen, Y., 2015. Simulation tool for assessing the release and environmental distribution of nanomaterials. Beilstein J. Nanotechnol. 6, 938–951.

Loosli, F., Yi, Z., Berti, D., Baalousha, M., 2018. Toward a better extraction of titanium dioxide engineered nanomaterials from complex environmental matrices. NanoImpact 11, 119–127.

Luo, Z.X., Wang, Z.H., Li, Q.Z., Pan, Q.K., Yan, C.Z., Liu, F., 2011. Spatial distribution, electron microscopy analysis of titanium and its correlation to heavy metals: occurrence and sources of titanium nanomaterials in surface sediments from Xiamen Bay. Chin. J. Environ. Monitor. 13, 1046–1052.

Luo, Z., Wang, Z., Yan, Y., Li, J., Yan, C., Xing, B., 2018. Titanium dioxide nanoparticles enhance inorganic arsenic bioavailability and methylation in two freshwater algae species. Environ. Pollut. 238, 631–637.

Luoma, S.N., Rainbow, P.S., 2008. Metal Contamination in Aquatic Environments. Science and Lateral Management. Cambridge University Press, New York.

Ma, C., Liu, H., Chen, G., Zhao, Q., Eitzer, B., Wang, Z., et al., 2017. Effects of titanium oxide nanoparticles on tetracycline accumulation and toxicity in *Oryza sativa* (L.). Environ. Sci. Nano 4, 1827–1839.

Ma, C.X., White, J.C., Zhao, J., Zhao, Q., Xing, B.S., 2018. Uptake of engineered nanoparticles by food crops: characterization, mechanisms, and implications. Annu. Rev. Food Sci. Technol. 9, 129–153.

Machado, A.A.S., Kloas, W., Zarfl, C., Hempel, S., Rillig, M.C., 2018. Microplastics as an emerging threat to terrestrial ecosystems. Glob. Change Biol. 24, 1405–1416.

Madigan, M.T., Martinko, J.M., Bender, K.S., Buckley, D.H., Stahl, D.A., 2015. Brock Biology of Microorganisms, 14th ed. Pearson, San Franscisco, CA.

Maes, H.M., Stibany, F., Giefers, S., Daniels, B., Deutschmann, B., Baumgartner, W., et al., 2014. Accumulation and distribution of multiwalled carbon nanotubes in zebrafish (*Danio rerio*). Environ. Sci. Technol. 48, 12256–12264.

Majedi, S.M., Kelly, B.C., Lee, H.K., 2014. Evaluation of a cloud point extraction approach for the preconcentration and quantification of trace CuO nanoparticles in environmental waters. Anal. Chim. Acta 814, 39–48.

McNaught, A.D., Wilkinson, A., 1997. IUPAC. Compendium of Chemical Terminology, 2nd ed. Blackwell Scientific Publications, Oxford (the "Gold Book").

Miao, L., Wang, C., Hou, J., Wang, P., Ao, Y., Li, Y., et al., 2015. Enhanced stability and dissolution of CuO nanoparticles by extracellular polymeric substances in aqueous environment. J. Nanopart. Res. 17, 404.

Mielke, R.E., Priester, J.H., Werlin, R.A., Gelb, J., Horst, A.M., Orias, E., et al., 2013. Differential growth of and nanoscale TiO_2 accumulation in *Tetrahymena thermophila* by direct feeding versus trophic transfer from *Pseudomonas aeruginosa*. Appl. Environ. Microbiol. 79, 5616–5624.

Misra, S.K., Dybowska, A., Berhanu, D., Luoma, S.N., Valsami-Jones, E., 2012. The complexity of nanoparticle dissolution and its importance in nanotoxicological studies. Sci. Total Environ. 438, 225–232.

Mitrano, D.M., Nowack, B., 2017. The need for a life-cycle based aging paradigm for nanomaterials: importance of real-world test systems to identify realistic particle transformations. Nanotechnology 28, 072001.

Mitrano, D.M., Ranville, J.F., Bednar, A., Kazor, K., Hering, A.S., Higgins, C.P., 2014a. Tracking dissolution of silver nanoparticles at environmentally relevant concentrations in laboratory, natural, and processed waters using single particle ICP-MS (spICP-MS). Environ. Sci. Nano 1, 248–259.

Mitrano, D.M., Rimmele, E., Wichser, A., Erni, R., Height, M., Nowack, B., 2014b. Presence of nanoparticles in wash water from conventional silver and nano-silver textiles. ACS Nano 8, 7208–7219.

Mitrano, D.M., Motellier, S., Clavaguera, S., Nowack, B., 2015. Review of nanomaterial aging and transformations through the life cycle of nano-enhanced products. Environ. Int. 77, 132–147.

Montaño, M.D., Badiei, H.R., Bazargan, S., Ranville, J.F., 2014. Improvements in the detection and characterization of engineered nanoparticles using spICP-MS with microsecond dwell times. Environ. Sci. Nano 1, 338–346.

Montoro Bustos, A.R., Winchester, M.R., 2016. Single-particle-ICP-MS advances. Anal. Bioanal. Chem. 408, 5051–5052.

Morachis, J.M., Mahmoud, E.A., Sankaranarayanan, J., Almutairi, A., 2012. Triggered rapid degradation of nanoparticles for gene delivery. J. Drug Deliv. 2012, 1–7.

Mortimer, M., Gogos, A., Bartolome, N., Kahru, A., Bucheli, T.D., Slaveykova, V.I., 2014a. Potential of hyperspectral imaging microscopy for semi-quantitative analysis of nanoparticle uptake by protozoa. Environ. Sci. Technol. 48, 8760–8767.

Mortimer, M., Kahru, A., Slaveykova, V.I., 2014b. Uptake, localization and clearance of quantum dots in ciliated protozoa *Tetrahymena thermophila*. Environ. Pollut. 190, 58–64.

Mortimer, M., Petersen, E., Buchholz, B., Holden, P., 2016a. Separation of bacteria, protozoa and carbon nanotubes by density gradient centrifugation. Nanomaterials 6, 181.

Mortimer, M., Petersen, E.J., Buchholz, B.A., Orias, E., Holden, P.A., 2016b. Bioaccumulation of multiwall carbon nanotubes in *Tetrahymena thermophila* by direct feeding or trophic transfer. Environ. Sci. Technol. 50, 8876–8885.

Mudunkotuwa, I.A., Pettibone, J.M., Grassian, V.H., 2012. Environmental implications of nanoparticle aging in the processing and fate of copper-based nanomaterials. Environ. Sci. Technol. 46, 7001–7010.

Mukherjee, A., Majumdar, S., Servin, A.D., Pagano, L., Dhankher, O.P., White, J.C., 2016. Carbon nanomaterials in agriculture: a critical review. Front. Plant Sci. 7, 172.

Nam, K.T., Lee, Y.J., Krauland, E.M., Kottmann, S.T., Belcher, A.M., 2008. Peptide-mediated reduction of silver ions on engineered biological scaffolds. ACS Nano 2, 1480–1486.

National Research Council, 2012. A Research Strategy for Environmental, Health, and Safety Aspects of Engineered Nanomaterials. The National Academies Press, Washington, DC.

Nowack, B., Mitrano, D.M., 2018. Procedures for the production and use of synthetically aged and product released nanomaterials for further environmental and ecotoxicity testing. NanoImpact 10, 70–80.

Nowack, B., David, R.M., Fissan, H., Morris, H., Shatkin, J.A., Stintz, M., et al., 2013. Potential release scenarios for carbon nanotubes used in composites. Environ. Int. 59, 1–11.

Nowack, B., Baalousha, M., Bornhöft, N., Chaudhry, Q., Cornelis, G., Cotterill, J., et al., 2015. Progress towards the validation of modeled environmental concentrations of engineered nanomaterials by analytical measurements. Environ. Sci. Nano 2, 421–428.

Odzak, N., Kistler, D., Behra, R., Sigg, L., 2014. Dissolution of metal and metal oxide nanoparticles in aqueous media. Environ. Pollut. 191, 132–138.

Pédrot, M., Boudec, A.L., Davranche, M., Dia, A., Henin, O., 2011. How does organic matter constrain the nature, size and availability of Fe nanoparticles for biological reduction? J. Colloid Interface Sci. 359, 75–85.

Peralta-Videa, J.R., Hernandez-Viezcas, J.A., Zhao, L., Diaz, B.C., Ge, Y., Priester, J.H., et al., 2014. Cerium dioxide and zinc oxide nanoparticles alter the nutritional value of soil cultivated soybean plants. Plant Physiol. Biochem. 80, 128–135.

Perstrimaux, C., Le Faucheur, S., Mortimer, M., Stoll, S., Aubry, F.B., Botter, M., et al., 2015. Stability of titanium dioxide nanoparticle agglomerates in transitional waters and their effects towards plankton from lagoon of Venice (Italy). Aquat. Geochem. 21, 343–362.

Peters, R.J.B., van Bemmel, G., Milani, N.B.L., den Hertog, G.C.T., Undas, A.K., van der Lee, M., et al., 2018. Detection of nanoparticles in Dutch surface waters. Sci. Total Environ. 621, 210–218.

Petersen, E.J., Akkanen, J., Kukkonen, J.V.K., Weber Jr., W.J., 2009. Biological uptake and depuration of carbon nano-tubes by *Daphnia magna*. Environ. Sci. Technol. 43, 2969–2975.

Petersen, E.J., Pinto, R.A., Mai, D.J., Landrum, P.F., Weber, W.J., 2011. Influence of polyethyleneimine graftings of multi-walled carbon nanotubes on their accumulation and elimination by and toxicity to *Daphnia magna*. Environ. Sci. Technol. 45, 1133–1138.

Petersen, E.J., Flores-Cervantes, D.X., Bucheli, T.D., Elliott, L.C.C., Fagan, J.A., Gogos, A., et al., 2016. Quantification of carbon nanotubes in environmental matrices: current capabilities, case studies, and future prospects. Environ. Sci. Technol. 50, 4587–4605.

Plata, D.L., Reddy, C.M., Gschwend, P.M., 2012. Thermogravimetry–mass spectrometry for carbon nanotube detection in complex mixtures. Environ. Sci. Technol. 46, 12254–12261.

Prasad, A., Lead, J.R., Baalousha, M., 2015. An electron microscopy based method for the detection and quantification of nanomaterial number concentration in environmentally relevant media. Sci. Total Environ. 537, 479–486.

Priester, J.H., Stoimenov, P.K., Mielke, R.E., Webb, S.M., Ehrhardt, C., Zhang, J.P., et al., 2009. Effects of soluble cadmium salts versus CdSe quantum dots on the growth of planktonic *Pseudomonas aeruginosa*. Environ. Sci. Technol. 43, 2589–2594.

Priester, J.H., Ge, Y., Mielke, R.E., Horst, A.M., Moritz, S.C., Espinosa, K., et al., 2012. Soybean susceptibility to manufactured nanomaterials with evidence for food quality and soil fertility interruption. Proc. Natl. Acad. Sci. U.S.A. 109, E2451–E2456.

Priester, J.H., Ge, Y., Chang, V., Stoimenov, P.K., Schimel, J.P., Stucky, G.D., et al., 2013. Assessing interactions of hydrophilic nanoscale TiO_2 with soil water. J. Nanopart. Res. 15, 1899.

Priester, J.H., Singhal, A., Wu, B., Stucky, G.D., Holden, P.A., 2014. Integrated approach to evaluating the toxicity of novel cysteine-capped silver nanoparticles to *Escherichia coli* and *Pseudomonas aeruginosa*. Analyst 139, 954–963.

Priester, J.H., Moritz, S.C., Espinosa, K., Ge, Y., Wang, Y., Nisbet, R.M., et al., 2017. Damage assessment for soybean cultivated in soil with either CeO_2 or ZnO manufactured nanomaterials. Sci. Total Environ. 579, 1756–1768.

Reddy, P.V.L., Hernandez-Viezcas, J.A., Peralta-Videa, J.R., Gardea-Torresdey, J.L., 2016. Lessons learned: are engineered nanomaterials toxic to terrestrial plants? Sci. Total Environ. 568, 470–479.

Reddy Pullagurala, V.L., Adisa, I.O., Rawat, S., Kim, B., Barrios, A.C., Medina-Velo, I.A., et al., 2018. Finding the conditions for the beneficial use of ZnO nanoparticles towards plants—a review. Environ. Pollut. 241, 1175–1181.

Rico, C.M., Johnson, M.G., Marcus, M.A., 2018. Cerium oxide nanoparticles transformation at the root–soil interface of barley (*Hordeum vulgare* L.). Environ. Sci. Nano 5, 1807–1812.

Roco, M.C., 2011. The long view of nanotechnology development: the National Nanotechnology Initiative at 10 years. J. Nanopart. Res. 13, 427–445.

Rohr, J.R., Salice, C.J., Nisbet, R.M., 2017. Chemical safety must extend to ecosystems. Science 356, 917.

Salvati, A., Nelissen, I., Haase, A., Åberg, C., Moya, S., Jacobs, A., et al., 2018. Quantitative measurement of nanoparticle uptake by flow cytometry illustrated by an interlaboratory comparison of the uptake of labelled polystyrene nanoparticles. NanoImpact 9, 42–50.

Sanchís, J., Olmos, M., Vincent, P., Farré, M., Barceló, D., 2016. New insights on the influence of organic co-contaminants on the aquatic toxicology of carbon nanomaterials. Environ. Sci. Technol. 50, 961–969.

Sauer, U.G., Aumann, A., Ma-Hock, L., Landsiedel, R., Wohlleben, W., 2015. Influence of dispersive agent on nanomaterial agglomeration and implications for biological effects *in vivo* or *in vitro*. Toxicol. in Vitro 29, 182–186.

Schaumann, G.E., Philippe, A., Bundschuh, M., Metreveli, G., Klitzke, S., Rakcheev, D., et al., 2015. Understanding the fate and biological effects of Ag- and TiO_2-nanoparticles in the environment: the quest for advanced analytics and interdisciplinary concepts. Sci. Total Environ. 535, 3–19.

Schierz, A., Parks, A.N., Washburn, K.M., Chandler, G.T., Ferguson, P.L., 2012. Characterization and quantitative analysis of single-walled carbon nanotubes in the aquatic environment using near-infrared fluorescence spectroscopy. Environ. Sci. Technol. 46, 12262–12271.

Schlich, K., Hund-Rinke, K., 2015. Influence of soil properties on the effect of silver nanomaterials on microbial activity in five soils. Environ. Pollut. 196, 321–330.

Schlich, K., Hoppe, M., Kraas, M., Schubert, J., Chanana, M., Hund-Rinke, K., 2018. Long-term effects of three different silver sulfide nanomaterials, silver nitrate and bulk silver sulfide on soil microorganisms and plants. Environ. Pollut.

Seltenrich, N., 2018. Down to Earth: the emerging field of planetary health. Environ. Health Perspect. 126, 072001.

Servin, A.D., Pagano, L., Castillo-Michel, H., De la Torre-Roche, R., Hawthorne, J., Hernandez-Viezcas, J.A., et al., 2017. Weathering in soil increases nanoparticle CuO bioaccumulation within a terrestrial food chain. Nanotoxicology 11, 98–111.

Song, R., Qin, Y., Suh, S., Keller, A.A., 2017. Dynamic model for the stocks and release flows of engineered nanomaterials. Environ. Sci. Technol. 51, 12424–12433.

Steinmetz, Z., Wollmann, C., Schaefer, M., Buchmann, C., David, J., Tröger, J., et al., 2016. Plastic mulching in agriculture. Trading short-term agronomic benefits for long-term soil degradation? Sci. Total Environ. 550, 690–705.

Stevenson, L.M., Dickson, H., Klanjscek, T., Keller, A.A., McCauley, E., Nisbet, R.M., 2013. Environmental feedbacks and engineered nanoparticles: mitigation of silver nanoparticle toxicity to *Chlamydomonas reinhardtii* by algal-produced organic compounds. PLoS One 8, e74456.

Sun, T.Y., Gottschalk, F., Hungerbühler, K., Nowack, B., 2014. Comprehensive probabilistic modelling of environmental emissions of engineered nanomaterials. Environ. Pollut. 185, 69–76.

Sun, T.Y., Bornhöft, N.A., Hungerbühler, K., Nowack, B., 2016. Dynamic probabilistic modeling of environmental emissions of engineered nanomaterials. Environ. Sci. Technol. 50, 4701–4711.

Tangaa, S.R., Selck, H., Winther-Nielsen, M., Khan, F.R., 2016. Trophic transfer of metal-based nanoparticles in aquatic environments: a review and recommendations for future research focus. Environ. Sci. Nano 3, 966–981.

Tella, M., Auffan, M., Brousset, L., Issartel, J., Kieffer, I., Pailles, C., et al., 2014. Transfer, transformation, and impacts of ceria nanomaterials in aquatic mesocosms simulating a pond ecosystem. Environ. Sci. Technol. 48, 9004–9013.

Terraube, J., Fernández-Llamazares, Á., Cabeza, M., 2017. The role of protected areas in supporting human health: a call to broaden the assessment of conservation outcomes. Curr. Opin. Environ. Sustain. 25, 50–58.

Therezien, M., Thill, A., Wiesner, M.R., 2014. Importance of heterogeneous aggregation for NP fate in natural and engineered systems. Sci. Total Environ. 485–486, 309–318.

Tong, Z., Bischoff, M., Nies, L., Applegate, B., Turco, R.F., 2007. Impact of fullerene (C_{60}) on a soil microbial community. Environ. Sci. Technol. 41, 2985–2991.

Vance, M.E., Kuiken, T., Vejerano, E.P., McGinnis, S.P., Hochella, M.F., Rejeski, D., et al., 2015. Nanotechnology in the real world: redeveloping the nanomaterial consumer products inventory. Beilstein J. Nanotechnol. 6, 1769–1780.

Vaseghi, Z., Nematollahzadeh, A., Tavakoli, O., 2018. Green methods for the synthesis of metal nanoparticles using biogenic reducing agents: a review. Rev. Chem. Eng. 34, 529–559.

von der Kammer, F., Ferguson, P.L., Holden, P.A., Masion, A., Rogers, K.R., Klaine, S.J., et al., 2012. Analysis of engineered nanomaterials in complex matrices (environment and biota): general considerations and conceptual case studies. Environ. Toxicol. Chem. 31, 32–49.

Wang, Y., Miao, A.J., Luo, J., Wei, Z.B., Zhu, J.J., Yang, L.Y., 2013. Bioaccumulation of CdTe quantum dots in a freshwater alga *Ochromonas danica*: a kinetics study. Environ. Sci. Technol. 47, 10601–10610.

Wang, H., Adeleye, A.S., Huang, Y., Li, F., Keller, A.A., 2015. Heteroaggregation of nanoparticles with biocolloids and geocolloids. Adv. Colloid Interface Sci. 226, 24–36.

Wang, J., Lankone, R.S., Reed, R.B., Fairbrother, D.H., Ranville, J.F., 2016a. Analysis of single-walled carbon nanotubes using spICP-MS with microsecond dwell time. NanoImpact 1, 65–72.

Wang, Z., Zhang, L., Zhao, J., Xing, B., 2016b. Environmental processes and toxicity of metallic nanoparticles in aquatic systems as affected by natural organic matter. Environ. Sci. Nano 3, 240–255.

Wang, Y., Chang, C.H., Ji, Z., Bouchard, D.C., Nisbet, R.M., Schimel, J.P., et al., 2017. Agglomeration determines effects of carbonaceous nanomaterials on soybean nodulation, dinitrogen fixation potential, and growth in soil. ACS Nano 11, 5753–5765.

Wang, P., Menzies, N.W., Chen, H., Yang, X., McGrath, S.P., Zhao, F.-J., et al., 2018a. Risk of silver transfer from soil to the food chain is low after long-term (20 years) field applications of sewage sludge. Environ. Sci. Technol. 52, 4901–4909.

Wang, Y., Mortimer, M., Chang, C., Holden, P., 2018b. Alginic acid-aided dispersion of carbon nanotubes, graphene, and boron nitride nanomaterials for microbial toxicity testing. Nanomaterials 8, 76.

Weir, A., Westerhoff, P., Fabricius, L., Hristovski, K., von Goetz, N., 2012. Titanium dioxide nanoparticles in food and personal care products. Environ. Sci. Technol. 46, 2242–2250.

Werlin, R., Priester, J.H., Mielke, R.E., Kramer, S., Jackson, S., Stoimenov, P.K., et al., 2011. Biomagnification of cadmium selenide quantum dots in a simple experimental microbial food chain. Nat. Nanotechnol. 6, 65–71.

Wigger, H., Wohlleben, W., Nowack, B., 2018. Redefining environmental nanomaterial flows: consequences of the regulatory nanomaterial definition on the results of environmental exposure models. Environ. Sci. Nano 5, 1372–1385.

Wilkinson, J., Hooda, P.S., Barker, J., Barton, S., Swinden, J., 2017. Occurrence, fate and transformation of emerging contaminants in water: an overarching review of the field. Environ. Pollut. 231, 954–970.

Wohlleben, W., Neubauer, N., 2016. Quantitative rates of release from weathered nanocomposites are determined across 5 orders of magnitude by the matrix, modulated by the embedded nanomaterial. NanoImpact 1, 39–45.

Yin, Y., Tan, Z., Hu, L., Yu, S., Liu, J., Jiang, G., 2017. Isotope tracers to study the environmental fate and bioaccumulation of metal-containing engineered nanoparticles: techniques and applications. Chem. Rev. 117, 4462–4487.

Yu, S., Yin, Y., Zhou, X., Dong, L., Liu, J., 2016. Transformation kinetics of silver nanoparticles and silver ions in aquatic environments revealed by double stable isotope labeling. Environ. Sci. Nano 3, 883–893.

Zhang, L.W., Petersen, E.J., Zhang, W., Chen, Y.S., Cabrera, M., Huang, Q.G., 2012. Interactions of C-14-labeled multi-walled carbon nanotubes with soil minerals in water. Environ. Pollut. 166, 75–81.

Zhang, L., Petersen, E.J., Habteselassie, M.Y., Mao, L., Huang, Q., 2013. Degradation of multiwall carbon nanotubes by bacteria. Environ. Pollut. 181, 335–339.

Zhao, Y., Allen, B.L., Star, A., 2011. Enzymatic degradation of multiwalled carbon nanotubes. J. Phys. Chem. A 115, 9536–9544.

Zhu, X.S., Wang, J.X., Zhang, X.Z., Chang, Y., Chen, Y.S., 2010. Trophic transfer of TiO_2 nanoparticles from daphnia to zebrafish in a simplified freshwater food chain. Chemosphere 79, 928–933.

Zuverza-Mena, N., Martínez-Fernández, D., Du, W., Hernandez-Viezcas, J.A., Bonilla-Bird, N., López-Moreno, M.L., et al., 2017. Exposure of engineered nanomaterials to plants: insights into the physiological and biochemical responses—a review. Plant Physiol. Biochem. 110, 236–264.

Zweers, M.L.T., Engbers, G.H.M., Grijpma, D.W., Feijen, J., 2004. In vitro degradation of nanoparticles prepared from polymers based on DL-lactide, glycolide and poly(ethylene oxide). J. Controlled Release 100, 347–356.

CHAPTER

Fate of engineered nanomaterials in agroenvironments and impacts on agroecosystems

4

Venkata L. Reddy Pullagurala[1,4], Ishaq O. Adisa[1,5], Swati Rawat[1,4], Jason C. White[3,5], Nubia Zuverza-Mena[3], Jose A. Hernandez-Viezcas[2,4], Jose R. Peralta-Videa[1,2,4] and Jorge L. Gardea-Torresdey[1,2,4,5]

[1]*Environmental Science and Engineering PhD Program, The University of Texas at El Paso, El Paso, TX, United States*

[2]*Department of Chemistry and Biochemistry, The University of Texas at El Paso, El Paso, TX, United States*

[3]*Department of Analytical Chemistry, The Connecticut Agricultural Experiment Station, New Haven, CT, United States*

[4]*University of California Center for Environmental Implications of Nanotechnology (UC CEIN), The University of Texas at El Paso, El Paso, TX, United States*

[5]*The Center for Nanotechnology and Agricultural Pathogen Suppression, New Haven, CT, United States*

4.1 Introduction

The expansion of nanotechnologies has resulted in the large-scale production of engineered nanomaterials (ENMs). A wide range of the ENMs including ZnO, TiO_2, CuO, and carbon nanotubes (CNTs) are actively employed in various fields (Reddy et al., 2014, 2017; Palombo et al., 2014). However, the environmental fate and transport of these emerging contaminants are widely debated. A lack of clear understanding about their environmental impact may lead to significant ecological damage (Guinée et al., 2017). Thus clear knowledge about the compartmentalization of ENMs in air, water, and soil is vital. Keller et al. (2014) estimated that 63%–91% of the total ENM waste end up in landfills, whereas 8%–28% are disposed into soil (Keller et al., 2014). It is clear that the soil compartment is more vulnerable to the negative effects of ENM exposure (Pullagurala et al., 2018a). Therefore it is essential to have a life cycle assessment of ENMs in agricultural environments. This life cycle assessment must include the fate and transport of these ENMs in the soil upon disposal, as well as the impacts on all the major agriculturally relevant biota (Kavitha et al., 2018).

Exposure to Engineered Nanomaterials in the Environment. DOI: https://doi.org/10.1016/B978-0-12-814835-8.00004-2

Based on the literature, we know that ENMs behave differently as a function of their physical and chemical properties. These materials undergo transformation in soil (Gardea-Torresdey et al., 2014) and in plants (Wang et al., 2012). In addition, their interaction with soil components and other cocontaminants has also been reported (Servin and White, 2016; Deng et al., 2017). Consequently, increased knowledge about material retention and bioavailability in different agricultural environments is crucial (Cornelis et al., 2014). Another equally important aspect of the studies is the impact of the ENM exposure on biota within agricultural ecosystems (Pradhan and Mailapalli, 2017). There are many experimental studies in the literature investigating ENM exposure; however, little consensus exists over whether ENM exposure is either beneficial or detrimental (Reddy et al., 2016). The probable reason for the contradictory findings in the literature is attributable to multiple factors associated with the study conditions, such as species used, dosing regime, and endpoints measured. The direct uptake or toxicity of ENMs may not be the only factor that could have an impact on plant growth. Plants form a large number of complex symbiotic relationships with microbes in the rhizosphere (i.e., nitrogen fixation), and impacts of ENMs on these species could also indirectly impact plant health (Saia et al., 2014). Disturbances in the microbial population can lead to nitrogen starvation, which could result in stunted growth. In addition, invertebrates in soil habitats also have a significant role in agricultural productivity as they form burrows, which are very important in water infiltration and assist in biogeochemical cycling. Thus it is important to have a comprehensive understanding of all the factors involved in the interaction of ENMs with environmental components. In this chapter, we review the current literature regarding the fate of the various ENMs in soil, as well as the factors that influence their bioavailability and transport. We also evaluate the impacts of ENMs exposure on plants and other relevant biota.

4.2 Factors influencing the fate, transport, and retention of engineered nanomaterials in soil

Soil is one of the primary sinks for ENMs. Upon the ENM entry into soil, their fate and transport are determined by several factors that include physicochemical properties of soil [such as cation exchange capacity (CEC), soil composition, and pH] and that of the ENMs (such as morphology, size, zeta potential, surface coating, and chemical composition, among other things) (Rawat et al., 2018a) (Fig. 4.1). Understanding the dynamics of ENMs in the soil medium is critical to evaluating their fate, transport, and interactions of ENMs in terrestrial habitats.

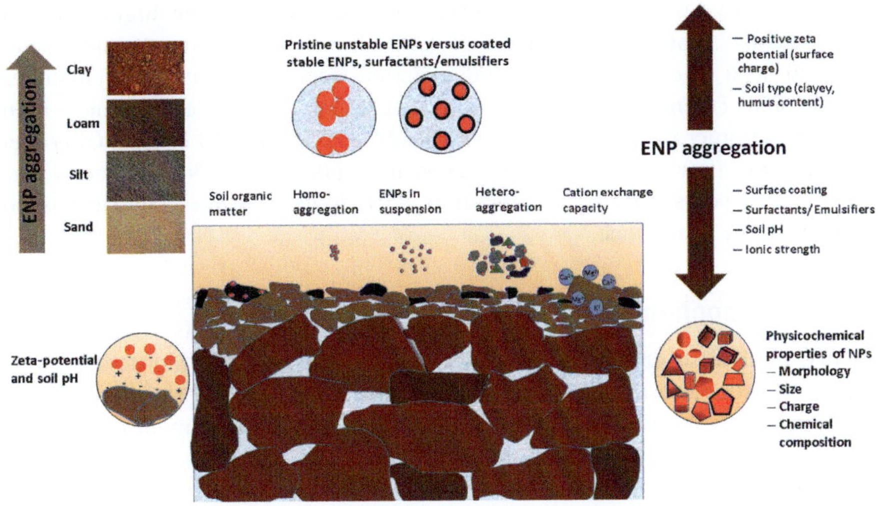

FIGURE 4.1

Factors influencing the fate, transport, and retention of ENMs in soil. *ENM*, Engineered nanomaterial.

4.2.1 Soil type

Soil characteristics (percentage of silt, clay, loam; grain size; porosity; and dissolved organic matter) are important to determining the fate and transport of ENMs (Park et al., 2016). Soil particles predominantly carry a negative charge, which greatly influences ENM's transport and retention (Lin et al., 2010a). Soils with higher clay content or higher organic matter tend to hetero-aggregate with the nanoparticles (NPs), which limit particle transport. In addition, ENMs' surface coating and associated zeta potential (ζ) determine their movement through soil, although these features will be dynamically modified by soil biota being degraded and/or deposited over time.

Darlington et al. (2009) observed differential transport and retention of phosphate-treated aluminum (Al) NPs, as well as untreated Al NPs, in soil and sand. The authors reported more retention and consequently, less mobility, in soil than sand. The untreated Al NPs, suspended in water, were transported much less in soil [39% of the initial concentration (C_o)] than in sand (68% of C_o). The treated Al NPs had breakthrough at six pour volumes and a low transport (71% of C_o) in soil, compared to breakthrough at five pour volumes. In addition, they had improved transport (95% of C_o) in sand column. Zhao et al. (2012) reported that ZnO NPs were preferentially retained in sandy loam soil (99%) as compared to sandy soil (68%–99%), with column travel distances of 5.3 and 19.2 cm in sandy loam and sandy soil, respectively. This was likely due to the higher clay or organic matter content in former soil type. In a comparative column study conducted with three kinds of coated quantum dots in two types of matrices, Quevedo and Tufenkji (2012) observed attachment efficiencies in sandy loam greater by an order of magnitude than in sandy quartz. Sharma et al. (2014) tried to analyze the flow of multiwalled carbon nanotubes (MWCNTs) through fine, medium, and coarse sand, and reported 85% lower retention in finer sand, compared to the coarse one.

4.2.2 Physicochemical properties of soil

Soil pH, total dissolved solids, organic matter content, CEC, moisture content, texture, and structure all have been shown to influence ENM's transformation, aggregation, and plausible transport (Dwivedi et al., 2015; Majumdar et al., 2016; Rodrigues et al., 2016). In addition, ENM's physicochemical properties such as size, charge, chemical composition, morphology, and surface characteristics, along with the solution chemistry [ionic strength (IS) and pH of the soil solution] will also significantly impact ENM's transport in soil (Dror et al., 2015; Park et al., 2016). The valence of the primary salt cation in solution could also have an effect on particle stability, for example, the divalent cation calcium (Ca^{2+}) gives a greater stability than the monovalent potassium (K^{+}) (Emerson et al., 2014; Park et al., 2016). Quevedo and Tufenkji (2012) studied the transport of three types of coated quantum dots in sand columns and observed that Ca^{2+} promoted greater retention than K^{+}. For instance, the particle size of CdSe quantum dots at 0.1 mM ionic concentration was 76 nm in KCl versus 257 nm in $CaCl_2$ solution, which is a 70.42% increase.

In the soil environment, ENMs may become coated with the organic matter (humic and/or fulvic acids), thus yielding an increase in particle size (Lowry et al., 2012). The ENMs may also undergo a change in the zeta potential during this transformation process, with positively charged materials becoming negative because of the soil-derived coating (Cornelis et al., 2014). ENM can also get attached to clay particles and/or natural organic matter in the native or dissolved ionic state (given the CEC of soil), resulting in hetero-aggregation. This reduces the available surface area of the particles, as well as the surface-to-volume ratio for any chemical interaction. These ENM transformations also affect material bioavailability and possible toxicity, since the particles are no longer in the nanoscale form. Some particles may even undergo pseudo-homo-aggregation with similar NPs (Lowry et al., 2012). Between homo and hetero-aggregation, the latter appears to be the dominant occurrence, leading to particle size increases (Quik et al., 2012). Emerson et al. (2014) confirmed this with column experiments in natural soil that was designed to simulate the vadose zone. More than 99% of the added [bare and Suwannee River–natural organic matter (SR-NOM) coated] iron oxide and (SR-NOM, citrate, and dodecanethiol coated) silver NPs were retained within the first 5 cm of the column due to hetero-aggregation.

Aggregation with the natural organic matter generally limits the movement of ENMs in soil. However, a study conducted by Sagee et al. (2012) with Ag NP transport in sandy clay soil illustrated contrary results. The mobility of the NM increased as the concentration of humic acid increased. This is an indication of very strong soil and NP interactions. In addition, the humic acid layer around the Ag NPs was apparently limiting mechanical straining and enhancing the transport. Likewise, the transport was improved in the presence of larger soil aggregates. The NPs and soil interact at the soil aggregate surface; smaller aggregates provide much larger surface area, hence lesser transport. This unconventional behavior of the NM was apparently due to the natural matrix used in this study, compared to pure quartz sand, bromide tracer soil, or even glass beads.

In a study conducted with nano-sized iron (II) oxide, the NPs underwent extensive aggregation within at pH 5–8, with maximum aggregation at pH 8.5. However, in the presence of Suwannee River humic acid, the same NPs aggregated at a pH range of 4–5. Transmission electron microscopy (TEM) and field flow fractionation demonstrated the formation of a humic acid layer on the FeO NPs, further promoting aggregation. Particle aggregation continues to increase with increasing humic acid concentration in the pH range of 2–6 (Baalousha et al., 2008). With metal-based NPs the pH of the soil controls the dissolution and oxidation state of the metal ion. The soil pH also determines the electrostatic interactions among NPs and the porous media (Lin et al., 2010b). However, in most studies, IS appears to be more important than pH in determining flow through the matrix (Ben-Moshe et al., 2010). In aquatic systems/solutions the decreasing pH normally promotes dissolution of metal ions. In a soil system, on the contrary, many factors related with soil chemistry affect the simple equation (Anderson et al., 2017; Hounslow, 2018).

Emulsifiers or surfactants are commonly used to stabilize NMs and limit homo- and hetero-aggregations with nearby chemical species. These capping agents limit the impact of environmental factors such as air, water, and pH variations and restrict transformations, including oxidation, dissolution, or aggregation (Dwivedi et al., 2015; Kanel and Choi, 2007). These stabilizing agents also improve NM's transport through the soil (Chekli et al., 2016). Lin et al. (2010b) compared the mobility and retention of two differently stabilized nano–zero valent iron (NZVI) particles in porous media. Polyacrylic acid (PAA) and carboxymethyl cellulose (CMC) were used to produce the two kinds of NZVI; these particles had limited aggregation because of steric and electrostatic repulsion. The former (PAA coated) was found to transport more effectively than the latter (CMC coated) in the matrix. This was because the chemistry of formation of PAA coated NZVI enabled much more steric hindrance, compared to the CMC coated NZVI. In a study conducted by Lu et al. (2014), positively charged CNTs were generated after stabilization with cetylpyridinium chloride; this modification enhanced material retention to negatively charged soil particles. Conversely, negatively charged CNTs were generated by stabilization with sodium dodecylbenzenesulfonate and octylphenol ethoxylate (TX-100); here, improved transport was evident throughout the breakthrough curve.

The critical coagulation coefficient (CCC), the attachment efficiency (α), and the related electrolyte IS also influence NPs aggregation in soil. Attachment efficiency (α) ranges from 0 to 1, where 0 indicates slow and unfavorable association to soil and 1 indicates a rapid and extensive attachment. CCC signifies the lowest amount of electrolyte needed to disrupt the colloidal suspension (Huynh and Chen, 2011). A lower CCC value indicates strong aggregation possibilities and a higher CCC range suggests strong colloidal nature and stability (Dwivedi et al., 2015). The solution IS will play an important role in determining the balance between NPs aggregation and a stable colloidal suspension (Lin et al., 2010a). The critical deposition constant (CDC) is another indicator IS reactivity similar to the CCC; the difference being that the CDC applies to a particle–surface interface and the CCC applies to homo-aggregation (Cornelis et al., 2014). Sharma et al. (2014) conducted a deposition and transport study with MWCNT in saturated sand columns with variable IS, pH, grain size, and flow velocity. IS was found to be the most important factor; the CDC of 4 mM gave maximum MWCNT retention.

For graphene oxide (GO) the IS of the soil solution was more important than pH in ENM retention in an aqueous saturated porous medium (Lanphere et al., 2013). The hydrodynamic diameter and electrophoretic mobility measurements indicated that GO transport decreased with increasing IS of the KCl solution, likely resulting from greater aggregation. An amount of 0.01 and 0.1 M KCl in the packed bed column yielded a 5% and 95% retention, respectively (Lanphere et al., 2013). Qi et al. (2014) reported similar findings with GO in natural (Lula) soil columns. The higher IS of the soil solution (hence reduced zeta potential of the ENM) leads to increased aggregation and reduced stability in the colloidal state. Looking at the breakthrough curves of GO NPs in Lula soil, it was observed that the gradation from 0 to 50 mM NaCl dropped the NP transport from 100% to 60%, indicating increased aggregation. Stebounova et al. (2011) illustrated the tendency of Ag NPs to aggregate in biological fluids of high IS, remaining in suspension as single NP in IS < 10 mg/L. Similar findings of lower transport, improved aggregation, and higher hydrodynamic diameter have been reinforced by other column studies with C_{60}, MWCNT, and $n\mathrm{TiO_2}$, respectively (Hedayati et al., 2016; Xu et al., 2017).

4.3 Impact of the engineered nanomaterials exposure on terrestrial plants

The potential interactions of ENMs with plants are numerous, ranging from enhancing plant growth/productivity and controlling diseases to overt phytotoxicity (De la Rosa et al., 2017; Zuverza-Mena et al., 2017; Reddy et al., 2016; Du et al., 2017; Ma et al., 2015a; Gardea-Torresdey et al., 2014). However, there are a large number of important knowledge gaps in understanding the effects of ENMs in agroecosystems and in their potential accumulation in the food chain (De la Rosa et al., 2017). Fig. 4.2 shows a schematic diagram ENMs exposure to plants, with potential subsequent biochemical and physiological impacts. In the environment, nanomaterials (NMs) tend to agglomerate, dissolve, precipitate, and biotransform as described above (Reddy et al., 2016; Markus et al., 2015). However, depending on the ENM, its properties, and the dose; the effects on biota will be different, even within the same environment. It has been reported that some ENMs have beneficial impacts on plants, such as nanofertilizers and nano-pesticides (Servin et al., 2015). Elmer and White (2016) reported that NPs (CuO, MnO, and ZnO) may enhance plant growth and nutrition, effectively suppressing disease in tomato and eggplant. However, in other systems, these same ENMs have shown detrimental effects (Verma et al., 2018; Zuverza-Mena et al., 2017; Servin and White, 2016; Gardea-Torresdey et al., 2014). The measured endpoints of toxicity include seed germination, photosynthetic rate, plant growth and development, fruit production, reactive oxygen species (ROS) generation, and synthesis of biomolecules such as carbohydrate, proteins, lipids, and hormones (Zuverza-Mena et al., 2017; Reddy et al., 2016; Mahmoodzadeh and Aghili, 2014; Rico et al., 2014; Siddiqui and Al-Whaibi, 2014). The impacts vary with the exposure conditions, ENM characteristics, and species (Tan et al., 2018; De la Rosa et al., 2017). Table 4.1 shows the summary of biochemical and physiological responses of plants to selected ENMs.

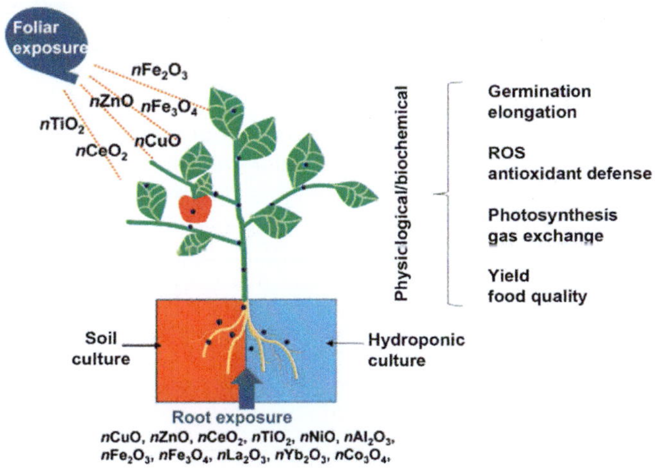

FIGURE 4.2

Annotated diagram showing biochemical and physiological impact of ENMs in plants. *ENM*, Engineered nanomaterial.

Adapted from Du, W., Tan, W., Peralta-Videa, J.R., Gardea-Torresdey, J.L., Ji, R., Yin, Y., et al., 2017. Interaction of metal oxide nanoparticles with higher terrestrial plants: physiological and biochemical aspects. Plant Physiol. Biochem. 110, 210–225.

Table 4.1 Summary of biochemical and physiological impact of selected ENMs on plants.

S. No.	ENM	Size	Concentration	Plant	Application mode	Biochemical/ physiological response	Reference
1	SWCNHs	50–100 nm	0.025–0.10 mg/L	Rice, soybean, tomato, corn and tobacco cell culture	Seed exposure to ENM in magenta boxes	Seed germination activated, seedling growth enhanced and increased tobacco cell growth	Lahiani et al. (2015)
2	MWCNTs	Inner diameter: 5–10 nm	500–5000 mg/kg	Corn, tomato, soybean, and zucchini	Soil	No effect on tomato and zucchini growth; decreased biomass in soybean and corn at 500 mg/kg	De La Torre-Roche et al. (2013)
	C_{60} fullerenes	Outer diameter: 13–18 nm					
3	Fullerol	Hydrodynamic size: 1.5 ± 0.2 nm	0.9–47.2 nM	Bitter melon	Seed pretreatment	Promoted growth and yield; fullerol found in the plant tissues	Kole et al. (2013)
	$[C_{60}(OH)_{20}]$	5.0 ± 0.7 nm					
4	Single-bilayer graphene oxide sheet	0.5–5 μm	100–1600 mg/L	Faba bean	Seed presoak	Stressed growth and development; reduction in POD enzyme activity	Anjum et al. (2014)
5	CeO_2 NP		500 mg/kg	Radish	Soil	Reduced seed germination and no effect after 15 days	Corral-Diaz et al. (2014)
6	Citric acid–coated/ uncoated CeO_2 NPs		0–500 mg/L	Radish	Hydroponics	No effect on germination, stem biomass and root and shoot length; CA coated increased root biomass and water content at 200 mg/L	Trujillo-Reyes et al. (2013)
7	CeO_2 NPs		2000 mg/L	Cucumber	Hydroponics	No effect on root and shoot biomass	Rui et al. (2015)
8	Citric acid–coated/ uncoated CeO_2 NPs		62.5–500 mg/kg	Tomato	Soil	62.5 mg/kg increased total fruit number; increased shoot length at 500 mg/kg	Barrios et al. (2016)

9	Citric acid–coated/ uncoated CeO_2 NPs		62.5–500 mg/kg	Tomato	Soil	CA-coated $nCeO_2$ reduced the dry weight	Barrios et al. (2017)
10	*n*Cu, *n*CuO, Cu $(OH)_2$	100–1000 nm	5–20 mg/L	Lettuce and alfalfa	Hydroponics	Reduced root growth, reduced shoot P and Fe contents and increased root APX activity in lettuce; decreased root length, increased shoot Cu, P, and S contents, reduced root and shoot CAT activity and increased root APX activity in alfalfa	Hong et al. (2015)
11	CuO NPs	10–100 nm $\geq 10^4$ nm $\leq 50^4$ nm	40–120 mg/L	Rice	Seed exposed to NP suspension	Decreased germination weight and yield, and root and shoot growth	Shaw and Hossain (2013)
12	CuO NPs		0–500 mg/kg	Bell pepper	Soil	No effect on stem elongation, dry biomass, leaf size and fruit production	Rawat et al. (2018b)
13	Cu NPs, CuO NPs, $Cu(OH)_2$	100–1000 nm	20, 80 mg/kg	Cilantro	Soil	*n*Cu (80 mg/kg) and *n*CuO (20 and 80 mg/kg) reduced germination and shoot elongation	Zuverza-Mena et al. (2015)
14	CuO, CuO: ZnO NPs	10–100 nm $\geq 10^4$ nm ≤ 50	100–500 ppm	Bean	Sand	At 500 ppm, inhibited growth, shoot nutritional element imbalance; increased Na, and decreased Fe, Mn and Zn contents	Dimkpa et al. (2015)

(Continued)

Table 4.1 Summary of biochemical and physiological impact of selected ENMs on plants. *Continued*

S. No.	ENM	Size	Concentration	Plant	Application mode	Biochemical/ physiological response	Reference
15	ZVI NPs	25 nm	10 g/L	Garden cress, White mustard and broom corn	Petri dish	Increased seedling length and biomass production in all plants	Libralato et al. (2016)
16	Fe/Fe_3O_4 NPs	50–60 nm	10, 20 mg/L	Lettuce	Hydroponics	Reduced root size, reduced chlorophyll content, particle aggregation on the root surface and increased antioxidant enzyme activities related to mineral element composition	Trijullo-Reyes et al. (2014)
17	Fe_2O_3 NPs	10 nm		Soybean	Foliar	Increased leaf and pod dry weights, and grain yield	Sheykhbaglou et al. (2010)
18	ZnO NPs	50 nm	0–1 g/L	Mesquite, Palo verde and tumbleweed	In vitro	No effect on germination of all plants	De la Rosa et al. (2011)
19	ZnO NPs	44.4 nm	0–4000 mg/L	Onion	Hydroponics	Reduced root growth	Ghodake et al. (2011)
20	ZnO NPs	10 nm	5, 10, 20 mg/L	Arabidopsis	Agar	Inhibited seed germination, root elongation and leaf total number	Lee et al. (2010)
21	ZnO NPs	1.2–6.8 nm	400–4000 mg/L	Alfalfa	Soil	Reduced biomass	Bandyopadhyay et al. (2015)
22	ZnO NPs		250–750 mg/kg	Guan bean		Increased biomass, chlorophyll and total soluble protein	Raliya and Tarafdar (2013)

Common plant species and biological name in the order of appearance in the table: *Alfalfa:* Medicago sativa, *bell pepper:* Capsicum annuum, *bitter melon:* Momordica charantia, *broom corn:* Sorghum saccharatum, *corn:* Zea mays L, *cilantro:* Coriandrum sativum, *faba bean:* Vicia faba, *garden cress:* Lepidium savitum, *lettuce:* Lactuca sativa, *onion:* Allium cepa, *palo verde:* Parkinsonia, *radish:* Raphanus sativus, *rice:* Oryza sativa, *tomato:* Solanum lycopersicum/Lycopersicon esculentum, *tumble weed:* Salsola, *tobbaco:* Nicotiana tabacum, *white mustard:* Sinapis alba, *zucchini:* Cucurbita pepo ssp. Pepo.

APX, *Ascorbate peroxidase;* CA, *citric acid;* CAT, *catalase;* MWCNT, *multiwall carbon nanotube;* NP, *nanoparticle;* POD, *peroxidase;* SWCNH, *single-walled carbon nanohorn;* ZVI, *zero valent iron.*

4.3.1 Engineered nanomaterials uptake and translocation in edible plant species

Plant–ENM interactions can be illustrated by using a model, which includes the source of ENMs, transport media, and the point contacts with the plant (soil, water, or air) (De la Rosa et al., 2017). An important model proposed by Dietz and Herth (2011) elaborated pathways in which NMs can translocate and accumulate within plants. Upon release into the soil, ENMs can bind to the soil particles and/or soil organic matter, being biotransformed prior to accumulation by plants. The root uptake and bioavailability of ENMs in plants depend on the particle retention/movement dynamics, which are influenced by both biotic and abiotic factors (Reddy et al., 2016; Jaisi and Elimelech, 2009). The model also highlighted ENM interactions with plant-root structures and plant-shoot compartments (stem and foliar tissues) (De la Rosa et al., 2017). The authors note that NPs exhibit differential transport in plants, which depends not only on particle size and surface properties but also on species-specific physiology and morphology. Furthermore, it is evident that during in planta transport, ENM biotransformation may also occur (De la Rosa et al., 2017; Zhang et al., 2012a). Generally, there are two main exposure routes for plants regarding the uptake and translocation of ENMs; root or foliar pathways (Du et al., 2017). Studies focused on the root pathway have used either soilless (Hernandez-Viezcas et al., 2016) or soil media (Rico et al., 2015). ENMs can be taken up through apoplastic pathways via roots, and then translocated to the vascular system via the symplastic pathways (Zhao et al., 2014). Apoplasm involves the movement of fluids through adjacent cell walls of epidermis and cortex without penetrating the cytoplasm, while symplasm is the cytoplasmic movement of fluids within plant cells. Several factors affect the absorption including root exudates (Huang et al., 2017), surface chemistry of the ENMs (Avellan et al., 2017), and microorganisms in the rhizosphere. After absorption the particles can then move through the xylem and phloem to the aboveground tissues (Hernandez-Viezcas et al., 2013; Servin et al., 2013; Zhao et al., 2014). Conversely, foliar exposure pathways involve basipetal movement of the ENMs from the leaves through the phloem to the stem and to the root (Elmer and White, 2016; Hong et al., 2014). Speciation and biotransformation of ENMs are critical to the understanding of the mechanisms of their uptake, translocation, and bioaccumulation in plant systems (Zhang et al., 2012b). Techniques such as TEM and scanning electron microscopy with energy dispersive X-ray spectroscopy, confocal scanning microscopy, and synchrotron radiation are effective tools for ENM characterization. These instruments reveal the forms in which ENMs can be taken up and eventually transformed within the plant, which may have significant impact on the physiological and biochemical effects. The uptake and subsequent biotransformation of ENMs in plants differ across plant type and species. For example, previous results have revealed that metallic oxides of

Zn and Cu are transformed to dissolved ions rather than maintaining their nanoforms (De la Rosa et al., 2017). For example, it was reported that ZnO NPs biotransformed rapidly in soybean (Hernandez-Viezcas et al., 2013), wheat (Dimkpa et al., 2013), and cowpea (Wang et al., 2013). Speciation may also affect ENM mobilization in the soil. Dimkpa et al. (2013) revealed that CuO NPs agglomerate first in the soil and then build up as larger complexes as CuO NPs and Cu(I)-sulfur. In addition, CeO_2 NPs can also exist in the nanoform and as $CePO_4$, $Ce(OH)_4$, or carboxylates in plant tissues (Zhang et al., 2012a; Hernandez-Viezcas et al., 2016). Some ENMs such as TiO_2 NPs have been reported to be taken up and translocated in plants without conformational changes in their nanoform (Larue et al., 2012; Servin et al., 2013; Tan et al., 2018).

4.3.2 Physiological responses

4.3.2.1 *Engineered nanomaterials influence germination, growth, and yield*

The interaction between ENMs and plants can be described in terms of the physicochemical properties of specific ENMs (Verma et al., 2018). Binding of organic ligands in the soil, and eventual biotransformation, may also influence the physiological response of plants to ENM exposure (De la Rosa et al., 2017). The impact of a number of NMs on plants was recently reviewed by Zuverza-Mena et al. (2017). The range of responses was found to vary across exposure time, concentration, NM types, and plant species.

Carbon-based NMs. Single-walled carbon nanohorns (SWCNHs), at 0.025–0.10 mg/L increased seed germination and enhanced the growth of corn (*Zea mays*), soybean (*Glycine max*), rice (*Oryza sativa*), and tomato (*Solanum lycopersicum*); the most effective concentration varied with plant species and the greatest effect was noted with tobacco (Lahiani et al., 2015). Interestingly, higher concentrations of SWCNHs had no significant negative effects on the plants. Similarly, MWCNTs and C_{60} fullerenes at 500 mg/kg had negligible effects on zucchini and tomato growth in soil after exposure for 28 days. Conversely, corn and soybean biomass were reduced under the same exposure concentrations (De La Torre-Roche et al., 2013). Pretreatment of bitter lemon seed with fullerol [$C_{60}(OH)_{20}$] solution for 48 hours prior to growth in potting mix significantly enhanced growth and fruit yield; however, fullerol was detected within the root and shoot components of the exposed plants (Kole et al., 2013). Anjum et al. (2014) conducted germination studies of single-bilayer GO in faba beans; the authors reported reduced plant growth and development. The mechanism of toxicity was largely attributed to the interference of the compound with the nutrient and water uptake in plant. However, their growth enhancement potentials were suggested to presumably be caused by cell division and development facilitated by favorable gene expression (De La Torre-Roche et al., 2013).

Cerium oxide NM. Many studies have examined the effect of cerium oxide NPs on different plant species. Corral-Diaz et al. (2014) reported that after 4 days of radish (*Raphanus sativus*) treatment with CeO_2 NPs at 500 mg/kg, the seed germination decreased but had no effect after 15 days. This is an indication that the CeO_2 NPs only have effect on radish at germination stage. Citric acid (CA) coated and uncoated CeO_2 NPs at 50, 100, and 200 mg/L showed no effect on germination, root and shoot lengths, and stem biomass of radish (Trujillo-Reyes et al., 2013). However, at 200 mg/L, the CA-coated CeO_2 NPs (at 1:7 molar ratio) significantly increased the root biomass and water content in the plants (Trujillo-Reyes et al., 2013). Rui et al. (2015) showed that CeO_2 NPs at 2000 mg/L, with and without phosphate, did not affect the root and shoot biomass of cucumber grown in nutrient solution as compared with the control. Conversely, exposure of CA-coated and CA-uncoated CeO_2 NPs at 500 mg/kg, via soil, increased tomato shoot length by 9% and 13%, respectively (Barrios et al., 2016). However, the increase in length was not correlated with the increase in dry weight (Barrios et al., 2017). The above literatures indicate that the effect of CeO_2 NPs varies with plant species, concentrations, and exposure conditions. Ce compounds are reported to have both antioxidant characteristics and oxidative stress potential, largely depending on dose; this may be responsible for their contrasting effects on plant phenotype. Moreover, surface charge as influenced by surface coating agents can influence the impact of the NMs in plants (Medina-Velo et al., 2017; Barrios et al., 2016, 2017).

Copper-based NMs. Hydroponic exposure of lettuce and alfalfa to 5, 10, and 20 mg/L of Cu NPs, CuO NPs, and $Cu(OH)_2$ NPs for 15 days decreased the root growth by 49% in both plants (Hong et al., 2015). In rice, CuO NPs at 0.5, 1.0, and 1.5 mM decreased germination by 13%, 16%, and 20%, respectively, compared with the control (Shaw and Hossain, 2013). In addition, such treatments decreased the root length (74%, 82%, and 91%), shoot lengths (28%, 37%, and 45%), and shoot weight (21%, 31%, and 53%) after 14 days of growth (Shaw and Hossain, 2013). In addition, Dimkpa et al. (2015) reported the growth inhibition in the root by ~67% and the shoot by ~22% in bean exposed to 500 mg/kg CuO NPs in soil, compared with control. Similarly, the exposure of cilantro to soil amended with CuO NPs (at 20 and 80 mg/kg) decrease the germination rate by ~50%, while Cu NPs (80 mg/kg) reduced the shoot elongation by 12.4% (Zuverza-Mena, et al., 2015). In oregano, Du et al. (2017) reported that Cu NPs (at 0–200 mg/kg) reduced shoot biomass by 21.6%–58.5%, compared with control. Conversely, in soil, Rawat et al. (2018b) reported that CuO NPs at 125, 250, and 500 mg/kg had no effect on plant growth or fruit production in bell pepper. It has been largely reported that the growth inhibition effect of Cu-based ENMs could be attributed to effective Cu delivery capacity of these compounds.

Iron NMs. Trujillo-Reyes et al. (2014) revealed that lettuce exposed to 20 mg/L of Fe/Fe_3O_4 NPs in hydroponic solution exhibited reduced leave biomass by 55% (Trujillo-Reyes et al., 2014). In addition, *Arabidopsis thaliana* grown in an agar medium containing 4 mg/kg of Fe_2O_3 NPs showed a significant reduction in total biomass (Marusenko et al., 2013). However, toxic effect on plant growth and development could result from aggregation of NPs on the root surface, which may hinder water permeability and photosynthetic activities, decreasing plant growth (De la Rosa et al., 2017; Libralato et al., 2016; Trujillo-Reyes et al., 2014; Asli and Neumann, 2009). However, Mukherjee et al. (2014) revealed that 25 days exposure of iron doping ZnO NPs (Fe@ZnO NPs) had minimal toxicity effects in *A. thaliana* relative to the bare ZnO NPs in term of zinc uptake, chlorophyll content, and H_2O_2 production. It was also reported that foliar exposure of soybean to Fe_2O_3 NPs at 0.75 g/L increased the leaf and pod dry weight by 42%, and at 0.5 g/L, increased grain yield by 48% (Sheykhbaglou et al., 2010). Li et al. (2015) reported that nanoscale zerovalent iron at 40 and 80 μmol/L, biostimulates growth and development in peanut seedlings. These authors suggested that such NPs could permeate the seed coat enabling water absorption into the seeds, thus enhancing seed germination and seedling development.

Zinc NMs. An in vitro study on mesquite, palo verde, and tumbleweed revealed that the exposure to 0–4000 mg/L of ZnO NPs for 4, 6, and 7 days had no significant effect on seed germinations (De la Rosa et al., 2011). However, an inhibitory effect on *A. thaliana* seed germination was observed after 30 days' exposure to ZnO NPs at 400, 2000, and 4000 mg/L in agar medium (Lee et al., 2010). Ghodake et al. (2011) showed that the hydroponic exposure of onion to ZnO NPs at 5, 10, and 20 μg/mL significantly reduced the root length by 68%, 74%, and 74%, respectively. In addition, alfalfa grown in soil treated with 250, 500, and 750 mg/kg ZnO NPs exhibited an 80% reduction in root biomass (Bandyopadhyay et al., 2015). In contrast, Raliya and Tarafdar (2013) revealed that foliar exposure of cluster bean (*Cyamopsis tetragonoloba* L.) to 10 mg/L of ZnO NPs increased the plant biomass by 27% compared with the control. Medina-Velo et al. (2017) reported that the effects of ZnO NPs (0–500 mg/L) depend on the surface charge. These researchers reported that uncoated ZnO NPs (Z-COTE) did not produce phenotypic changes in *Phaseolus vulgaris*, while Z-COTE HP1 (triethoxycaprylylsilane-coated ZnO), at all concentrations, increased root length (~44%) and leaf length (~13%) compared with control. Disruption of cell morphology by Zn NPs at certain concentrations has been largely reported, which is possibly responsible for phytotoxic impact. Generally, zinc exists in plant primarily as Zn^{2+} ions or as complexes with organic acid chelates (Sturikova et al., 2018). Zinc ion serves as activator for several enzymes such superoxide dismutase (SOD), and carbonic anhydrase; in addition, it is important in macromolecule metabolism and synthesis of proteins (De la Rosa et al., 2011; Pullagurala et al., 2018b,c). The mechanism of Zn ions toxicity in plants is based on the competition with other ions at the binding sites (Sturikova et al., 2018).

4.3.2.2 Chlorophyll content and micro/macronutrients accumulation

Leaf pigments, such as chlorophylls, are widely used parameter to evaluate plant response to stress or contaminant (Du et al., 2017; De la Rosa et al., 2017). Impact of ENMs on chlorophyll contents in plants has been studied extensively. Rice seeds exposed to CeO_2 NPs at 125 mg/L for 10 days in aqueous media exhibited significant reduction in chlorophyll *a* and *b* by 27% and 60%, respectively (Rico et al., 2013). Ma et al. (2013) showed that *A. thaliana* exhibited a decreased chlorophyll content upon exposure to CeO_2 NPs at 250 mg/L in agar medium (Ma et al., 2013). Zhao et al. (2013) reported no changes in chlorophyll content and gaseous exchange when cucumber was exposed to CeO_2 NPs at 400 and 800 mg/kg for 53 days in the mixture of loam sand, sand soil, and organic soil. Similarly, corn grown in soil treated with CeO_2 NPs at 400 and 800 mg/kg for 84 days exhibited no significant impact on chlorophyll content and gaseous exchange in the plants (Zhao et al., 2015). Mung bean grown in agar treated with CuO NPs at 100, 200, and 500 mg/L showed reduction in chlorophyll content by 33%, 44%, and 71%, respectively, compared with the control (Nair et al., 2014). Chlorophyll accumulation increased in lettuce exposed to Fe/Fe_3O_4 NPs at 10 and 20 mg/L under hydroponic conditions by 12 and 7 soil-plant analyses development (SPAD) units (Trujillo-Reyes et al., 2014). Marusenko et al. (2013) also reported decreased chlorophyll content of *A. thaliana* grown in algal medium exposed to Fe_2O_3 NPs. In addition, an increase in chlorophyll content was observed in cluster bean (*C. tetragonoloba* L.) exposed to 10 mg/L of ZnO NPs through soil for 28 days (Marusenko et al., 2013). Du et al. (2017) revealed that Cu NPs at 50, 100, and 200 mg/kg did not alter the chlorophyll content of oregano plant (*Origanum vulgare*) grown for 60 days. Moreover, a study have shown that foliar exposure of basil (*Ocimum basilicum*) to $Cu(OH)_2$ NPs at 4.8 mg/pot did not altered chlorophyll in a low anthocyanin variety (LAV) but increased it in a high anthocyanin variety (HAV) (Tan et al., 2018). In LAV and HAV varieties, only Mn was affected by foliar exposure to $Cu(OH)_2$ NPs. The Mn content in LAV and HAV varieties was reduced in leaves (68% and 79%) and stems (75% and 75%), respectively. Barrios et al. (2017) reported that in tomato, uncoated CeO_2 NPs at 500 mg/kg decreased the B (28%), Fe (78%), Mn (33%), and Ca (59%), while CA-coated CeO_2 NPs, at 125 and 500 mg/kg, increased B by 33%. Rico et al. (2015) reported increased Ca, K, Zn, Mg, Cu, Al, Fe, and P uptake in barley grown in soil amended with CeO_2 NPs at 125, 250, and 500 mg/kg. However, transgenic cotton showed reduction in Zn, Mg, Fe, and P in xylem sap on exposure to CeO_2 NPs at 100 and 500 mg/L (Ma et al., 2015b). In addition, in soil CeO_2 NPs at 0–1000 mg/kg altered the accumulation of essential elements in soybean (Peralta-Videa et al., 2014). Overall, the data corroborate that the effects depend on the type of ENMs and the species of plants. However, there are no studies shown results from a single plant species exposed to several NMs.

4.4 Oxidative stress responses

4.4.1 Enzyme assays

At the biomolecular level, plant defensive systems against environmental stress include several response mechanisms. These include the synthesis of oxidative stress enzymes such as catalase (CAT), SOD, and peroxidases, among others, and substrates including malondialdehyde (MDA) and a range of glycoproteins (De la Rosa et al., 2017). Peroxidation of polyunsaturated fatty acids with free radicals produces MDA, which can be used as a biomarker of oxidative stress (Placer et al., 1966; De la Rosa et al., 2017). Cilantro plants exposed to 0–500 mg/kg of $nCeO_2$ in soil showed an increase in CAT and ascorbate peroxidase activities at 62.5 and 125 mg/kg of $nCeO_2$. The authors suggested that the increased CAT activity corresponded with plant elongation, which, in turn, correlated with increases in cellular activity and H_2O_2 production (Morales et al., 2013). Short-term foliar exposure of *Physcomitrella patens* to nano-Fe at 5 ng to 50 mg for 7 days had no impact on ROS and reactive nitrogen species, whereas negative effects were reported under longer term studies. The negative effects were determined by MDA production and glutathione regulation (Canivet et al., 2015). Enzyme activities generally reveal catalytic interactions between NPs and plants and also demonstrate hormetic effect in plants. Hormesis effect in a plant can be defined as a dose–response activity wherein the chemical exposure at lower dose can induce beneficial effect while having an inhibitory effect at higher doses (Calabrese and Blain, 2009). The measurement of H_2O_2, membrane leakage, and lipid peroxidation (MDA) has proven to be very successful indicators in the oxidative stress assessment of plants.

4.4.2 Omics

Genomics (structure and function of DNA), transcriptomics (study of mRNA), proteomics (structure and study of proteins), and metabolomics (metabolite profiles of a unit sample) are well-known omics-based approaches in system biology. In general, the genomics and transcriptomic analyses work on the principle of microarray technologies, whereas the proteomic and metabolomic require mass spectrometry (Espín-Pérez et al., 2014).

A recent literature suggests that these analytical methods are good indicators of oxidative stress in ENMs—plants exposure studies (Ruotolo et al., 2018). For instance, Majumdar et al. (2015) performed a proteomic analysis in *P. vulgaris* seeds obtained from plants cultivated in soil amended with CeO_2 NPs. The authors reported upregulation of stress-related proteins in seeds of plants exposed to 62.5 and 125 mg/kg. Similar results were observed when the CeO_2 NPs were applied through the foliar route (Salehi et al., 2018). Besides proteomics, "metabolomics" has also been used to explore the effects of ENMs in plants. The gas chromatography–mass spectrometry (GC–MS)-based metabolomics study on

TiO_2 NPs exposed rice plants revealed increased levels of amino acids and palmitic acids in the grains (Zahra et al., 2017). Similar results were obtained when CuO NPs were exposed to *Cucumis sativus* plants. This NMR and GC–MS-based metabolomics approach reported alterations in sugar, amino acids, and fatty acid metabolites (Zhao et al., 2017). The data suggest that amino acids and fatty acids are most commonly affected by the ENMs exposure. Omics techniques have been used in very few studies; thus it is premature to conclude about the effects of a specific ENM in several plant species. The major setback about the omics-based approaches is the instrumentation, sample preparation, and data handling. Despite the complexities, omics-based studies are very essential for a better understanding of the interaction of ENM with plants. The omics approaches should be considered complementary rather than exclusive. An example of the differences encountered when studying gene regulation at transcriptomics and proteomics level is shown in Marmiroli et al. (2015).

4.5 Impact of the engineered nanomaterials exposure on other soil biota

In agroenvironments, plant growth and development depend on the association with a host of other organisms (Fig. 4.3). Thus the ENM's potential to affect other organisms, such as microbes and invertebrates, is worthy of study.

4.5.1 Bacteria

The soil microbiome is an essential part of the terrestrial ecosystem. These organisms have a pivotal role in plant rhizosphere, as well as in waste decomposition, nutrient cycling, and plant growth performance (Trivedi et al., 2017). A wide range of bacterial phyla are known to be important to plant performance in soil, including groups such as actinobacteria, proteobacteria, acidobacteria, and cyanobacteria. The relative abundances of these bacterial populations can be used as a good indicator of soil health (Vikram et al., 2016). Thus any disturbance to the bacterial community can have deleterious effects on agricultural productivity; a number of studies have been carried out on the impact of ENMs exposure toward soil bacterial communities. For instance, Sillen et al. (2015) reported that Ag NPs at 100 mg/kg caused inhibitory effects on the rhizobial bacterial community (Sillen et al., 2015); the microbiota exhibited reduced enzymatic activity and altered carbon usage. In a second study at a lower dose, Ag NP at 30 mg^{-1} reduced the nitrite production of ammonia oxidizing by 90% (Michels et al., 2017). In both of these previous studies the toxicity was largely attributed to ionic dissolution of the Ag metal. The role of the Ag^+ ion in bacterial inhibition has been characterized in a number of studies. For instance, Yang et al. (2013) exposed a number of bacterial strains to carbon-coated Ag NPs and $AgNO_3$; the

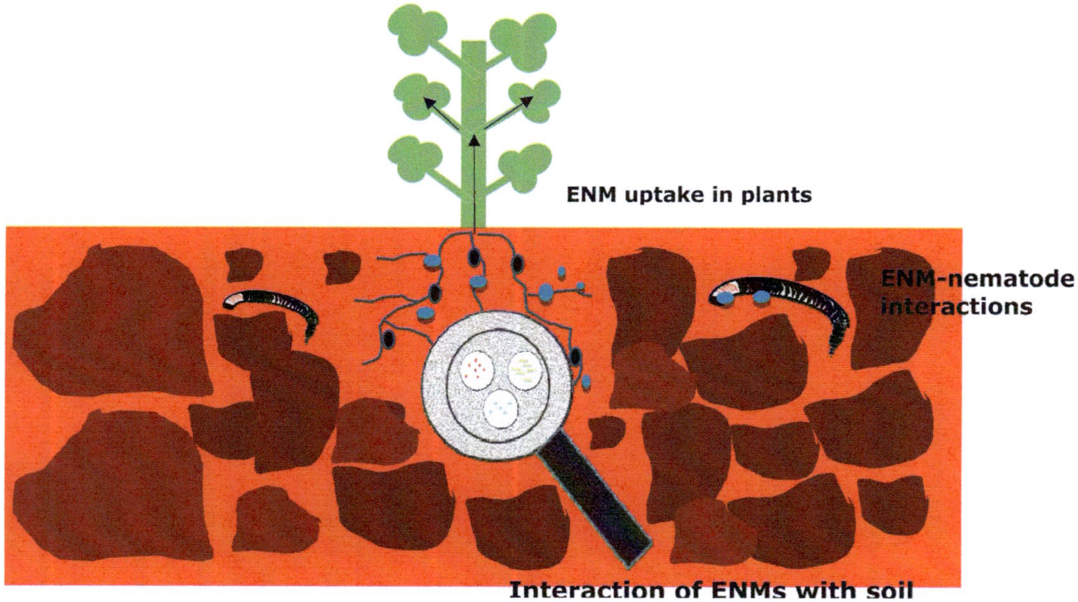

FIGURE 4.3

Impact of different ENMs on different biota. *ENM*, Engineered nanomaterial.

cellular and transcriptional response of the denitrifier *Pseudomonas stutzeri*, the nitrogen fixer *Azotobacter vinelandii*, and the nitrifier *Nitrosomonas europaea* were examined. The authors reported that Ag^+ was 20–48 times more toxic to the tested strains than was Ag NP. In addition to the Ag, other NPs such as ZnO and TiO_2 have been linked to significant compositional changes in soil microbial communities. According to Ge et al. (2011), ZnO NPs exposure reduced the soil microbial biomass at concentrations as low as 0.5 mg/kg, with endpoints including substrate-induced respiration and terminal restriction fragment length polymorphism analysis. Carbon-based ENMs have also been shown to have an impact on soil bacterial species. For instance, GO NP exposure under in vitro conditions was shown to negatively affect a wide range of bacteria, including *Bacillus marisflavi*, *Bacillus cereus*, *Bacillus subtilis*, *Bacillus megaterium*, and *Bacillus mycoides* (Gurunathan, 2015). Soil properties can have a significant impact of ENM fate and toxicity to resident bacterial populations. For instance, the CuO NP exposure had a strong effect on bacterial hydrolytic activity, oxidative potential, and composition of bacterial communities in a sandy loam soil but the effects were more modest in a sandy clay loam soil (Frenk et al., 2013; Adhikari et al., 2018). There are obviously also instances of species-specific response. For example, ZnO NP exposure had little overt bactericidal impact on Gram-negative bacteria such as *Pseudomonas aeruginosa*, whereas negative effects were clearly evident on the Gram-positive *B. subtilis* (Azam et al., 2012). It is possible that Gram-negative bacteria may be more tolerant to ENM exposure (Abbaszadegan et al., 2015). The impact of ENMs on the bacterial community tends may also be influenced by the presence of plant species and even the growth stage of those plants. For example, Liu et al. (2017) demonstrated that Ag NPs at concentration of 1 mg/kg had more inhibitory effect on soil bacterial community during the transition phase of wheat plant from seedling to vegetative stage. However, it must also be noted that there are published studies where ENM exposure had no impact on bacterial communities or in some cases, even exerted positive effects. For example, the CNTs exposure to soil grown tomato plants was shown to increase the relative abundance of *Bacteroidetes* and *Firmicutes* (Khodakovskaya et al., 2013). In a similar study carried out by Hamidat et al. (2016), nanoceria addition to soil favored the growth of hydrocarbon-degrading bacteria but the growth and development of bacteria that are related to disease suppressive activity of plant pathogens were negatively impacted. In another study, Au NP was shown to increase the population of the plant growth-promoting rhizobacteria (PGPR) such as *Pseudomonas fluorescens*, *B. subtilis*, *Paenibacillus elgii*, and *Pseudomonas putida* (Shukla et al., 2015). Conversely, ZrO_2 NPs at 1000 mg/L exposure had no impact on PGPR such as *B. megaterium*, *P. fluorescens*, *A. vinelandii*, and *Brevibacillus brevis* (Karunakaran et al., 2013). Thus the contradictory reports suggest that the bacterial species and type of ENM play a significant role on the overall effect. More molecular-based approaches are essential to characterize the ENMs effect on bacteria.

4.5.2 Mycorrhiza fungi

Arbuscular mycorrhizal fungi (AMF) are symbiotic microorganisms that have relationships with the majority of terrestrial plant species. The fungi offer range of benefits for the growth and development of plants. Importantly, there is evidence that AMF colonization can be inhibited by exposure to ENMs, although the effects may be complex. The diversity of AMF as a whole was decreased by the exposure of Fe_3O_4 NPs at the concentration of 10 mg/kg, whereas the number of species within the community was increased (Cao et al., 2017). On the other hand, when the AMF are inoculated into soil, their presence may ameliorate ENMs toxicity to plants. In one such study the AMF inoculation alleviated the negative impact of NP ZnO exposure effect on maize; fungal presence increased growth, nutrient uptake, and photosynthetic output (Li et al., 2015). Similar effects were observed in tomato plants inoculated with AMF and exposed Ag NP (Noori et al., 2017). The authors reported that AMF reduced by 14% the Ag uptake. This beneficial effect was attributed to the mycorrhizal colonization of the plant roots, which led to direct reductions in Zn and Ag uptake respectively. Specifically, AMF restricting ENM availability and uptake could be possibly through the secretion of glycoprotein called glomalin (Siani et al., 2017) (Fig. 4.4), which acts as a metal chelator in the rhizosphere.

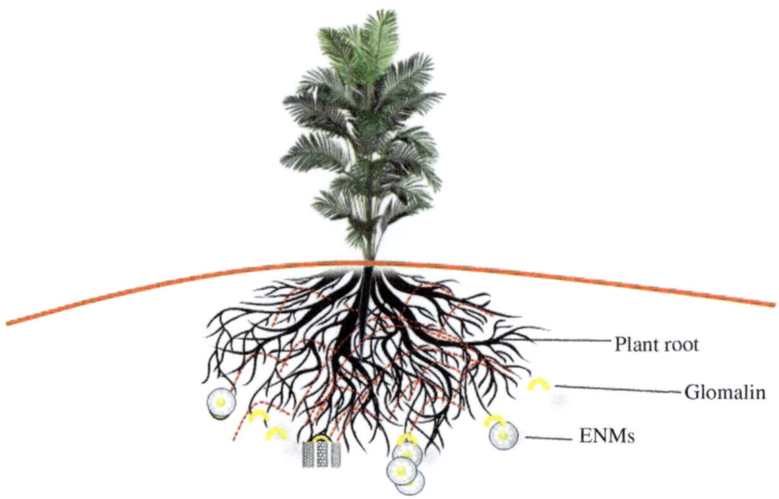

FIGURE 4.4

The AMF-based glomalin impact on ENMs uptake. *AMF*, Arbuscular mycorrhizal fungi; *ENM*, engineered nanomaterial.

4.5.3 Invertebrates

Soil invertebrates such as earthworms play a key role in soil and agroecosystem health through facilitated nutrient cycling and the formation of burrows in the soil that improve nutrient availability and drainage. Considering the significance of this activity, the impact of ENM exposure on these species is an important topic of study, though a number of studies have explored this topic (Table 4.2). For instance, the CuO NP exposure at 1000 mg/kg led to immune suppression and mortality of the earthworm species *Metaphire posthuma*. This was determined by the analysis of phagocytic response, generation of cytotoxic molecules, stress enzymes, and total protein of coelomocytes (Gautam et al., 2018). In another study, CuO NPs exposure on another invertebrate species, *Enchytraeus crypticus*, caused a shorter life span for adults (Gonçalves et al., 2017). Importantly, the juvenile earthworms were more sensitive to the harmful impacts of ENM exposure. Similar age-dependent effects were reported for *Lumbricus rubellus* populations. This upon C_{60} exposure, the juveniles experienced higher mortality than the adults (Van Der Ploeg et al., 2011). The primary mode of ENM's uptake from most soil invertebrates is through soil ingestion (Garcia-Velasco et al., 2016). The ENMs are attached to soil colloids, internalized upon the ingestion, and eventually some fractions are transferred to the digestive gut epithelium. The major reported mechanisms of ENM toxicity toward soil invertebrates include affecting pathways related to ribosome function, sugar/protein metabolism, as well as disruption of energy production and histone activity (Novo et al., 2015). Furthermore, nuclear magnetic resonate–based metabolomics has shown that the amino acids leucine, valine, isoleucine, and sugars such as glucose and maltose are potential bioindicators of ENM toxicity for invertebrates (Lankadurai et al., 2015).

There are a number of factors that can influence ENM based toward soil invertebrates. For instance, the bioavailability of metal-based ENMs is known to vary based on a number of factors. In a study carried out by Coutris et al. (2012) the accumulation of Co and Ag NPs in worms was 69% and 0.4%, respectively, over a period of 4 weeks. Thus it can be inferred that some metal-based NPs such as Ag are either less bioavailable or more rapidly excreted. Furthermore, the growth media may also play a role. For example, the exposure of ZnO and TiO_2 NPs caused minimal toxicity to *Eisenia fetida* in sand; effects were much more pronounced in soil. The enhanced effect was attributed to the soil-based effects such as dissolution of the ZnO NPs in the soil media (Cañas et al., 2011). ENM's size is also an important factor. Smaller size (10 nm) ZnO NPs were shown to cause more reproductive toxicity in *Caenorhabditis elegans* than larger 50 and 100 nm particles (Gupta et al., 2015). However, it must be noted that invertebrates may develop a capacity to avoid the effects of ENM exposure. For instance, some studies have reported that earthworms can differentiate and avoid

Table 4.2 Summary of ENMs impact on terrestrial invertebrates.

S. no.	Type of engineered nanomaterials	Organism type	Dosage	Effect	Reference
1	Ag NPs	*Eisenia andrei*	833 mg/g	Reproduction	Jesmer et al. (2017)
2	CuO NPs	*Enchytraeus crypticus*	1400 mg/kg	Shorter life span	Gonçalves et al. (2017)
3	CeO_2 NPs	*Porcellio scaber*	5000 mg	Increased lipid peroxidation	Malev et al. (2017)
4	ZnO NPs	*Caenorhabditis elegans*	500 mg	ATP content reduced and ROS is induced	Huang et al. (2017)
5	Au NPs	*C. elegans*	50×10^{10} particles/mL	Multigeneration effects	Moon et al. (2017)
6	Ag NPs	*Eisenia fetida*	500 mg/kg	Weight loss and mortality	Garcia-Velasco et al. (2016)
7	Ag NPs	*Lumbricus rubellus*	250 mg	Reproduction	Makama et al. (2016)
8	Ag NPs	*E. crypticus*	170 mg	Embryo development and hatchability	Bicho et al. (2016)
9	MoO_3	*E. fetida*	40 mg/kg	Increased antioxidant enzymes	Lebedev et al. (2016)
10	TiO_2 NP	*C. elegans*	38.5 mg/mL	Affects TCA cycle	Ratnasekhar et al. (2015)
11	ZnO NPs	*C. elegans*	0.7 mg/L	Survivability of worms	Gupta et al. (2015)
12	C_{60} NPs	*E. fetida*	7 $\mu g/cm^2$	Amino acids and sugars	Lankadurai et al. (2015)
13	MWCNTs	*E. fetida*	1000 mg/L	Antioxidant enzymes	Zhang et al. (2014)
14	Ag NP	*E. fetida*	500 mg Ag/kg	Oxidative stress and immune gene regulation	Hayashi et al. (2013)
15	C_{60}	*L. rubellus*	154 mg/kg	Juvenile mortality	Van Der Ploeg et al. (2013)
16	Ag NP	*E. fetida*	500 mg/kg	Gene expression	Tsyusko et al. (2012)
17	TiO_2 NPs	*E. andrei*	10,000 mg	Avoidance response	McShane et al. (2012)
18	ZnO NPs	*E. fetida*	10,000 mg/L	Reproductive toxicity	Cañas et al. (2011)
19	Al_2O_3 NPs	*E. fetida*	5000 mg/kg	Reproduction	Coleman et al. (2010)
20	Ag NPs	*E. andrei*	1758 mg Ag/kg	Gene expression related to protein metabolism	Novo et al. (2015)

ATP, Adenosine triphosphate; *ENM*, Engineered nanomaterial; *MWCNT*, multiwall carbon nanotube; *NP*, nanoparticle; *ROS*, reactive oxygen species; *TCA*, tricarboxylic acid cycle.

ENM ingestion (Coleman et al., 2010). For instance, Shoults-Wilson et al. (2011) demonstrated that *E. fetida* were able to sense the presence of Ag NPs and exhibited avoidance behavior. In conclusion, unlike studies on plants or bacteria, the exposure effects on invertebrates are not conflicting. There is a clear evidence that ENMs have strong negative effect on the soil invertebrates, irrespective of the genus and species. This negative effect is attributed to the direct oral ingestion of the ENMs.

4.6 Trophic transfer in terrestrial food chain

One important question regarding the fate of engineered ENMs is the potential transfer between organisms through the food chain, including humans (Gardea-Torresdey et al., 2014). A limited number of studies have addressed the issue. Some early experiments revealed that *Manduca sexta* bioaccumulate higher concentrations of Au from indirect NP exposure than from direct exposure to gold NPs (nano-Au). Judy et al. (2011) fed caterpillars with tobacco leaves that contained gold within their tissues. More specifically, these tissues were from 4-week-old tobacco plants exposed for 7 days to Au NPs at 100 mg/L under hydroponic conditions. This indirect exposure resulted in bioaccumulation factors (BAF) of 6.2–11.6. In a subsequent experiment, Judy et al. (2012) fed *M. sexta* with Au NP deposited on the surface of tomato leaves. The BAF in these hornworms under this more direct exposure to Au NPs was 0.16. This finding biomagnification in a primary consumer from indirect exposure to Au NPs (i.e., the food supply) is important. Unrine et al. (2012) evaluated the bioavailability of Au NP from soil to secondary consumers. Bullfrogs (*Rana catesbeiana*) ingested earthworms (*E. fetida*) whose soil environment contained nano-Au at 200 mg/kg. The metal concentration in the tissues decreased by a 100-fold from one compartment to the next in the chain (soil–worm–frog). Bullfrogs were also given Au NPs directly (orally) for comparison purposes. Gold accumulated in the liver, kidney, spleen, muscles, stomach, and intestinal tissues, regardless of the exposure method. However, the kidney and liver of frogs fed with contaminated worms had higher concentrations of Au than those of bullfrogs who received that Au NP by oral gavage. This study again suggests higher bioavailability of Au NP from an indirect route (feeding from a lower trophic level organism) than from more direct exposure.

Researchers have also examined the trophic transfer of NPs of rare earth elements in terrestrial systems. Hawthorne et al. (2014) grew zucchini plants in soil with NPs or bulk-sized ceria (nano-CeO_2 or bulk-CeO_2) at ~1200 mg/kg. Leaves from the exposed plants were supplied to crickets (*Acheta domesticus*), and crickets subsequently to wolf spiders. The authors reported that in comparison to the control, stems and leaves of plants treated with CeO_2 NPs had greater biomass (higher wet weight) but that the reproductive tissues (i.e., flowers) had reduced mass. Importantly, there was greater Ce in the tissues of plants exposed to nano-CeO_2 than those exposed to the bulk material. Consequently, crickets fed with plants grown in the presence of CeO_2 NPs contained 56% more Ce than those fed from control or bulk-CeO_2 (34 vs 15 μg/kg). Importantly, more than 90% of the ingested Ce was excreted by the crickets, regardless of particle size. Subsequently, there was only measurable Ce in the wolf spiders' tissues from the NP treatments; the controls and bulk exposures had nondetectable levels of the element.

Ceria has also been investigated in a trophic pathway from soil to kidney bean plants (*P. vulgaris*), to Mexican bean beetles (MBBs) (*Epilachna varivestis*), and to spined soldier bugs (SSBs) (*Podisus maculiventris*) (Majumdar et al., 2016). Soil was amended with CeO_2 NPs at 1000–2000 mg/kg or bulk-CeO_2 at 1000 mg/kg. In the first trophic level (plant) the accumulation of Ce in shoots ranged between 1 and 1.3 μg/kg dry biomass but, unlike the Hawthorne study, did not differ greatly as a function of particle size. The reasons for this difference are unknown but could be a function of the different exposure conditions (different soil and plant species). In the Majumdar et al. (2016) study the root-to-shoot translocation factor from the bulk treatment (0.07 ± 0.01) was twice than that of the nanotreatment (0.04 and 0.03 for nano-CeO_2 at 1000 and 2000 mg/kg, respectively), although translocation was statistically equivalent among treatments. At the second trophic level the MBB larvae excreted almost all the Ce that was ingested (88%–98%), whereas the adults retained a greater amount and only excreted ~32%–36% Ce. This finding of increased analyte retention with age/physiological status is interesting and is worthy of significant additional investigation. At the last trophic level (SSB), higher biomagnification factors (BMFs) were found for the nano-CeO_2 fed animals than for the bulk-CeO_2; the BMF from the MBB to SSB was 5.32–6.7 for the nano-Ce treatment and 1.62 ± 0.12 for the bulk exposure.

In another study, De La Torre-Roche et al. (2015) grew lettuce (*Lactuca sativa*) in soil amended with 500 mg/kg of La_2O_3 (bulk or nano-sized). The leaves were fed to two herbivores: darkling beetles (*Tenebrionoidea*) or crickets (*A. domesticus*). Subsequently, crickets were depurated (removed from a La-based diet) and supplied to mantises. Lanthanum content in lettuce leaves ranged from 12.5 to 23.5 mg/kg, with the plants exposed to bulk-La_2O_3 containing the highest amounts of La. After a 2-day depuration period, La in darkling beetles was 0.18 and 0.085 mg/kg from the bulk or nano-La_2O_3 treatments, respectively. After a 2-day depuration time, crickets that consumed leaves from the bulk-La_2O_3 treatment contained more La than those who fed from the nanotreatment. However, crickets that were depurated for 7 days contained statistically equivalent amounts of La (0.18 and 0.16 mg/kg), regardless of particle size. The concentrations of La in mantises that were fed to 7-day depurated crickets were 0.020 and 0.025 mg/kg. The BMF in darkling beetles was 0.007 for both the bulk or nanotreatment; the values were 0.007 and 0.012 for crickets under a bulk or nano-La_2O_3 consumption, respectively. The authors also highlighted BMFs below 1 from crickets to manties, indicating that lanthanum is mobile through the food chain, regardless of the particle size (bulk or nanobiomagnification does not occur).

It is clear that NP transformation in soil and during the transfer process is poorly understood. The NPs may be aggregating, dissolving into ions or reforming within the matrix that contains them or may be transformed to other species through oxidation/reduction reactions (Lowry et al., 2012; Gao and Lowry, 2018). Along those lines, Servin et al. (2017) studied the role of transformation of CuO NPs in a terrestrial system with a trophic study involving lettuce, crickets, and lizards (*Anolis carolinensis*). Different forms of Cu (bulk-CuO, nano-CuO, or $CuSO_4$) at 400 mg/kg were introduced into the soil matrix. Lettuce seedlings were transplanted into the soil the same day as the Cu-based compound was introduced or was planted after the CuO had been aged for 70 days. Aging or weathering in the soil had no effect on the accumulation of Cu in the lettuce aerial tissues; Cu in the plant leaves ranged from 6.0 to 9.0 mg/kg and was statistically equivalent among the untreated, treated, and aged-growing conditions. However, in the lettuce roots, the amount of Cu increased by 53% on plants grown in aged soil with nano-CuO as compared to the aged-bulk particle treatment. Interestingly, synchrotron data demonstrated that weathering promoted the reduction of Cu from nano-CuO to Cu_2O and Cu_2S, which correlated well with the increased availability and uptake. In the animal species the amount of Cu in control crickets was found to be comparable to crickets previously fed for 15 days from the different Cu-treated plants. Overall, Cu ranged from 16.6 to 22.1 mg/kg. Similarly, no significant differences were observed in the lizards' Cu content after ingesting for 20 days' crickets on the Cu-based diet. Importantly, no biomagnification was observed. Given the findings on CuO fate in soil during weathering, follow-up studies are needed that track these processes and to evaluate their role on exposure and risk. Thus it can be concluded that biomagnification and trophic transfer depend on various factors. Most importantly, the mode of exposure and age of the organism have to be considered. Irrespective of the bioavailability at different trophic levels, excretory ability of the organisms will play a crucial role in the overall impact.

4.7 Knowledge gaps, limitations, and conclusion

The impact of ENMs on agroecosystems is predicted to be significant, but there are still many research questions. The impact of weather/season, geography, climate change, and environmental factors such as soil properties, wind, temperature, and precipitation is still unknown. Particle properties and species-specific responses will also be important but their impact is poorly understood. As such, it is important to conduct studies under environmentally relevant conditions whenever possible, although doing so often raises issues of experimental design, complexity, and confounding factors that may compromise the mechanistic understanding.

In the current plant exposure literature the concentrations used are typically not environmentally relevant and often occur over short exposure times. Chronic exposures to low-dose scenarios are far more relevant but require more sensitive endpoints. To this end, we emphasize new work on omics-based platforms, such as transcriptomics, metabolomics, and proteomics. These approaches, when applied appropriately, are much more likely to provide a mechanistic understanding of the biotic responses. It is also notable that the majority of the literature involves conventional ENMs including TiO_2, CeO_2, Ag, Au, and ZnO. However, there is little or no research on other emerging ENMs such as quantum dots, graphene, or on complex composite materials that appear in actual consumer products. In realistic situations the ENM will not occur as a single toxicant but likely be a part of a much more complex exposure scenario. Thus more research should be focused on ENM exposure in the presence of other ENMs, as well as conventional soil organic and heavy metal contaminants. The impact of ENM exposure to soil and plant microbiomes is an area where investigation is critically needed. The impact of ENMs on soil beneficial bacteria cannot be underestimated. In the absence of proper precautions, decreased nitrogen fixation and poor soil health might be the potential consequences. Furthermore, the negative effects on the soil invertebrates are very clear. On the positive side, there are also findings about the prospects of ENMs as potential plant growth promoters and nanopesticides.

As noted above, the impact of ENMs on agricultural ecosystems is complex. Multiple biotic and abiotic factors play key roles. Various soil, plant, and other biotic factors exert influence on the overall outcome of exposure. However, based on the majority of the findings in the literature, one can predict a number of realistic exposure scenarios where ENMs can be potentially hazardous within agricultural ecosystems. As such, intensive investigation in the area is warranted. As of now, we can conclude that ENMs' impact on agroenvironments is still far from conclusive. The impact appears to be multifaceted; thus we suggest ENM-based assessments on case–case scenarios since it seems impossible to find fixed commonalities.

Acknowledgments

This material is based upon work supported by the National Science Foundation and the Environmental Protection Agency under Cooperative Agreement Number DBI-1266377. Any opinions, findings, and conclusions or recommendations expressed in this material are those of the author(s) and do not necessarily reflect the views of the National Science Foundation or the Environmental Protection Agency. This work has not been subjected to EPA review and no official endorsement should be inferred. The authors also acknowledge the USDA grant 2016-67021-24985 and the NSF grants EEC-1449500, CHE-0840525, and DBI-1429708. Partial funding was provided by the NSF ERC on Nanotechnology-Enabled Water Treatment (EEC-1449500). This work was also supported by grant 2G12MD007592 from the National Institutes on Minority Health and Health Disparities (NIMHD), a component of the National Institutes of Health (NIH) and by the grant 1000001931 from the ConTex program. J.L.G.-T. acknowledges the Dudley family for the Endowed Research

Professorship, the Academy of Applied Science/US Army Research Office, Research and Engineering Apprenticeship program (REAP) at UTEP, grant no. W11NF-10-2-0076, subgrant 13-7, and the STARs retention award from the University of Texas System.

References

Abbaszadegan, A., Ghahramani, Y., Gholami, A., Hemmateenejad, B., Dorostkar, S., Nabavizadeh, M., et al., 2015. The effect of charge at the surface of silver nanoparticles on antimicrobial activity against Gram-positive and Gram-negative bacteria: a preliminary study. J. Nanomater. 16 (1), 53.

Adhikari, T., Dube, G., Kundu, S., Patra, A.K., 2018. Impact of copper oxide nanoparticles on growth of different bacterial species. Energy and Environment. Springer, Singapore, pp. 47–55.

Anderson, A., McLean, J., McManus, P., Britt, D., 2017. Soil chemistry influences the phytotoxicity of metal oxide nanoparticles. Int. J. Nanotechnol. 14 (1–6), 15–21.

Anjum, N.A., Singh, N., Singh, M.K., Sayeed, I., Duarte, A.C., Pereira, E., et al., 2014. Single-bilayer graphene oxide sheet impacts and underlying potential mechanism assessment in germinating faba bean (*Vicia faba* L.). Sci. Total Environ. 472, 834–841.

Asli, S., Neumann, P.M., 2009. Colloidal suspensions of clay or titanium dioxide nanoparticles can inhibit leaf growth and transpiration via physical effects on root water transport. Plant Cell Environ. 32 (5), 577–584.

Avellan, A., Schwab, F., Masion, A., Chaurand, P., Borschneck, D., Vidal, V., et al., 2017. Nanoparticle uptake in plants: gold nanomaterial localized in roots of *Arabidopsis thaliana* by X-ray computed nanotomography and hyperspectral imaging. Environ. Sci. Technol. 51 (15), 8682–8691.

Azam, A., Ahmed, A.S., Oves, M., Khan, M.S., Habib, S.S., Memic, A., 2012. Antimicrobial activity of metal oxide nanoparticles against Gram-positive and Gram-negative bacteria: a comparative study. Int. J. Nanomed. 7, 6003.

Baalousha, M., Manciulea, A., Cumberland, S., Kendall, K., Lead, J.R., 2008. Aggregation and surface properties of iron oxide nanoparticles: influence of pH and natural organic matter. Environ. Toxicol. Chem. 27 (9), 1875–1882.

Bandyopadhyay, S., Plascencia-Villa, G., Mukherjee, A., Rico, C.M., José-Yacamán, M., Peralta-Videa, J.R., et al., 2015. Comparative phytotoxicity of ZnO NPs, bulk ZnO, and ionic zinc onto the alfalfa plants symbiotically associated with *Sinorhizobium meliloti* in soil. Sci. Total Environ. 515, 60–69.

Barrios, A.C., Rico, C.M., Trujillo-Reyes, J., Medina-Velo, I.A., Peralta-Videa, J.R., Gardea-Torresdey, J.L., 2016. Effects of uncoated and citric acid coated cerium oxide nanoparticles, bulk cerium oxide, cerium acetate, and citric acid on tomato plants. Sci. Total Environ. 563, 956–964.

Barrios, A.C., Medina-Velo, I.A., Zuverza-Mena, N., Dominguez, O.E., Peralta-Videa, J. R., Gardea-Torresdey, J.L., 2017. Nutritional quality assessment of tomato fruits after exposure to uncoated and citric acid coated cerium oxide nanoparticles, bulk cerium oxide, cerium acetate and citric acid. Plant Physiol. Biochem. 110, 100–107.

Ben-Moshe, T., Dror, I., Berkowitz, B., 2010. Transport of metal oxide nanoparticles in saturated porous media. Chemosphere 81 (3), 387–393.

Bicho, R.C., Ribeiro, T., Rodrigues, N.P., Scott-Fordsmand, J.J., Amorim, M.J., 2016. Effects of Ag nanomaterials (NM300K) and Ag salt ($AgNO_3$) can be discriminated in a full life cycle long term test with *Enchytraeus crypticus*. J. Hazard. Mater. 318, 608–614.

Calabrese, E.J., Blain, R.B., 2009. Hormesis and plant biology. Environ. Pollut. 167 (1), 42–48.

Cañas, J.E., Qi, B., Li, S., Maul, J.D., Cox, S.B., Das, S., et al., 2011. Acute and reproductive toxicity of nano-sized metal oxides (ZnO and TiO_2) to earthworms (*Eisenia fetida*). J. Environ. Monit. 13 (12), 3351–3357.

Canivet, L., Dubot, P., Garçon, G., Denayer, F.O., 2015. Effects of engineered iron nanoparticles on the bryophyte, *Physcomitrella patens* (Hedw.) Bruch & Schimp, after foliar exposure. Ecotoxicol. Environ. Saf. 113, 499–505.

Cao, J., Feng, Y., Lin, X., Wang, J., Xie, X., 2017. Iron oxide magnetic nanoparticles deteriorate the mutual interaction between arbuscular mycorrhizal fungi and plant. J. Soils Sediments 17 (3), 841–851.

Chekli, L., Brunetti, G., Marzouk, E., Maoz-Shen, A., Smith, E., Naidu, R., et al., 2016. Evaluating the mobility of polymer-stabilised zero-valent iron nanoparticles and their potential to co-transport contaminants in intact soil cores. Environ. Pollut. 216, 636–645.

Coleman, J.G., Johnson, D.R., Stanley, J.K., Bednar, A.J., Weiss, C.A., Boyd, R.E., et al., 2010. Assessing the fate and effects of nano aluminum oxide in the terrestrial earthworm, *Eisenia fetida*. Environ. Toxicol. Chem. 29 (7), 1575–1580.

Cornelis, G., Hund-Rinke, K., Kuhlbusch, T., Van den Brink, N., Nickel, C., 2014. Fate and bioavailability of engineered nanoparticles in soils: a review. Crit. Rev. Environ. Sci. Technol. 44 (24), 2720–2764.

Corral-Diaz, B., Peralta-Videa, J.R., Alvarez-Parrilla, E., Rodrigo-García, J., Morales, M.I., Osuna-Avila, P., et al., 2014. Cerium oxide nanoparticles alter the antioxidant capacity but do not impact tuber ionome in *Raphanus sativus* (L). Plant Physiol. Biochem. 84, 277–278.

Coutris, C., Hertel-Aas, T., Lapied, E., Joner, E.J., Oughton, D.H., 2012. Bioavailability of cobalt and silver nanoparticles to the earthworm *Eisenia fetida*. Nanotoxicology 6 (2), 186–195.

Darlington, T.K., Neigh, A.M., Spencer, M.T., Guyen, O.T., Oldenburg, S.J., 2009. Nanoparticle characteristics affecting environmental fate and transport through soil. Environ. Toxicol. Chem. 28 (6), 1191–1199.

De la Rosa, G., López-Moreno, M.L., Hernandez-Viezcas, J.A., Montes, M.O., Peralta-Videa, J., Gardea-Torresdey, J., 2011. Toxicity and biotransformation of ZnO nanoparticles in the desert plants *Prosopis juliflora-velutina*, *Salsola tragus* and *Parkinsonia florida*. Int. J. Nanotechnol. 8 (6–7), 492–506.

De la Rosa, G., García-Castañeda, C., Vázquez-Núñez, E., Alonso-Castro, Á.J., Basurto-Islas, G., Mendoza, Á., et al., 2017. Physiological and biochemical response of plants to engineered NMs: implications on future design. Plant Physiol. Biochem. 110, 226–235.

De La Torre-Roche, R., Hawthorne, J., Deng, Y., Xing, B., Cai, W., Newman, L.A., et al., 2013. Multiwalled carbon nanotubes and C_{60} fullerenes differentially impact the accumulation of weathered pesticides in four agricultural plants. Environ. Sci. Technol. 47 (21), 12539–12547.

De la Torre-Roche, R., Servin, A., Hawthorne, J., Xing, B., Newman, L.A., Ma, X., et al., 2015. Terrestrial trophic transfer of bulk and nanoparticle La_2O_3 does not depend on particle size. Environ. Sci. Technol. 49 (19), 11866–11874.

Deng, R., Lin, D., Zhu, L., Majumdar, S., White, J.C., Gardea-Torresdey, J.L., et al., 2017. Nanoparticle interactions with co-existing contaminants: joint toxicity, bioaccumulation and risk. Nanotoxicology 11 (5), 591–612.

Dietz, K.J., Herth, S., 2011. Plant nanotoxicology. Trends Plant Sci. 16 (11), 582–589.

Dimkpa, C.O., Latta, D.E., McLean, J.E., Britt, D.W., Boyanov, M.I., Anderson, A.J., 2013. Fate of CuO and ZnO nano-and microparticles in the plant environment. Environ. Sci. Technol. 47 (9), 4734–4742.

Dimkpa, C.O., McLean, J.E., Britt, D.W., Anderson, A.J., 2015. Nano-CuO and interaction with nano-ZnO or soil bacterium provide evidence for the interference of nanoparticles in metal nutrition of plants. Ecotoxicology 24 (1), 119–129.

Dror, I., Yaron, B., Berkowitz, B., 2015. Abiotic soil changes induced by engineered nanomaterials: a critical review. J. Contam. Hydrol. 181, 3–16.

Du, W., Tan, W., Peralta-Videa, J.R., Gardea-Torresdey, J.L., Ji, R., Yin, Y., et al., 2017. Interaction of metal oxide nanoparticles with higher terrestrial plants: physiological and biochemical aspects. Plant Physiol. Biochem. 110, 210–225.

Dwivedi, A.D., Dubey, S.P., Sillanpää, M., Kwon, Y., Lee, C., Varma, R.S., 2015. Fate of engineered nanoparticles: implications in the environment. Coord. Chem. Rev. 287, 64–78.

Elmer, W.H., White, J.C., 2016. The use of metallic oxide nanoparticles to enhance growth of tomatoes and eggplants in disease infested soil or soilless medium. Environ. Sci. Nano 3 (5), 1072–1079.

Emerson, H.P., Hart, A.E., Baldwin, J.A., Waterhouse, T.C., Kitchens, C.L., Mefford, O.T., 2014. Physical transformations of iron oxide and silver nanoparticles from an intermediate scale field transport study. J. Nanopart. Res. 16 (2), 2258.

Espín-Pérez, A., Krauskopf, J., de Kok, T.M., Kleinjans, J.C., 2014. "OMICS-based" biomarkers for environmental health studies. Curr. Environ. Health Rep. 1, 353–362.

Frenk, S., Ben-Moshe, T., Dror, I., Berkowitz, B., Min, D., 2013. Effect of metal oxide nanoparticles on microbial community structure and function in two different soil types. PLoS One 8 (12), e84441.

Gao, X., Lowry, G.V., 2018. Progress towards standardized and validated characterizations for measuring physicochemical properties of manufactured nanomaterials relevant to nano health and safety risks. Nanoimpact 9, 14–30.

Garcia-Velasco, N., Gandariasbeitia, M., Irizar, A., Soto, M., 2016. Uptake route and resulting toxicity of silver nanoparticles in *Eisenia fetida* earthworm exposed through Standard OECD Tests. Ecotoxicology 25 (8), 1543–1555.

Gardea-Torresdey, J.L., Rico, C.M., White, J.C., 2014. Trophic transfer, transformation, and impact of engineered nanomaterials in terrestrial environments. Environ. Sci. Technol. 48 (5), 2526–2540.

Gautam, A., Ray, A., Mukherjee, S., Das, S., Pal, K., Das, S., et al., 2018. Immunotoxicity of copper nanoparticle and copper sulfate in a common Indian earthworm. Ecotoxicol. Environ. Saf. 148, 620–631.

Ge, Y., Schimel, J.P., Holden, P.A., 2011. Evidence for negative effects of TiO_2 and ZnO nanoparticles on soil bacterial communities. Environ. Sci. Technol. 45 (4), 1659–1664.

Ghodake, G., Seo, Y.D., Le, D.S., 2011. Hazardous phytotoxic nature of cobalt and zinc oxide nanoparticles assessed using *Allium cepa*. J. Hazard. Mater. 186 (1), 952–955.

Gonçalves, M.F., Gomes, S.I., Scott-Fordsmand, J.J., Amorim, M.J., 2017. Shorter lifetime of a soil invertebrate species when exposed to copper oxide nanoparticles in a full life-span exposure test. Sci. Rep. 7, 1355.

Guinée, J.B., Heijungs, R., Vijver, M.G., Peijnenburg, W.J., 2017. Setting the stage for debating the roles of risk assessment and life-cycle assessment of engineered nanomaterials. Nat. Nanotechnol. 12 (8), 727.

Gupta, S., Kushwah, T., Vishwakarma, A., Yadav, S., 2015. Optimization of ZnO-NPs to investigate their safe application by assessing their effect on soil nematode *Caenorhabditis elegans*. Nanoscale Res. Lett. 10 (1), 303.

Gurunathan, S., 2015. Cytotoxicity of graphene oxide nanoparticles on plant growth promoting rhizobacteria. J. Ind. Eng. Chem. 32, 282–291.

Hamidat, M., Barakat, M., Ortet, P., Chanéac, C., Rose, J., Bottero, J.Y., et al., 2016. Design defines the effects of nanoceria at a low dose on soil microbiota and the potentiation of impacts by the canola plant. Environ. Sci. Technol. 50 (13), 6892–6901.

Hawthorne, J., De la Torre Roche, R., Xing, B., Newman, L.A., Ma, X., Majumdar, S., et al., 2014. Particle-size dependent accumulation and trophic transfer of cerium oxide through a terrestrial food chain. Environ. Sci. Technol. 48 (22), 13102–13109.

Hayashi, Y., Heckmann, L.H., Simonsen, V., Scott-Fordsmand, J.J., 2013. Time-course profiling of molecular stress responses to silver nanoparticles in the earthworm *Eisenia fetida*. Ecotoxicol. Environ. Saf. 98, 219–226.

Hedayati, M., Sharma, P., Katyal, D., Fagerlund, F., 2016. Transport and retention of carbon-based engineered and natural nanoparticles through saturated porous media. J. Nanopart. Res. 18 (3), 57.

Hernandez-Viezcas, J.A., Castillo-Michel, H., Andrews, J.C., Cotte, M., Rico, C., Peralta-Videa, J.R., et al., 2013. In situ synchrotron X-ray fluorescence mapping and speciation of CeO_2 and ZnO nanoparticles in soil cultivated soybean (*Glycine max*). ACS Nano 7, 1415–1423.

Hernandez-Viezcas, J.A., Castillo-Michel, H.A., Lopez-Moreno, M., Peralta-Videa, J.R., Gardea-Torresdey, J.L., 2016. Interactions between CeO_2 nanoparticles and the desert plant mesquite: a spectroscopy approach. ACS Sustain. Chem. Eng. 4 (3), 1187e1192.

Hong, J., Peralta-Videa, J.R., Rico, C., Sahi, S., Viveros, M.N., Bartonjo, J., et al., 2014. Evidence of translocation and physiological impacts of foliar applied CeO_2 nanoparticles on cucumber (*Cucumis sativus*) plants. Environ. Sci. Technol. 48, 4376e4385.

Hong, J., Rico, C.M., Zhao, L., Adeleye, A.S., Keller, A.A., Peralta-Videa, J.R., et al., 2015. Toxic effects of copper-based nanoparticles or compounds to lettuce (*Lactuca sativa*) and alfalfa (*Medicago sativa*). Environ. Sci.: Process. Impacts 17 (1), 177–185.

Hounslow, A., 2018. Water quality data: analysis and interpretation. CRC press.

Huang, Y., Zhao, L., Keller, A.A., 2017. Interactions, transformations, and bioavailability of nano-copper exposed to root exudates. Environ. Sci. Technol. 51 (17), 9774–9783.

Huynh, K.A., Chen, K.L., 2011. Aggregation kinetics of citrate and polyvinylpyrrolidone coated silver nanoparticles in monovalent and divalent electrolyte solutions. Environ. Sci. Technol. 45 (13), 5564–5571.

Jaisi, D.P., Elimelech, M., 2009. Single-walled carbon nanotubes exhibit limited transport in soil columns. Environ. Sci. Technol. 43, 9161–9166.

Jesmer, A.H., Velicogna, J.R., Schwertfeger, D.M., Scroggins, R.P., Princz, J.I., 2017. The toxicity of silver to soil organisms exposed to silver nanoparticles and silver nitrate in biosolids-amended field soil. Environ. Toxicol. Chem. 36 (10), 2756–2765.

Judy, J.D., Unrine, J.M., Bertsch, P.M., 2011. Evidence for biomagnification of gold nanoparticles within a terrestrial food chain. Environ. Sci. Technol. 45 (2), 776–781.

Judy, J.D., Unrine, J.M., Rao, W., Bertsch, P.M., 2012. Bioaccumulation of gold nanomaterials by *Manduca sexta* through dietary uptake of surface contaminated plant tissue. Environ. Sci. Technol. 46 (22), 12672–12678.

Kanel, S., Choi, H., 2007. Transport characteristics of surface-modified nanoscale zero-valent iron in porous media. Water Sci. Technol. 55 (1–2), 157–162.

Karunakaran, G., Suriyaprabha, R., Manivasakan, P., Yuvakkumar, R., Rajendran, V., Kannan, N., 2013. Impact of nano and bulk ZrO_2, TiO_2 particles on soil nutrient contents and PGPR. J. Nanosci. Nanotechnol. 13 (1), 678–685.

Kavitha, B., Reddy, P.V.L., Kim, B., Lee, S.S., Pandey, S.K., Kim, K.H., 2018. Benefits and limitations of biochar amendment in agricultural soils: a review. J. Environ. Manage. 227, 146–154.

Keller, A.A., Vosti, W., Wang, H., Lazareva, A., 2014. Release of engineered nanomaterials from personal care products throughout their life cycle. J. Nanopart. Res. 16 (7), 2489.

Khodakovskaya, M.V., Kim, B.S., Kim, J.N., Alimohammadi, M., Dervishi, E., Mustafa, T., et al., 2013. Carbon nanotubes as plant growth regulators: effects on tomato growth, reproductive system, and soil microbial community. Small 9 (1), 115–123.

Kole, C., Kole, P., Randunu, K.M., Choudhary, P., Podila, R., Ke, P.C., et al., 2013. Nanobiotechnology can boost crop production and quality: first evidence from increased plant biomass, fruit yield and phytomedicine content in bitter melon (*Momordica charantia*). BMC Biotechnol. 13 (1), 37.

Lahiani, M.H., Chen, J., Irin, F., Puretzky, A.A., Green, M.J., Khodakovskaya, M.V., 2015. Interaction of carbon nanohorns with plants: uptake and biological effects. Carbon 81, 607–619.

Lankadurai, B.P., Nagato, E.G., Simpson, A.J., Simpson, M.J., 2015. Analysis of *Eisenia fetida* earthworm responses to sub-lethal C_{60} nanoparticle exposure using 1H-NMR based metabolomics. Ecotoxicol. Environ. Saf. 120, 48–58.

Lanphere, J.D., Luth, C.J., Walker, S.L., 2013. Effects of solution chemistry on the transport of graphene oxide in saturated porous media. Environ. Sci. Technol. 47 (9), 4255–4261.

Larue, C., Laurette, J., Herlin-Boime, N., Khodja, H., Fayard, B., Flank, A.M., et al., 2012. Accumulation, translocation and impact of TiO_2 nanoparticles in wheat (*Triticum aestivum* spp.): influence of diameter and crystal phase. Sci. Total Environ. 431, 197–208.

Lebedev, S., Yausheva, E., Galaktionova, L., Sizova, E., 2016. Impact of molybdenum nanoparticles on survival, activity of enzymes, and chemical elements in *Eisenia fetida* using test on artificial substrata. Environ. Sci. Pollut. Res. 23 (18), 18099–18110.

Lee, C.W., Mahendra, S., Zodrow, K., Li, D., Tsai, Y.C., Braam, J., et al., 2010. Developmental phytotoxicity of metal oxide nanoparticles to *Arabidopsis thaliana*. Environ. Toxicol. Chem. 29 (3), 669–675.

Li, X., Yang, Y., Gao, B., Zhang, M., 2015. Stimulation of peanut seedling development and growth by zero-valent iron nanoparticles at low concentrations. PLoS One 10 (4), e0122884.

Libralato, G., Devoti, A.C., Zanella, M., Sabbioni, E., Mičetić, I., Manodori, L., et al., 2016. Phytotoxicity of ionic, micro-and nano-sized iron in three plant species. Ecotoxicol. Environ. Saf. 123, 81–88.

Lin, D., Tian, X., Wu, F., Xing, B., 2010a. Fate and transport of engineered nanomaterials in the environment. J. Environ. Qual. 39 (6), 1896–1908.

Lin, Y., Tseng, H., Wey, M., Lin, M., 2010b. Characteristics of two types of stabilized nano zero-valent iron and transport in porous media. Sci. Total Environ. 408 (10), 2260–2267.

Liu, G., Zhang, M., Jin, Y., Fan, X., Xu, J., Zhu, Y., et al., 2017. The effects of low concentrations of silver nanoparticles on wheat growth, seed quality, and soil microbial communities. Water Air Soil Pollut. 228 (9), 348.

Lowry, G.V., Gregory, K.B., Apte, S.C., Lead, J.R., 2012. Transformations of nanomaterials in the environment. Environ. Sci. Technol. 46 (13), 6893–6899.

Lu, Y., Yang, K., Lin, D., 2014. Transport of surfactant-facilitated multiwalled carbon nanotube suspensions in columns packed with sized soil particles. Environ. Pollut. 192, 36–43.

Ma, C., Chhikara, S., Xing, B., Musante, C., White, J.C., Dhankher, O.P., 2013. Physiological and molecular response of *Arabidopsis thaliana* (L.) to nanoparticle cerium and indium oxide exposure. ACS Sustain. Chem. Eng. 1 (7), 768–778.

Ma, C., Rui, Y., Liu, S., Li, X., Xing, B., Liu, L., 2015b. Phytotoxic mechanism of nanoparticles: destruction of chloroplasts and vascular bundles and alteration of nutrient absorption. Sci. Rep. 5, 11618.

Ma, C., White, J.C., Dhankher, O.P., Xing, B., 2015a. Metal-based nanotoxicity and detoxification pathways in higher plants. Environ. Sci. Technol. 49 (12), 7109–7122.

Mahmoodzadeh, H., Aghili, R., 2014. Effect on germination and early growth characteristics in wheat plants (*Triticum aestivum* L.) seeds exposed to TiO_2 nanoparticles. J. Chem. Health Risk 4 (1), 29–36.

Majumdar, S., Almeida, I.C., Arigi, E.A., Choi, H., VerBerkmoes, N.C., Trujillo-Reyes, J., et al., 2015. Environmental effects of nanoceria on seed production of common bean (*Phaseolus vulgaris*): a proteomic analysis. Environ. Sci. Technol. 49 (22), 13283–13293.

Majumdar, S., Trujillo-Reyes, J., Hernandez-Viezcas, J.A., White, J.C., Peralta-Videa, J.R., Gardea-Torresdey, J.L., 2016. Cerium biomagnification in a terrestrial food chain: influence of particle size and growth stage. Environ. Sci. Technol. 50 (13), 6782–6792.

Makama, S., Piella, J., Undas, A., Dimmers, W.J., Peters, R., Puntes, V.F., et al., 2016. Properties of silver nanoparticles influencing their uptake in and toxicity to the earthworm *Lumbricus rubellus* following exposure in soil. Environ. Pollut. 218, 870–878.

Malev, O., Trebše, P., Piecha, M., Novak, S., Budič, B., Dramićanin, M.D., et al., 2017. Effects of CeO_2 nanoparticles on terrestrial isopod *Porcellio scaber*: comparison of CeO_2 biological potential with other nanoparticles. Arch. Environ. Contam. Toxicol. 72 (2), 303–311.

Markus, A.A., Parsons, J.R., Roex, E.W.M., De Voogt, P., Laane, R.P.M., 2015. Modeling aggregation and sedimentation of nanoparticles in the aquatic environment. Sci. Total Environ. 506, 323–329.

Marmiroli, M., Imperiale, D., Pagano, L., Villani, M., Zappettini, A., Marmiroli, N., 2015. The proteomic response of *Arabidopsis thaliana* to cadmium sulfide quantum dots, and its correlation with the transcriptomic response. Front. Plant Sci. 6, 1104. Available from: https://doi.org/10.3389/fpls.2015.01104.

Marusenko, Y., Shipp, J., Hamilton, G.A., Morgan, J.L., Keebaugh, M., Hill, H., et al., 2013. Bioavailability of nanoparticulate hematite to *Arabidopsis thaliana*. Environ. Pollut. 174, 150–156.

McShane, H., Sarrazin, M., Whalen, J.K., Hendershot, W.H., Sunahara, G.I., 2012. Reproductive and behavioral responses of earthworms exposed to nano-sized titanium dioxide in soil. Environ. Toxicol. Chem. 31 (1), 184–193.

Medina-Velo, I.A., Adisa, I., Tamez, C., Peralta-Videa, J.R., Gardea-Torresdey, J.L., 2017. Effects of surface coating on the bioactivity of metal-based engineered nanoparticles: lessons learned from higher plants. In: Yan, B., Zhou, H., Gardea-Torresdey, J.L. (Eds.), Bioactivity of Engineered Nanoparticles Singapore. Springer, pp. 43–61.

Michels, C., Perazzoli, S., Soares, H.M., 2017. Inhibition of an enriched culture of ammonia oxidizing bacteria by two different nanoparticles: silver and magnetite. Sci. Total Environ. 586, 995–1002.

Moon, J., Kwak, J.I., Kim, S.W., An, Y.J., 2017. Multigenerational effects of gold nanoparticles in *Caenorhabditis elegans*: continuous versus intermittent exposures. Environ. Pollut. 220, 46–52.

Morales, M.I., Rico, C.M., Hernandez-Viezcas, J.A., Nunez, J.E., Barrios, A.C., Tafoya, A., et al., 2013. Toxicity assessment of cerium oxide nanoparticles in cilantro (*Coriandrum sativum* L.) plants grown in organic soil. J. Agric. Food Chem. 61 (26), 6224–6230.

Mukherjee, A., Pokhrel, S., Bandyopadhyay, S., Mädler, L., Peralta-Videa, J.R., Gardea-Torresdey, J.L., 2014. A soil mediated phyto-toxicological study of iron doped zinc oxide nanoparticles (Fe@ZnO) in green peas (*Pisum sativum* L.). Chem. Eng. J. 258, 394–401.

Nair, P.M.G., Kim, S.H., Chung, I.M., 2014. Copper oxide nanoparticle toxicity in mung bean (*Vigna radiata* L.) seedlings: physiological and molecular level responses of in vitro grown plants. Acta Physiol. Plant. 36 (11), 2947–2958.

Noori, A., White, J.C., Newman, L.A., 2017. Mycorrhizal fungi influence on silver uptake and membrane protein gene expression following silver nanoparticle exposure. J. Nanopart. Res. 19 (2), 66.

Novo, M., Lahive, E., Díez-Ortiz, M., Matzke, M., Morgan, A.J., Spurgeon, D.J., et al., 2015. Different routes, same pathways: molecular mechanisms under silver ion and nanoparticle exposures in the soil sentinel *Eisenia fetida*. Environ. Pollut. 205, 385–393.

Palombo, M., Deshmukh, M., Myers, D., Gao, J., Szekely, Z., Sinko, P.J., 2014. Pharmaceutical and toxicological properties of engineered nanomaterials for drug delivery. Annu. Rev. Pharmacol. Toxicol. 54, 581–598.

Park, C.M., Chu, K.H., Heo, J., Her, N., Jang, M., Son, A., et al., 2016. Environmental behavior of engineered nanomaterials in porous media: a review. J. Hazard. Mater. 309, 133–150.

Peralta-Videa, J.R., Hernandez-Viezcas, J.A., Zhao, L., Diaz, B.C., Ge, Y., Priester, J.H., et al., 2014. Cerium dioxide and zinc oxide nanoparticles alter the nutritional value of soil cultivated soybean plants. Plant Physiol. Biochem. 80, 128–135.

Placer, Z.A., Cushman, L.L., Johnson, B.C., 1966. Estimation of product of lipid peroxidation (malonyl dialdehyde) in biochemical systems. Anal. Biochem. 16 (2), 359–364.

Pradhan, S., Mailapalli, D.R., 2017. Interaction of engineered nanoparticles with the agri-environment. J. Agric. Food Chem. 65 (38), 8279–8294.

Pullagurala, V.L.R., Rawat, S., Adisa, I.O., Hernandez-Viezcas, J.A., Peralta-Videa, J.R., Gardea-Torresdey, J.L., 2018a. Plant uptake and translocation of contaminants of emerging concern in soil. Sci. Total Environ. 636, 1585–1596.

Pullagurala, V.L.R., Adisa, I.O., Rawat, S., Kim, B., Barrios, A.C., Medina-Velo, I.A., et al., 2018b. Finding the conditions for the beneficial use of ZnO nanoparticles towards plants—a review. Environ. Pollut. 241, 1175–1181.

Pullagurala, V.L.R., Adisa, I.O., Rawat, S., Kalagara, S., Hernandez-Viezcas, J.A., Peralta-Videa, J.R., et al., 2018c. ZnO nanoparticles increase photosynthetic pigments and decrease lipid peroxidation in soil grown cilantro (*Coriandrum sativum*). Plant Physiol. Biochem. 132, 120–127.

Qi, Z., Zhang, L., Wang, F., Hou, L., Chen, W., 2014. Factors controlling transport of graphene oxide nanoparticles in saturated sand columns. Environ. Toxicol. Chem. 33 (5), 998–1004.

Quevedo, I.R., Tufenkji, N., 2012. Mobility of functionalized quantum dots and a model polystyrene nanoparticle in saturated quartz sand and loamy sand. Environ. Sci. Technol. 46 (8), 4449–4457.

Quik, J.T., Stuart, M.C., Wouterse, M., Peijnenburg, W., Hendriks, A.J., van de Meent, D., 2012. Natural colloids are the dominant factor in the sedimentation of nanoparticles. Environ. Toxicol. Chem. 31 (5), 1019–1022.

Raliya, R., Tarafdar, J.C., 2013. ZnO nanoparticle biosynthesis and its effect on phosphorous-mobilizing enzyme secretion and gum contents in clusterbean (*Cyamopsis tetragonoloba* L.). Agric. Res. 2 (1), 48–57.

Ratnasekhar, C., Sonane, M., Satish, A., Mudiam, M.K.R., 2015. Metabolomics reveals the perturbations in the metabolome of *Caenorhabditis elegans* exposed to titanium dioxide nanoparticles. Nanotoxicology 9 (8), 994–1004.

Rawat, S., Pullagurala, V.L., Adisa, I.O., Wang, Y., Peralta-Videa, J.R., Gardea-Torresdey, J.L., 2018a. Factors affecting fate and transport of engineered nanomaterials in a terrestrial environment. Curr. Opin. Environ. Sci. Health 6, 47–63.

Rawat, S., Pullagurala, V.L., Hernandez-Molina, M., Sun, Y., Niu, G., Hernandez-Viezcas, J.A., et al., 2018b. Impacts of copper oxide nanoparticles on bell pepper (*Capsicum annum* L.) plants: a full life cycle study. Environ. Sci. Nano 5, 83–95.

Reddy, J.K., Lalitha, K., Reddy, P.V.L., Sadanandam, G., Subrahmanyam, M., Kumari, V.D., 2014. Fe/TiO_2: a visible light active photocatalyst for the continuous production of hydrogen from water splitting under solar irradiation. Catal. Lett. 144 (2), 340–346.

Reddy, P.V.L., Hernandez-Viezcas, J.A., Peralta-Videa, J.R., Gardea-Torresdey, J.L., 2016. Lessons learned: are engineered nanomaterials toxic to terrestrial plants? Sci. Total Environ. 568, 470–479.

Reddy, P.A.K., Reddy, P.V.L., Kim, K.H., Kumar, M.K., Manvitha, C., Shim, J.J., 2017. Novel approach for the synthesis of nitrogen-doped titania with variable phase composition and enhanced production of hydrogen under solar irradiation. J. Ind. Eng. Chem. 53, 253–260.

Rico, C.M., Hong, J., Morales, M.I., Zhao, L., Barrios, A.C., Zhang, J.Y., et al., 2013. Effect of cerium oxide nanoparticles on rice: a study involving the antioxidant defense system and in vivo fluorescence imaging. Environ. Sci. Technol. 47 (11), 5635–5642.

Rico, C.M., Lee, S.C., Rubenecia, R., Mukherjee, A., Hong, J., Peralta-Videa, J.R., et al., 2014. Cerium oxide nanoparticles impact yield and modify nutritional parameters in wheat (*Triticum aestivum* L.). J. Agric. Food Chem. 62, 9669–9675.

Rico, C.M., Barrios, A.C., Tan, W., Rubenecia, R., Lee, S.C., Varela-Ramirez, A., et al., 2015. Physiological and biochemical response of soil-grown barley (*Hordeum vulgare* L.) to cerium oxide nanoparticles. Environ. Sci. Pollut. Res. 22 (14), 10551–10558.

Rodrigues, S.M., Trindade, T., Duarte, A.C., Pereira, E., Koopmans, G.F., Römkens, P.F. A.M., 2016. A framework to measure the availability of engineered nanoparticles in soils: trends in soil tests and analytical tools. TrAC Trends Anal. Chem. 75, 129–140.

Rui, Y., Zhang, P., Zhang, Y., Ma, Y., He, X., Gui, X., et al., 2015. Transformation of ceria nanoparticles in cucumber plants is influenced by phosphate. Environ. Pollut. 198, 8–14.

Ruotolo, R., Maestri, E., Pagano, L., Marmiroli, M., White, J.C., Marmiroli, N., 2018. Plant response to metal-containing engineered nanomaterials: an omics-based perspective. Environ. Sci. Technol. 52 (5), 2451–2467.

Sagee, O., Dror, I., Berkowitz, B., 2012. Transport of silver nanoparticles (AgNPs) in soil. Chemosphere 88 (5), 670–675.

Saia, S., Benítez, E., García-Garrido, J.M., Settanni, L., Amato, G., Giambalvo, D., 2014. The effect of arbuscular mycorrhizal fungi on total plant nitrogen uptake and nitrogen recovery from soil organic material. J. Agric. Sci. 152 (3), 370–378.

Salehi, H., Chehregani, A., Lucini, L., Majd, A., Gholami, M., 2018. Morphological, proteomic and metabolomic insight into the effect of cerium dioxide nanoparticles to *Phaseolus vulgaris* L. under soil or foliar application. Sci. Total Environ. 616, 1540–1551.

Servin, A.D., White, J.C., 2016. Nanotechnology in agriculture: next steps for understanding engineered nanoparticle exposure and risk. Nanoimpact 1, 9–12.

Servin, A.D., Morales, M.I., Castillo-Michel, H., Hernandez-Viezcas, J.A., Munoz, B., Zhao, L.J., et al., 2013. Synchrotron verification of TiO_2 accumulation in cucumber fruit: a possible pathway of TiO_2 nanoparticle transfer from soil into the food chain. Environ. Sci. Technol. 47 (20), 11592e11598.

Servin, A., Elmer, W., Mukherjee, A., De la Torre-Roche, R., Hamdi, H., White, J.C., et al., 2015. A review of the use of engineered nanomaterials to suppress plant disease and enhance crop yield. J. Nanopart. Res. 17 (2), 92.

Servin, A.D., Pagano, L., Castillo-Michel, H., De la Torre-Roche, R., Hawthorne, J., Hernandez-Viezcas, J.A., et al., 2017. Weathering in soil increases nanoparticle CuO bioaccumulation within a terrestrial food chain. Nanotoxicology 11 (1), 98–111.

Sharma, P., Bao, D., Fagerlund, F., 2014. Deposition and mobilization of functionalized multiwall carbon nanotubes in saturated porous media: effect of grain size, flow velocity and solution chemistry. Environ. Earth Sci. 72 (8), 3025–3035.

Shaw, A.K., Hossain, Z., 2013. Impact of nano-CuO stress on rice (*Oryza sativa* L.) seedlings. Chemosphere 93 (6), 906–915.

Sheykhbaglou, R., Sedghi, M., Shishevan, M.T., Sharifi, R.S., 2010. Effects of nano-iron oxide particles on agronomic traits of soybean. Not. Sci. Biol. 2 (2), 112.

Shoults-Wilson, W.A., Reinsch, B.C., Tsyusko, O.V., Bertsch, P.M., Lowry, G.V., Unrine, J.M., 2011. Effect of silver nanoparticle surface coating on bioaccumulation and reproductive toxicity in earthworms (*Eisenia fetida*). Nanotoxicology 5 (3), 432–444.

Shukla, S.K., Kumar, R., Mishra, R.K., Pandey, A., Pathak, A., Zaidi, M.G.H., et al., 2015. Prediction and validation of gold nanoparticles (GNPs) on plant growth promoting rhizobacteria (PGPR): a step toward development of nano-biofertilizers. Nanotechnol. Rev. 4 (5), 439–448.

Siani, N.G., Fallah, S., Pokhrel, L.R., Rostamnejadi, A., 2017. Natural amelioration of zinc oxide nanoparticle toxicity in fenugreek (*Trigonella foenum-gracum*) by arbuscular mycorrhizal (*Glomus intraradices*) secretion of glomalin. Plant Physiol. Biochem. 112, 227–238.

Siddiqui, M.H., Al-Whaibi, M.H., 2014. Role of nano-SiO_2 in germination of tomato (*Lycopersicum esculentum* seeds mill). Saudi J. Biol. Sci. 21, 13–17.

Sillen, W.M., Thijs, S., Abbamondi, G.R., Janssen, J., Weyens, N., White, J.C., et al., 2015. Effects of silver nanoparticles on soil microorganisms and maize biomass are linked in the rhizosphere. Soil Biol. Biochem. 91, 14–22.

Stebounova, L.V., Guio, E., Grassian, V.H., 2011. Silver nanoparticles in simulated biological media: a study of aggregation, sedimentation, and dissolution. J. Nanopart. Res. 13 (1), 233–244.

Sturikova, H., Krystofova, O., Huska, D., Adam, V., 2018. Zinc, zinc nanoparticles and plants. J. Hazard Mater. 349, 101–110.

Tan, W., Peralta-Videa, J.R., Gardea-Torresdey, J.L., 2018. Interaction of titanium dioxide nanoparticles with soil components and plants: current knowledge and future research needs—a critical review. Environ. Sci. Nano 5, 257–278.

Trivedi, P., Delgado-Baquerizo, M., Trivedi, C., Hamonts, K., Anderson, I.C., Singh, B.K., 2017. Keystone microbial taxa regulate the invasion of a fungal pathogen in agroecosystems. Soil Biol. Biochem. 111, 10–14.

Trujillo-Reyes, J., Vilchis-Nestor, A.R., Majumdar, S., Peralta-Videa, J.R., Gardea-Torresdey, J.L., 2013. Citric acid modifies surface properties of commercial CeO_2 nanoparticles reducing their toxicity and cerium uptake in radish (*Raphanus sativus*) seedlings. J. Hazard. Mater. 263, 677–684.

Trujillo-Reyes, J., Majumdar, S., Botez, C.E., Peralta-Videa, J.R., Gardea-Torresdey, J.L., 2014. Exposure studies of core–shell Fe/Fe_3O_4 and Cu/CuO NPs to lettuce (*Lactuca sativa*) plants: are they a potential physiological and nutritional hazard? J. Hazard. Mater. 267, 255–263.

Tsyusko, O.V., Hardas, S., Shoults-Wilson, W.A., Starnes, C.P., Joice, G., Butterfield, D. A., et al., 2012. Short-term molecular-level effects of silver nanoparticle exposure on the earthworm, *Eisenia fetida*. Environ. Pollut. 171, 249–255.

Unrine, J.M., Shoults-Wilson, W.A., Zhurbich, O., Bertsch, P.M., Tsyusko, O.V., 2012. Trophic transfer of Au nanoparticles from soil along a simulated terrestrial food chain. Environ. Sci. tech. 46 (17), 9753–9760.

Van Der Ploeg, M.J.C., Baveco, J.M., Van der Hout, A., Bakker, R., Rietjens, I.M.C.M., Van den Brink, N.W., 2011. Effects of C_{60} nanoparticle exposure on earthworms (*Lumbricus rubellus*) and implications for population dynamics. Environ. Pollut. 159 (1), 198–203.

Van Der Ploeg, M.J.C., Handy, R.D., Heckmann, L.-H., Van Der Hout, A., Van Den Brink, N.W., 2013. C_{60} exposure induced tissue damage and gene expression alterations in the earthworm *Lumbricus rubellus*. Nanotoxicology 7 (4), 432–440.

Verma, S.K., Das, A.K., Patel, M.K., Shah, A., Kumar, V., Gantait, S., 2018. Engineered nanomaterials for plant growth and development: a perspective analysis. Sci. Total Environ. 630, 1413–1435.

Vikram, S., Guerrero, L.D., Makhalanyane, T.P., Le, P.T., Seely, M., Cowan, D.A., 2016. Metagenomic analysis provides insights into functional capacity in a hyper arid desert soil niche community. Environ. Microbial. 18 (6), 1875–1888.

Wang, Z., Xie, X., Zhao, J., Liu, X., Feng, W., White, J.C., et al., 2012. Xylem- and phloem-based transport of CuO nanoparticles in maize (*Zea mays* L.). Environ. Sci. Technol. 46, 4434–4441.

Wang, P., Menzies, N.W., Lombi, E., McKenna, B.A., Johannessen, B., Glover, C.J., et al., 2013. Fate of ZnO nanoparticles in soils and cowpea (*Vigna unguiculata*). Environ. Sci. Technol. 47 (23), 13822–13830.

Xu, X., Xu, N., Cheng, X., Guo, P., Chen, Z., Wang, D., 2017. Transport and aggregation of rutile titanium dioxide nanoparticles in saturated porous media in the presence of ammonium. Chemosphere 169, 9–17.

Yang, Y., Wang, J., Xiu, Z., Alvarez, P.J., 2013. Impacts of silver nanoparticles on cellular and transcriptional activity of nitrogen-cycling bacteria. Environ. Toxicol. Chem. 32 (7), 1488–1494.

Zahra, Z., Waseem, N., Zahra, R., Lee, H., Badshah, M.A., Mehmood, A., et al., 2017. Growth and metabolic responses of rice (*Oryza sativa* L.) cultivated in phosphorus-deficient soil amended with TiO_2 nanoparticles. J. Agric. Food Chem. 65 (28), 5598–5606.

Zhang, L., Hou, L., Wang, L., Kan, A.T., Chen, W., Tomson, M.B., 2012a. Transport of fullerene nanoparticles (*n*C60) in saturated sand and sandy soil: controlling factors and modeling. Environ. Sci. Technol. 46, 7230–7238.

Zhang, P., Ma, Y., Zhang, Z., He, X., Zhang, J., Guo, Z., et al., 2012b. Biotransformation of ceria nanoparticles in cucumber plants. ACS Nano 6 (11), 9943–9950.

Zhang, L., Hu, C., Wang, W., Ji, F., Cui, Y., Li, M., 2014. Acute toxicity of multi-walled carbon nanotubes, sodium pentachlorophenate, and their complex on earthworm *Eisenia fetida*. Ecotoxicol. Environ. Saf. 103, 29–35.

Zhao, L., Peralta-Videa, J.R., Ren, M., Varela-Ramirez, A., Li, C., Hernandez-Viezcas, J. A., et al., 2012. Transport of Zn in a sandy loam soil treated with ZnO NPs and uptake by corn plants: electron microprobe and confocal microscopy studies. Chem. Eng. J. 184, 1–8.

Zhao, L., Sun, Y., Hernandez-Viezcas, J.A., Servin, A.D., Hong, J., Niu, G., et al., 2013. Influence of CeO_2 and ZnO nanoparticles on cucumber physiological markers and bioaccumulation of Ce and Zn: a life cycle study. J. Agric. Food Chem. 61 (49), 11945–11951.

Zhao, L.J., Peralta-Videa, J.R., Rico, C.M., Hernandez-Viezcas, J.A., Sun, Y.P., Niu, G.H., et al., 2014. CeO_2 and ZnO nanoparticles change the nutritional qualities of cucumber (*Cucumis sativus*). J. Agric. Food Chem. 62 (13), 2752e2759.

Zhao, L., Sun, Y., Hernandez-Viezcas, J.A., Hong, J., Majumdar, S., Niu, G., et al., 2015. Monitoring the environmental effects of CeO_2 and ZnO nanoparticles through the life cycle of corn (*Zea mays*) plants and in situ μ-XRF mapping of nutrients in kernels. Environ. Sci. Technol. 49 (5), 2921–2928.

Zhao, L., Hu, J., Huang, Y., Wang, H., Adeley, A., Ortiz, C., et al., 2017. 1H NMR and GC–MS based metabolomics reveal nano-cu altered cucumber (*Cucumis sativus*) fruit nutritional supply. Plant Physiol. Biochem. 110, 138–146.

Zuverza-Mena, N., Medina-Velo, I.A., Barrios, A.C., Tan, W., Peralta-Videa, J.R., Gardea-Torresdey, J., 2015. Copper nanoparticles/compounds impact agronomic and physiological parameters in cilantro (*Coriandrum sativum*). Environ. Sci.: Process. Impacts 10, 1783–1793.

Zuverza-Mena, N., Martínez-Fernández, D., Du, W., Hernandez-Viezcas, J.A., Bonilla-Bird, N., López-Moreno, M.L., et al., 2017. Exposure of engineered nanomaterials to plants: insights into the physiological and biochemical responses—a review. Plant Physiol. Biochem. 110, 236–264.

CHAPTER

Fate of engineered nanomaterials in urban and work environments

5

Guodong Yuan[1], Benny K.G Theng[2], Lirong Feng[3] and Dongxue Bi[3]

[1]*School of Environmental and Chemical Engineering, Zhaoqing University, Zhaoqing, P.R. China*

[2]*Manaaki Whenua—Landcare Research, Palmerston North, New Zealand*

[3]*Yantai Institute of Coastal Zone Research, University of Chinese Academy of Sciences, Yantai, P.R. China*

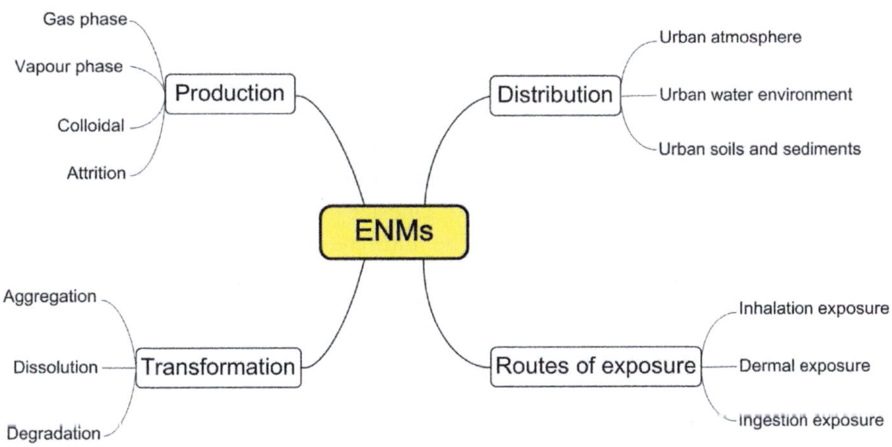

ENMs in urban and work environments.

5.1 Introduction

Engineered nanomaterials (ENMs) may be defined as intentionally produced materials with at least one dimension in the 1–100 nm size range. There are also unintentionally generated nanomaterials with a diameter of less than 100 nm, commonly referred to as "ultrafine particles" (UFPs) or incidental nanomaterials (INMs). In urban areas the majority of UFPs derive from fossil fuel combustion,

Exposure to Engineered Nanomaterials in the Environment. DOI: https://doi.org/10.1016/B978-0-12-814835-8.00005-4

cooking, coal burning, and welding operations (Kumar et al., 2014a; Li et al., 2016). According to Aitken et al. (2004), more than 1 million workers in the United Kingdom may be exposed to nanoparticles through incidental production. Baalousha et al. (2016) have described the release of ENMs into the outdoor urban environment, while Kumar et al. (2014b) have indicated that motor traffic is the major source of UFPs in the outdoor environment of large cities.

Worldwide production of nanomaterials is projected to increase to 58,000 t by 2020 (Maynard, 2006), and up to 6 million workers may be involved in the production of an estimated US$3.3 trillion worth of consumer and household goods incorporating ENMs (Brenner and Neu-Baker, 2015; Roco, 2011). Among these articles are cleaning and coating products, cosmetics, lubricants, plastics, printer ink, sporting goods, sunscreens, and textiles, while the associated ENMs include metal oxides (Al_2O_3, CeO_2, SiO_2, TiO_2, ZnO), metals (Ag, Au, Fe), carbon (single- and multiwalled carbon nanotubes, fullerene, carbon black), nanoclays, and organic polymers (Brar et al., 2010; Goswami et al., 2017; Hansen et al., 2016; Munafò et al., 2015; Nel et al., 2006; Pietroiusti and Magrini, 2014; Theng, 2012).

The incorporated ENMs would inevitably find their way into the environment during the transportation and disposal of such goods (Bystrzejewska-Piotrowska et al., 2009; Caballero-Guzman et al., 2015; Gottschalk et al., 2009, 2013; Part et al., 2018). Wastewater, waste products, and sewage sludge are the major routes by which ENMs finish up in the urban environment (Giese et al., 2018; Gottschalk and Nowack, 2011; Gottschalk et al., 2013; Keller and Lazareva, 2014). The application of sewage sludge to agricultural lands, for example, would not only release ENMs to surface and groundwater but also lead to their accumulation in soil biota, fruits, and cereal grains (Brar et al., 2010; Duester et al., 2014; Gardea-Torresdey et al., 2014; Whiteley et al., 2013). The ultimate environmental recipients of ENMs, however, are sediments and soils (Joner et al., 2008; Peijnenburg et al., 2016; Peralta-Videa et al., 2011) although these nanomaterials would by then have been significantly transformed (Batley et al., 2013; Dwivedi et al., 2015; Garner and Keller, 2014; Goswami et al., 2017; Hartmann et al., 2014). Similarly, most nanoparticles in technical applications occur as functionalized materials rather than in their pristine state (Mitrano et al., 2015; Nowack and Bucheli, 2007).

Besides raising questions about the occurrence, fate, and potential toxicity of ENMs (Colvin, 2003; Nel et al., 2006), a robust methodology for their detection and characterization in food, water, and soil needs to be developed and tested (Tiede et al., 2008). Furthermore, the behavior, bioavailability, and environmental fate of ENMs are not well understood (Lin et al., 2010). As the biological activity of nanoparticles is generally greater than that of their micro-size counterparts, ENMs pose a potential health risk to workers in nanotechnology industries (Hristozov and Malsch, 2009; Schulte et al., 2008).

This chapter is an attempt at assessing the effect of ENMs on, and their fate in, the urban and work environments where 68% of the world's population is projected to live and move by 2050 (UN Department of Economic and Social Affairs, 2018).

5.2 Urbanization and exposure to engineered nanomaterials in the urban workplace

Schulte et al. (2008) have suggested that the workplace is the environment in which the earliest and most extensive exposure to ENMs is likely to occur. Thus workers in the construction, health care, energy, automobile, aerospace, chemical, electronic, and communication industries would be highly vulnerable (Pietroiusti and Magrini, 2014; Woskie, 2010). The production of ENMs involves (1) gas phase processes (flame pyrolysis, high temperature evaporation, plasma synthesis); (2) vapor deposition synthesis (widely used for the manufacture of semiconductors and carbon nanotubes); (3) colloidal or liquid phase methods in which chemical reactions in solvents lead to the formation of nanoparticles in liquid suspension for distribution and use; and (4) mechanical processes (wet grinding, milling, alloying). Although all the four processes have potential risks of exposure, the nature, likely level, and probability of exposure will vary with the nature and stage of the process (Table 5.1).

Brenner and Neu-Baker (2015) have reported the occurrence on work surfaces of metal oxide nanoparticles, used or generated by chemical mechanical planarization. As a result, the skin of workers in semiconductor facilities would be exposed to such nanoparticles. It is not yet possible, however, to make any collective judgment about the potential risks of ENMs to human health (Hristozov and Malsch, 2009). In discussing this topic, Pietroiusti and Magrini (2014),

Table 5.1 Risks of exposure in the production processes of engineered nanomaterial (ENMs).

Processes	ENM formation	Inhalation risks	Dermal and ingestion risks
Gas phase	In air	Leakage from reactor, product recovery, postrecovery, processing, and packing	Airborne contamination of workplace, handling of product, plant cleaning and maintenance
Vapor phase	On substrate	Product recovery, postrecovery, processing, and packing	Dry contamination of workplace, handling of product, plant cleaning, and maintenance
Colloidal	Liquid suspension	Product drying, processing, and spillage	Spillage contamination of workplace, handling of product, plant cleaning/maintenance
Attrition	Liquid suspension	Product drying, processing and spillage	Spillage at workplace, handling of product, plant cleaning, and maintenance

Source: *Modified from Aitken, R.J., Creely, K.S., Tran, C.L., 2004. Nanoparticles: an occupational hygiene review. In: Research Report 274. HSE Books, Sudbury, UK.*

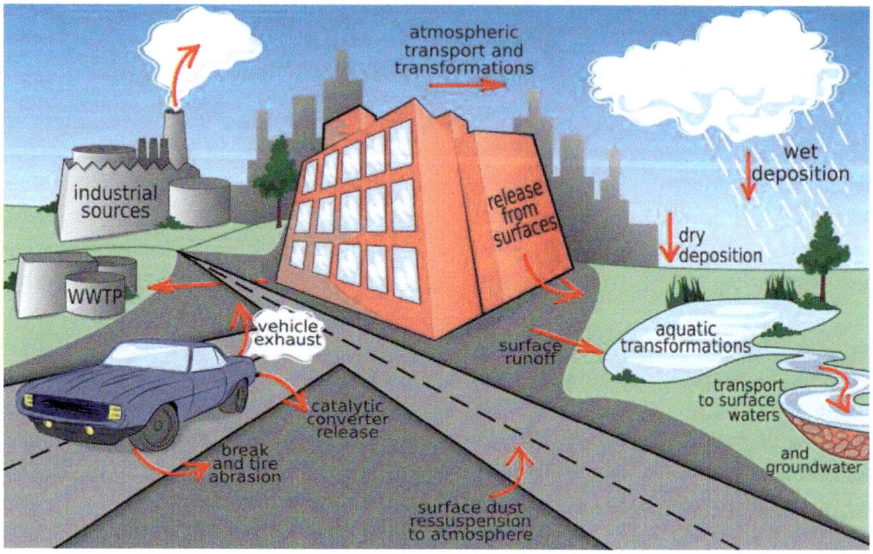

FIGURE 5.1

Sources and pathways of nanomaterials in urban environment.

After Baalousha, M., Yang, Y., Vance, M.E., Colman, B.P., McNeal, S., Xu, J., et al., 2016. Outdoor urban nanomaterials: the emergence of a new, integrated, and critical field of study. Sci. Total Environ. 557–558, 740–753, with permission.

Woskie (2010), and Yokel and MacPhail (2011) have recommended the installation of good workplace practices, the adoption of a precautionary approach, and the development of a comprehensive health and safety program until health-based occupational exposure limits are officially developed and released.

ENMs have been incorporated into a wide range and variety of commercial products, and their release into urban environment is inevitable throughout the life cycle of production, use, and disposal. The release may be intentional (e.g., when ENMs are used for environmental remediation), nonintentional (e.g., due to wear and tear of materials containing ENMs), or by accidental spills (Fig. 5.1). When released into the urban environment, ENMs would interact with gases, water, natural organic matter (NOM), and clay materials. Thus the behavior and distribution of ENMs will be determined by the specific environmental conditions as well as their intrinsic properties, such as size, form, and reactivity.

Urbanization affects the fate of ENMs in two ways. First, it leads to the widespread and increased use of products containing ENMs (Table 5.2), such as scratch-resistant surface coatings, quality fuels, superior paints, water- and dust-repelling covers, self-cleaning surfaces, photocatalytic pavements, photovoltaic materials, moisture absorbents, high performance tires, antireflection layers for road signs and markings, not to mention various household appliances, including washing machines and refrigerators (Baalousha et al., 2016). The revised

Table 5.2 Engineered nanomaterials and their applications.

Class	Component	Applications
Metal oxides	Zinc oxide	Cosmetic sunscreens and UV coatings; paints, plastics, and packaging
	Titanium dioxide	Cosmetics
	Cerium dioxide	Automobile catalyst
	Mixed oxides	Cosmetics
Carbon products	Fullerenes	Plastics, catalysts, battery and fuel cell electrodes, super-capacitors, water purification systems, cosmetics, orthopedic implants, conductive coatings, adhesives and composites, sensors, and components in electronics, aircraft, aerospace, and automotive industries
	Single-walled and multiwalled carbon nanotubes	
	Amorphous carbon	Inks, photocopier toner, automobile tires
Metals	Silver	Bactericide in wound dressings, socks and other textiles, air filters, toothpaste, baby products, vacuum cleaners, and washing machines
	Iron	Remediation of groundwater, sediments, soils
	Gold	Electronics in flexible conducting inks or films, and as catalysts
	Bimetallic nanoparticles Fe–Pd, Fe–Ni, Fe–Ag	Remediation of organics in waters; usually supported nanoparticles
Quantum dots semiconductors	CdTe, CdSe/ZnS, CdSe, PbSe, and InP	Medical applications, photovoltaics, security inks, and photonics and telecommunications

inventory was released in October 2013. The Nanotechnology Consumer Products Inventory of the Woodrow Wilson International Center for Scholars currently lists 1814 consumer products of ENMs (Vance et al., 2015). The health and fitness category contains the most products (762, or 42% of the total). Silver is the most frequently used nanomaterial (435 products, or 24%). Second, urbanization increases ENM deposition on impervious cover, reduces storm water infiltration, and enhances surface runoff. The resultant increase in ENM transport to surface water would lead to a deterioration of water quality.

5.3 Processes controlling the fate of engineered nanomaterials in urban environment

The transformation of ENMs in the urban environment through aggregation, dissolution, and degradation (Fig. 5.2) can dramatically affect their behavior, fate,

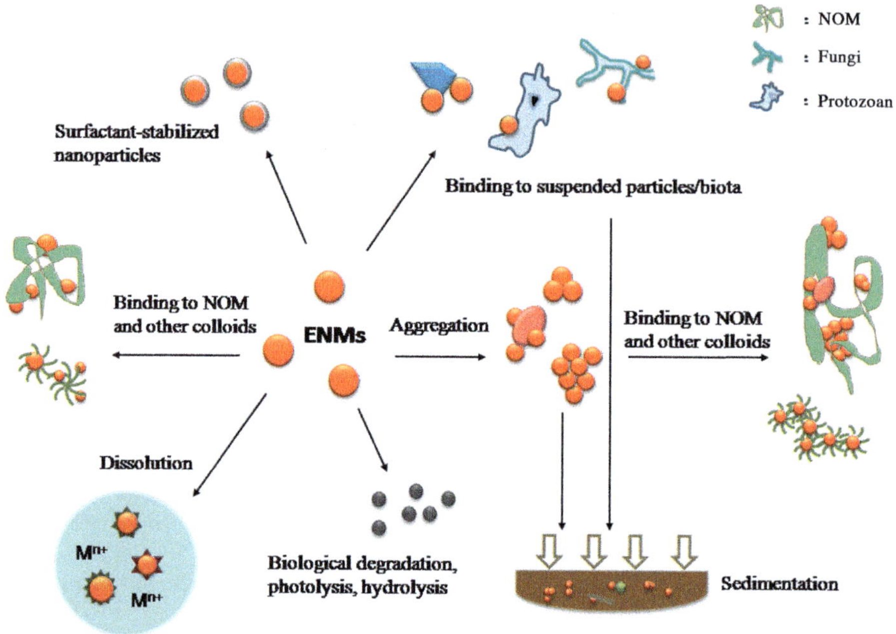

FIGURE 5.2

Processes that affect the fate of ENMs in the urban environment. *ENM*, Engineered nanomaterial.

and toxicity. Biodegradation is not relevant to inorganic ENMs, nor is it important for carbon-based ENMs. Table 5.3 summarizes the relative importance of transformation processes in compartments.

Photochemical transformation, induced by light, affects the fate and behavior of ENMs by influencing aggregation/agglomeration, adsorption to other surfaces, contaminant binding, and degradation of surface coatings. Some metal ENMs can act as photocatalysts upon excitation, producing reactive species that can degrade organic compounds and cause an ecotoxic effect. Nanoparticles of TiO_2, for example, are 2–4 orders of magnitude more toxic to a Japanese fish under simulated solar radiation than under ambient laboratory light (Ma et al., 2012). Like those of TiO_2, nanoparticles of ZnO display a high photocatalytic activity under UV illumination, whereas those of CeO_2 and CuO are much less active. Oxidation of C_{60} by exposure to UV light increases the negative surface charge and hydrophilicity of the molecule (Qu et al., 2012).

Because of their small particle size, large surface area, and high surface reactivity, there is a natural propensity for ENMs to grow in size. The surface of ENMs can be modified through contact with surfactants or bulky polymeric additives to slow down nanoparticle aggregation during storage and transportation.

Table 5.3 Relevance of environmental transformation processes of engineered nanomaterials (ENMs) in different physical compartments.

Environmental transformation processes		Compartments			
		Air	Water	Sediment	Soil
Chemical and photochemical processes	Photochemical reactions	++	+	−	−
	Redox reactions	−	++[a]	++[a]	+[a]
	Dissolution/speciation	−	−/++[a]	−/++[a]	−/++[a]
Physical processes	Aggregation/agglomeration	+	++	+	+
	Sedimentation/deposition	+	++	−	−
Interactions with surfaces/substances	NOM adsorption	−	+	++	+
	Sorption to other surfaces	−	+	++	++
Biologically mediated processes	Biodegradation	−	−/++[a]	−/++[a]	−/++[a]
	Biomodification	−	−/+	−/+	−/+

++, Highly relevant for inclusion in ENM fate modeling. +, relevant for inclusion in ENM fate modeling. −, low/no for inclusion in ENM fate modeling.
[a]*Highly dependent on the ENM chemical composition.*
Source: *After Hartmann, N.B., Skjolding, L.M., Hansen, S.F., Kjølholt, J., Gottschalck, F., Baun, A., 2014. Environmental Fate and Behaviour of Nanomaterials. The Danish Environmental Protection Agency.*

Ke and Qiao (2007) described three mechanisms by which surfactants help disperse single-walled carbon nanotubes: encapsulation, hemi-micellular adsorption, and random adsorption. The slower the aggregation the greater is the potential for interaction with biota. Phenrat et al. (2007), for example, have found that zerovalent iron nanoparticles in water could rapidly form micro-size aggregates of low reactivity and settled out of solution. ENMs with a surface charge can interact with opposite-charge ions (as in seawater), natural colloids (humic substances), clay minerals, and (Fe, Al, and Mn) hydroxides. The resultant decrease in interparticle electrostatic repulsion is conducive to aggregate formation.

The environmental fate and toxicity of ENMs are influenced by various intrinsic and external factors. Chemical compositions, surface area, particle size/shape/curvature/morphology, crystal phase, and surface coating/chemistry are intrinsic properties of relevance for ENM dissolution (Hartmann et al., 2014). They can affect dissolution directly, for example, by determining the ability of surface molecules to undergo reactions with water, or indirectly via affecting aggregation, which in turn influences the surface available for dissolution. In general the equilibrium solubility of particles increases as the particle size decreases, as expressed by the Ostwald–Freundlich equation (Hartmann et al., 2014). Size and surface area have been reported to substantially affect dissolution of metals and metal oxides. Bulk silver, for example, is considered insoluble in aqueous media. But Ag nanoparticles release free silver ions (Stone et al., 2010).

Most metal-based ENMs are hydrophilic and have a finite but often low solubility. Since the soluble ionic metal fraction is the most toxic to aquatic biota, it is desirable that the solubility of ENMs be determined. Although zinc oxide is commonly regarded as being insoluble, Franklin et al. (2007) have shown that its nanoparticles can rapidly dissolve up to 16 mg/L after 72 hours in a buffered solution (pH 7.5), which is well in excess of the 5 mg Zn/L toxicity level applicable to most aquatic biota.

Similarly, 250 mg/L solutions of CdSe semiconductor quantum dots can, upon oxidation, release as high as 80 mg Cd/L (Derfus et al., 2004). Carbon-based ENMs are typically lipophilic and virtually insoluble in natural waters. Their lipophilicity, however, can be altered by introducing substituents to the basic (fullerene or nanotube) structure. Sayes et al. (2004) reported that the cytotoxicity (to human liver cells) of fullerene derivatives was related to their solubility in that the least derivatized form was much more toxic than the highly soluble derivatives. The cytotoxic effects are related to the ability of fullerenes to generate oxygen free radicals that cause membrane damage.

Media composition and properties are external factors to influence the dissolution of ENMs and their fate and toxicity. They include pH, ionic strength, hardness, redox environment, inorganic ligands, and organic matter (Hartmann et al., 2014). Levard et al. (2012), for example, reported that the solubility of AgNPs increased as pH was lowered. The reaction of Ag nanoparticles with oxygen enhanced Ag dissolution, whereas removal of oxygen inhibited the dissolution (Liu and Hurt, 2010). The presence of S-containing cysteine increases the dissolution of Ag nanoparticles (Gondikas et al., 2012). While humic acid (100 mg/L) did not have a significant effect on ZnO dissolution at low pH (1, 3, and 6), the dissolution significantly increased at high pH (9 and 11) (Bian et al., 2011). Hyung et al. (2007) further showed that dissolved humic acid greatly enhanced the dispersion of multiwalled carbon nanotubes in deionized water, and much more so than that observed with the synthetic surfactant sodium dodecyl sulfate. Dissolved humic acids can also stabilize dispersions of iron oxide nanoparticles (Baalousha et al., 2008).

In investigating the behavior of cerium oxide nanoparticles in a model wastewater treatment system, Limbach et al. (2008) found that 6% of the nanoparticles was dispersed and released in the effluent (at 2–5 mg/L concentrations) as a result of stabilization by NOM and surfactants in the wastewater. The disaggregation of ENMs by household or industrial detergents has been known for years (Fernandes et al., 2006). By contrast, particulate NOM is likely to increase aggregation of ENMs. A high suspended sediment load in water, for example, can provide an effective removal mechanism for ENMs by enhancing their transport to, and accumulation in, bottom sediments.

5.4 Engineered nanomaterials in the urban atmosphere

Nanoparticles in the urban atmosphere mostly derive from combustion sources (traffic, forest fires), industrial and power plants using fossil fuels, and volcanic activity. An example of a nanoparticle-generating industrial activity is thermal spraying and coating. In this process a coating material (usually metal) is vaporized in a gas flame or plasma and deposited as a thin film onto a surface to improve its hardness or corrosion resistance. Nanoparticles can also form as an undesirable by-product of an industrial process. Welding, for example, can generate large quantities of nanoparticle aggregates in the form of a well-defined plume. Particles in the nanometer size range are also produced in large quantities from diesel engines and from domestic activities such as gas cooking.

Dall'Osto et al. (2011) have suggested that the emission of nanoparticles from road traffic (e.g., diesel engine exhaust) is the largest source of respiratory exposure for people in urban areas. The exposure to UFPs of different commuters (foot, bicycle, bus, tram, and car) in the city of Basel, Switzerland, has been reported by Ragetti et al. (2013). The closer the proximity to traffic the smaller is the average particle size and the higher the UFP concentration. Average UFP exposure concentrations decrease in the order of car > bicycle > foot > bus, tram.

The fate of UFPs in air is well documented. By comparison, the behavior, transport, and fate of airborne ENMs are less well understood. ENMs in air are highly mobile and they can mix rapidly. Their exposure to sunlight, especially UV light, would increase the probability for photochemical transformations. As airborne nanoparticles will settle out at a much slower rate than their larger counterparts, agglomeration will significantly increase the deposition of ENMs. Processes other than photochemical reactions, agglomeration, and deposition are not important or even irrelevant for airborne ENMs (Table 5.3).

For airborne ENMs and UFPs the single, most likely pathway of human exposure is through inhalation although dermal penetration and ingestion would contribute (Kumar et al., 2010; Maynard, 2006; Woskie, 2010; Yokel and MacPhail, 2011).

5.5 Engineered nanomaterials in the urban water environment

ENMs (and INMs) from industrial emissions, exterior façades, painted surfaces, and motor vehicle exhausts, enter the urban storm water ponds through runoff and atmospheric deposition (Baalousha et al., 2016). Keller and Lazareva (2014), for example, have estimated that as much as 66,000 tons of ENMs are released annually to surface waters. Among these, silver nanoparticles (AgNPs) have received much attention. With an estimated worldwide production between 500 and

1000 tons/year (Giese et al., 2018; Syafiuddin et al., 2018), it is inevitable that AgNPs will find their way into urban wastewater (Kaegi et al., 2013). Within the United Kingdom alone, 8.8 tons of silver nanoparticles enter the wastewater system every year (Whiteley et al., 2013). In a recent report, Giese et al. (2018) state that ENMs (CeO_2, SiO_2, Ag) pose a low risk for most environmental compartments although organisms living near ENM point sources, such as production plant outfalls and waste treatment plants, may be at increased risk. In developing a nanoFate model to simulate the time-dependent accumulation of metallic ENMs (CeO_2, CuO, TiO_2, ZnO) in the San Francisco Bay area, Garner et al. (2017) found that the accumulation of high-volume engineered TiO_2 and ZnO nanomaterials are likely to exceed the minimum toxic threshold in freshwater and some soils.

Because of their antimicrobial properties, AgNPs have been incorporated into a wide range of consumer products (creams, cosmetics, detergents, soaps, and textiles), home appliances (air and water filters, washing machines), and medical supplies (wound dressings) and textiles (Fabrega et al., 2011). Blaser et al. (2008) predicted that by 2015 biocidal plastics and textiles would account for up to 15% of the total silver released into water in the European Union, while Syafiuddin et al. (2018) stated that 5%–95% of the total amount of AgNPs in commercial products would end up in sewage treatment plants. In order to minimize the release of nanosilver from textiles, Voelker et al. (2015) have recommended the insertion of AgNPs into the textile fiber.

According to Kaegi et al. (2013), the discharged AgNPs are partially and rapidly sulfidized in the sewer system before being transported to the wastewater treatment plants, which constitutes their primary resting place in urban environments. Other chemical transformation routes for AgNPs include oxidative dissolution, chlorination, and photoreduction (Zhang et al., 2018). In vitro measurements by Larese et al. (2009) indicate low but detectable absorption of AgNPs through intact and damaged human skin. Sulfidation decreases the toxicity of AgNPs, and lowers their short-term environmental impact (Levard et al., 2012). The results of modeling studies by Gottschalk et al. (2009) indicate that nano-Ag in surface waters can pose a risk to aquatic organisms. Interestingly, biogenic AgNPs would appear to be less toxic than their chemically synthesized counterparts, and human cells are more resistant to the toxic effects of AgNPs than other organisms (de Lima et al., 2012). The mechanisms underlying the release, transformation, and toxicity of silver nanoparticles have been reviewed by Reidy et al. (2013).

5.6 Engineered nanomaterials in the urban soils and sediments

Like soils in general, urban soils contain many kinds of (natural) UFPs in the form of clay minerals, metal (hydr)oxides, and humic substances (Theng and Yuan, 2008).

By their nature and surface properties, soil nanoparticles are involved and participate in essential ecological services, ranging from regulating water storage and element cycling through sorbing and transporting chemical contaminants to serving as a source or sink of organic carbon and plant nutrients. For an informative and readable account of the application of nanotechnology to increase and sustain agricultural production, we refer to the report by Suppan (2013) and the review by Cornelis et al. (2014).

Urban soils are also a repository of various ENMs from cosmetic and pharmaceutical products, raising concerns about their potential adverse effects on soil biota and soil health. Mwaanga et al. (2017) have reported on the toxicity of various nanoparticles to earthworms in urban soils. Ingestion of Cu, CuO, and ZnO nanoparticles caused oxidative stress the intensity of which first increased with particle concentration and then declined. A similar but much less intense toxic effect was observed with the corresponding bulk materials.

The various transformations that ENMs may be subjected to in soils and sediments are less well understood than those occurring in water. The tendency of ENMs to aggregate in water would be conducive to their accumulation in sediments. In soil and sediment, ENMs are also likely to adsorb to, and interact with, clay mineral surfaces.

The large surface area and exchange capacity of soils would promote dissolution by providing protons to ENM with a pH-dependent solubility and act as a sink for ENM dissolution products (Batley and McLaughlin, 2010). Little is known, however, about the rate or extent of ENM dissolution in soils in relation to their bulk counterparts. Similarly, the aggregation behavior of ENMs in soils has not been examined in any detail. The relatively high ionic strength of soil pore waters would have a promoting effect on ENM aggregation. Aggregation of ENMs in soil would also lead to their entrapment and hence restrict their mobility (Wang et al., 2008).

As soil minerals and organic matter are predominantly negatively charged, positively charged hydrophilic ENMs would be retained strongly, while those with net negative charge will be highly mobile in most soils (Saleh et al., 2008). Binding of hydrophobic ENMs by soil organic matter may inhibit their mobility and availability to organisms. Given that contaminant partitioning (K_d, K_{oc}, or K_{ow}) is a key property used in risk assessment for a wide range of contaminants in terrestrial systems, this parameter of ENMs requires proper evaluation.

5.7 Routes of exposure to engineered nanomaterials

In contrast to the rapid advances in the production and applications of ENMs, their health and safety aspect is still in its formative stage (Yokel and MacPhail, 2011). Thus current recommendations to minimize exposure and hazards are largely based on common sense, knowledge by analogy to UFP toxicity, and general health and safety recommendations.

5.7.1 Inhalation exposure

Human exposures to ENMs during their activities of daily living occur through three major pathways: nasal, ocular, and dermal. It was confirmed that inhalation exposure is more fatal than dermal exposure (Manigrasso et al., 2017). Inhalation of fine particulate dusts is known to lead to pulmonary diseases. There is good evidence to indicate that the inflammatory response depends on the surface area of particles deposited in the proximal alveolar region of the lungs (Faux et al., 2003; Li et al., 2016; Maynard and Kuempel, 2005; Tran et al., 2000). Although the underlying mechanisms have yet to be elucidated, it seems likely that oxidative stress (through contact between particle and cell surface area), and the production of reactive oxygen species (ROS), play an important role in initiating the chain of events at the molecular and cellular level that lead to the observed inflammatory effects (Buzea et al., 2007; Hussain et al., 2009; Li et al., 2016).

The results of in vitro experiments by Faux et al. (2003) further indicate that highly reactive particles with a low surface area, and low toxicity particles with a high surface area, can exert the same oxidative stress level. The consensus is that surface area is a more appropriate dose metric than mass for the proinflammatory effects of (low toxicity, low solubility) nanoparticles under both in vitro and in vivo conditions (Donaldson and Tran 2002; Duffin et al., 2007; Monteiller et al., 2007). The concentration of nanoparticles might be very small in terms of mass but substantial in terms of surface area and huge in terms of particle number. Upon deposition in the pulmonary region, particles with different surface reactivity can exert an oxidative stress on the cells via contact between particle and cell surface due to imbalance between the production of ROS and their degradation (Limbach et al., 2007). Consequently, both highly reactive particles with a low surface area and low toxicity particles with a high surface area can exert the same oxidative stress level on the cells they come into contact with (Duffin et al., 2007). Using carbon black and titanium dioxide nanoparticles, Hussain et al. (2009) found that particle size differences of even a few nanometers could give rise to appreciable changes in inflammatory and oxidative stress responses. Accordingly, Xia et al. (2006) have proposed that ROS generation and oxidative stress can be used as a paradigm to compare the toxicity of nanoparticles. We might add that nanoparticles can potentially penetrate epithelial cells, enter the bloodstream from the lungs, and even translocate to the brain via the olfactory nerves (Buzea et al., 2007; Donaldson et al., 2005; Gilmour et al., 2004; Oberdörster et al., 2009). Thus the negative health effects of inhaled nanoparticles may not be confined to the lungs (Bakand et al., 2012).

In conclusion, scientific evidence, so far, has demonstrated that particle surface area and surface reactivity are likely to be the metric of choice to describe the inflammatory reaction to deposited particles in the proximal alveolar region of the lung. For nanoparticles, their potential dispersal to other organs as well as the possibility of exposure by other routes such as dermal or ingestion mean that possible health risks beyond the lung cannot be ruled out. Further research to generate vital data on the possible mode of action of nanoparticles in the extrapulmonary system is needed in order to assess realistically the health risks to nanoparticle exposure.

5.7.2 Dermal exposure

Research on dermal exposure to ENMs has largely focused on cosmetics, sunscreens, and other consumer health and personal care products that contain ENMs, notably TiO_2 and ZnO. Brenner and Neu-Baker (2015) have also reported the occurrence of metal (Si, Al, and Ce) oxide nanoparticles on working surfaces, used or generated by chemical mechanical planarization, and to which the skin of workers in semiconductor facilities would be exposed. Such particles may penetrate into hair follicles where some particle constituents can dissolve and enter the skin. Gulson et al. (2010), for example, have found that ZnO particles in sunscreens, applied outdoors, can be absorbed through human skin. Similarly, Hirai et al. (2012) reported that nanosilica particles could penetrate the skin of mice. Staroňová et al. (2012) also found that SiO_2 nanoparticles could penetrate into human skin but did not permeate through it. Calculations by Watkinson et al. (2013) further indicate that nanoparticles are too large to permeate healthy skin by passive diffusion. On the other hand, Sadrieh et al. (2010) failed to detect significant penetration of TiO_2 nanoparticles through the intact skin of minipigs. The apparent inconsistencies in results may be ascribed to the interplay of several factors, among which particle size and surface charge are major determinants (Carlson et al., 2008; Park et al., 2013), as well as to variations in experimental approaches and conditions (Baroli, 2010; Filon et al., 2015; Gulson et al., 2015).

Overall, there is no consistent evidence to indicate that specific health problems would arise from dermal penetration of UFPs (Nohynek and Dufour, 2012; Robertson et al., 2010). Similarly, Hristozov and Malsch (2009) have argued that it is not yet possible to make any collective judgment about the potential risks of ENMs to human health. Several investigators (Pietroiusti and Magrini, 2014; Woskie, 2010; Yokel and MacPhail, 2011) have recommended the installation of good workplace practices, the adoption of a precautionary approach, and the development of a comprehensive health and safety program until health-based occupational exposure limits are officially developed and released.

5.7.3 Ingestion exposure

So far, very little work has been done on ingestion exposure in occupational settings. Lead is one of the few materials the ingestion route of which has received some attention. Removing lead paint can give rise to high ingestion exposures through hand–mouth contact and food contamination. Workers involved in the supply and removal of scaffolding, for example, can have particularly high blood lead levels as a result of hand contamination and subsequent ingestion (Sen et al., 2002). In the first instance, mass concentration/uptake can usefully serve as a metric for assessing exposure (Abbott and Maynard, 2010). New and biologically relevant exposure metrics would be required to assess ingestion exposure to ENMs.

Merrifield et al. (2013) have shown that ingestion of food, containing copper and silver nanoparticles, can disrupt the endogenous microbiota in zebrafish. More recently, Guo et al. (2017) reported that ingested TiO_2 nanoparticles could alter nutrient absorption in human epithelial cells of the small intestine. Many in vitro investigations also indicate that silver nanoparticles are toxic to mammalian cells from skin, liver, lung, brain, and reproductive organs (Ahamed et al., 2010). As is the case with dermal exposure, data on the toxicological effects of ingested nanoparticles in terms of cell proliferation, oxidative stress, and organ dysfunction have at times been inconsistent, if not contradictory (McCracken et al., 2016).

Bergin and Witzmann (2013) have pointed out that the gastrointestinal tract is the likely route of entry for many nanoparticles, either directly through intentional ingestion, or indirectly through dissolution of nanoparticle-containing food. These workers also found that results on the toxicity of ingested nanoparticles were strikingly different between in vivo and in vitro studies. All the same, ingestion was unlikely to have acute or severe toxic effects at typical levels of exposure.

Highlights

- ENMs are released during life cycle via emission, deposition, runoff, and abrasion.
- Photochemical degradation, aggregation, and dissolution are the key transformation processes of ENMs in urban environment.
- Intrinsic (compositions, particle size, surface charge) and external (UV light, surfactant, humic substance, and pH) factors determine the fate of ENMs.
- Inhalation is more common than dermal and ingestion routes for human exposure to ENMs.

Acknowledgment

This work was supported by the Chinese National Key Research and Development Program (Grant No. 2016YFD0200303).

References

Abbott, L.C., Maynard, A.D., 2010. Exposure assessment approaches for engineered nanomaterials. Risk Anal. 30 (11), 1634–1644.

Ahamed, M., AlSalhi, M.S., Siddiqui, M.K.J., 2010. Silver nanoparticle applications and human health. Clin. Chim. Acta 411 (23–24), 1841–1848.

Aitken, R.J., Creely, K.S., Tran, C.L., 2004. Nanoparticles: An Occupational Hygiene Review. Research Report 274. HSE Books, Sudbury, UK.

Baalousha, M., Manciulea, A., Cumberland, S., Kendall, K., Lead, J.R., 2008. Aggregation and surface properties of iron oxide nanoparticles: influence of pH and natural organic matter. Environ. Toxicol. Chem. 27 (9), 1875–1882.

Baalousha, M., Yang, Y., Vance, M.E., Colman, B.P., McNeal, S., Xu, J., et al., 2016. Outdoor urban nanomaterials: the emergence of a new, integrated, and critical field of study. Sci. Total Environ. 557–558, 740–753.

Bakand, S., Hayes, A., Dechsakulthorn, F., 2012. Nanoparticles: a review of particle toxicology following inhalation exposure. Inhal. Toxicol. 24 (2), 125–135.

Baroli, B., 2010. Penetration of nanoparticles and nanomaterials in the skin: fiction or reality? J. Pharmaceut. Sci. 99 (1), 21–50.

Batley, G.E., McLaughlin, M.J., 2010. Fate of Manufactured Nanomaterials in the Australian Environment. Department of the Environment.

Batley, G.E., Kirby, J.K., McLaughlin, M.J., 2013. Fate and risks of nanomaterials in aquatic and terrestrial environments. Acc. Chem. Res. 46 (3), 854–862.

Bergin, I.L., Witzmann, F.A., 2013. Nanoparticle toxicity by the gastrointestinal rout: evidence and knowledge gaps. Int. J. Biomed. Nanosci. Nanotechnol. 3 (1–2), 163–210.

Bian, S., Mudunkotuwa, I.A., Rupasinghe, T., Grassian, V.H., 2011. Aggregation and dissolution of 4 nm ZnO nanoparticles in aquatic environments: influence of pH, ionic strength, size, and adsorption of humic acid. Langmuir 27 (10), 6059–6068.

Blaser, S.A., Scheringer, M., MacLeod, M., Hungerbühler, K., 2008. Estimation of cumulative aquatic exposure and risk due to silver: contribution of nano-functionalized plastics and textiles. Sci. Total Environ. 390 (2–3), 396–409.

Brar, S.K., Verma, M., Tyagi, R.D., Surampalli, R.Y., 2010. Engineered nanoparticles in wastewater and wastewater sludge—evidence and impacts. Waste Manage. 30 (3), 504–520.

Brenner, S.A., Neu-Baker, N.M., 2015. Occupational exposure to nanomaterials: assessing the potential for cutaneous exposure to metal oxide nanoparticles in a semiconductor facility. J. Chem. Health Saf. 22 (4), 10–19.

Buzea, C., Pacheco, I.I., Robbie, K., 2007. Nanomaterials and nanoparticles: sources and toxicity. Biointerphases 2 (4), MR17–MR71.

Bystrzejewska-Piotrowska, G., Golimowski, J., Urban, P.L., 2009. Nanoparticles: their potential toxicity, waste and environmental management. Waste Manage. 29 (9), 2587–2595.

Caballero-Guzman, A., Sun, T., Nowack, B., 2015. Flows of engineered nanomaterials through the recycling process in Switzerland. Waste Manage. 36, 33–43.

Carlson, C., Hussain, S.M., Schrand, A.M., Braydich-Stolle, L.K., Hess, K.L., Jones, R.L., et al., 2008. Unique cellular interaction of silver nanoparticles: size-dependent generation of reactive oxygen species. J. Phys. Chem. B 112 (43), 13608–13619.

Colvin, V.L., 2003. The potential environmental impact of engineered nanomaterials. Nat. Biotech. 21 (10), 1166–1170.

Cornelis, G., Hund-Rinke, K., Kuhlbusch, T., van der Brink, N., Nickel, C., 2014. Fate and bioavailability of engineered nanoparticles in soils; a review. Crit. Rev. Environ. Sci. Technol. 44 (24), 2720–2764.

Dall'Osto, M., Thorpe, A., Beddows, D.C.S., Harrison, R.M., Barlow, J.F., Dunbar, T., 2011. Remarkable dynamics of nanoparticles in the urban atmosphere. Atmos. Chem. Phys. 11 (13), 6623–6637.

de Lima, R., Seabra, A.B., Durán, N., 2012. Silver nanoparticles: a brief review of cytotoxicity and genotoxicity of chemically and biogenically synthesized nanoparticles. J. Appl. Toxicol. 32 (11), 867–879.

Derfus, A.M., Chan, W.C.W., Bhatia, S.N., 2004. Probing the nanotoxicity of semiconductor quantum dots. Nano Lett. 4 (1), 11–18.

Donaldson, K., Tran, C.L., 2002. Inflammation caused by particles and fibers. Inhal. Toxicol. 14 (1), 5–27.

Donaldson, K., Tran, L., Jimenez, L.A., Duffin, R., Newby, D.E., Mills, N., et al., 2005. Combustion-derived nanoparticles: a review of their toxicology following inhalation exposure. Particle Fibre Toxicol. 2, 10.

Duester, L., Burkhardt, M., Gutleb, A.C., Kaegi, R., Macken, A., Meermann, B., et al., 2014. Toward a comprehensive and realistic risk evaluation of engineered nanomaterials in the urban water system. Front. Chem. 2, 1–6.

Duffin, R., Tran, L., Brown, D., Stone, V., Donaldson, K., 2007. Proinflammogenic effects of low-toxicity and metal nanoparticles in vivo and in vitro: highlighting the role of particle surface area and surface reactivity. Inhal. Toxicol. 19 (10), 849–856.

Dwivedi, A.D., Dubey, S.P., Sillanpää, M., Kwon, Y.-N., Lee, C., Varma, R.S., 2015. Fate of engineered nanoparticles: implications in the environment. Coord. Chem. Rev. 287, 64–78.

Fabrega, J., Luoma, S.N., Tyler, C.R., Galloway, T.S., Lead, J.R., 2011. Silver nanoparticles: behaviour and effects in the aquatic environment. Environ. Int. 37 (2), 517–531.

Faux, S.P., Tran, C.L., Miller, B.G., Jones, A.D., Monteiller, C., Donaldson, K., 2003. In vitro determination of particulate toxicity: the dose-metric for poorly soluble dusts. Research Report 154. Institute of Occupational Medicine for the Health and Safety Executive, Norwich, UK.

Fernandes, M., Rosenkranz, P., Ford, A., Christofi, N., Stone, V., 2006. Ecotoxicology of Nanoparticles (NPs). SETAC Globe, pp. 43–45.

Filon, F.L., Mauro, M., Adami, G., Bovenzi, M., Crosera, M., 2015. Nanoparticles skin absorption: new aspects for a safety profile. Regul. Toxicol. Pharamcol. 72 (2), 301–322.

Franklin, N.M., Rogers, N.T., Apte, S.C., Batley, G.E., Casey, P.E., 2007. Comparative toxicity of nanoparticulate ZnO, bulk ZnO and $ZnCl_2$ to a freshwater microalga (*Pseudokirchneriella subcapitata*): the importance of particle solubility. Environ. Sci. Technol. 41 (24), 8484–8490.

Gardea-Torresdey, J.L., Ric, C.M., White, J.C., 2014. Trophic transfer, transformation, and impact of engineered nanomaterials in terrestrial environments. Environ. Sci. Technol. 48 (5), 2526–2540.

Garner, K.L., Keller, A.A., 2014. Emerging patterns for engineered nanomaterials in the environment: a review of fate and toxicity studies. J. Nanopart. Res. 16 (8), 2503.

Garner, K.L., Suh, S., Keller, A.A., 2017. Assessing the risk of engineered nanomaterials in the environment: development and application of the nanoFate model. Environ. Sci. Technol. 51 (10), 5541–5551.

Giese, B., Klaessig, F., Park, B., Kaegi, R., Steinfeldt, M., Wigger, H., et al., 2018. Risk, release and concentrations of engineered nanomaterial in the environment. Sci. Rep. 8 (1).

Gilmour, P.S., Ziesenis, A., Morrison, E.R., Vickers, M.A., Drost, E.M., Ford, I., et al., 2004. Pulmonary and systemic effects of short-term inhalation exposure to ultrafine carbon black particles. Toxicol. Appl. Pharmacol. 195 (1), 35–44.

Gondikas, A.P., Morris, A., Reinsch, B.C., Marinakos, S.M., Lowry, G.V., Hsu-Kim, H., 2012. Cysteine induced modifications of zero-valent silver nanomaterials: implications for particle surface chemistry, aggregation, dissolution, and silver speciation. Environ. Sci. Technol. 46, 7037–7045.

Goswami, L., Kim, K.-H., Deep, A., Das, P., Bhattacharya, S.S., Kumar, S., et al., 2017. Engineered nano particles: nature, behavior, and effect on the environment. J. Environ. Manage. 196, 297–315.

Gottschalk, F., Nowack, B., 2011. The release of engineered nanomaterials to the environment. J. Environ. Monit. 13 (5), 1145–1155.

Gottschalk, F., Sonderer, T., Scholz, R.W., Nowack, B., 2009. Modeled environmental concentrations of engineered nanomaterials (TiO_2, ZnO, Ag, CNT, fullerenes) for different regions. Environ. Sci. Technol. 43 (24), 9216–9220.

Gottschalk, F., Sun, T.Y., Nowack, B., 2013. Environmental concentrations of engineered nanomaterials: review of modeling and analytical studies. Environ. Pollut. 181 (6), 287–300.

Gulson, B., McCall, M., Korsch, M., Gomez, L., Casey, P., Oytam, Y., et al., 2010. Small amounts of zinc from zinc oxide particles in sunscreens applied outdoors are absorbed through human skin. Toxicol. Sci. 118 (1), 140–149.

Gulson, B., McCall, M.J., Bowman, D.M., Pinheiro, T., 2015. A review of critical factors for assessing the dermal absorption of metal oxide nanoparticles from sunscreens applied to humans, and a research strategy to address current deficiencies. Arch. Toxicol. 89 (11), 1909–1930.

Guo, Z., Martucci, N.J., Moreno-Olivas, F., Tako, E., Nahler, G.J., 2017. Titanium dioxide nanoparticle ingestion alters nutrient absorption in an in vivo model of the small intestine. NanoImpact 5, 70–82.

Hansen, S.F., Heggelund, L.R., Revilla Besora, P., Mackevica, A., Boldrin, A., Baun, A., 2016. Nanoproducts—what is actually available to European consumers? Environ. Sci. Nano 3 (1), 169–180.

Hartmann, N.B., Skjolding, L.M., Hansen, S.F., Kjølholt, J., Gottschalck, F., Baun, A., 2014. Environmental Fate and Behaviour of Nanomaterials. The Danish Environmental Protection Agency.

Hirai, T., Yoshikawa, T., Nabeshi, H., Yoshida, T., Akase, T., Yoshioka, Y., et al., 2012. Dermal absorption of amorphous nanosilica particles after topical exposure for three days. Pharmazie 67 (8), 742–743.

Hristozov, D., Malsch, I., 2009. Hazards and risks of engineered nanoparticles for the environment and human health. Sustainability 1 (4), 1161–1194.

Hussain, S., Boland, S., Baeza-Squiban, A., Hamel, R., Thomassen, L.C.J., Martens, J.A., et al., 2009. Oxidative stress and proinflammatory effects of carbon black and titanium dioxide nanoparticles: role of particle surface area and internalized amount. Toxicology 260 (1–3), 142–149.

Hyung, H., Fortner, J.D., Hughes, J.B., Kim, J.-H., 2007. Natural organic matter stabilization of multi-walled nanotubes in water. Environ. Sci. Technol. 41 (1), 179–184.

Joner, E., Hartnik, T., Amundsen, C., 2008. Environmental Fate and Ecotoxicity of Engineered Nanoparticles. Norwegian Pollution Control Authority.

Kaegi, R., Voegelin, A., Ort, C., Sinnet, B., Thalmann, B., Krismer, J., et al., 2013. Fate and transformation of silver nanoparticles in urban wastewater systems. Water Res. 47 (12), 3866–3877.

Ke, P.C., Qiao, R., 2007. Carbon nanomaterials in biological systems. J. Phys. Condens. Matt. 19 (37), 1501–1507.

Keller, A.A., Lazareva, A., 2014. Predicted releases of engineered nanomaterials: from global to regional to local. Environ. Sci. Technol. Lett. 1 (1), 65–70.

Kumar, P., Robins, A., Vardoulakis, S., Britter, R., 2010. A review of the characteristics of nanoparticles in the urban atmosphere and the prospects for developing regulatory controls. Atmos. Environ. 44 (39), 5035–5052.

Kumar, A., Kumar, P., Anandan, A., Fernandes, T.F., Ayoko, G.A., Biskos, G., 2014a. Engineered nanomaterials: knowledge gaps in fate, exposure, toxicity, and future directions. J. Nanomater. 2014 (7), 5.

Kumar, P., Morawska, L., Birmili, W., Paasonen, P., Hu, M., Kulmala, M., et al., 2014b. Ultrafine particles in cities. Environ. Int. 66 (2), 1–10.

Larese, F.F., D'Agostin, F., Crosera, M., Adami, G., Renzi, N., Bovenzi, M., et al., 2009. Human skin penetration of silver nanoparticles through intact and damaged skin. Toxicology 255 (1–2), 33–37.

Levard, C., Hotze, E.M., Lowry, G.V., Brown Jr., G.E., 2012. Environmental transformations of silver nanoparticles: impact on stability and toxicity. Environ. Sci. Technol. 46 (13), 6900–6914.

Li, N., Georas, S., Alexis, N., Fritz, P., Xia, T., Williams, M.A., et al., 2016. A work group report on ultrafine particles (American Academy of Allergy, Asthma & Immunology): why ambient ultrafine and engineered nanoparticles should receive special attention for possible adverse health outcomes in human subjects. J. Allergy Clin. Immunol 138 (2), 386–396.

Limbach, L.K., Wick, P., Manser, P., Grass, R.N., Bruinink, A., Stark, W.J., 2007. Exposure of engineered nanoparticles to human lung epithelial cells: influence of chemical composition and catalytic activity on oxidative stress. Environ. Sci. Technol. 41 (11), 4158–4163.

Limbach, L.K., Bereiter, R., Müller, E., Krebs, R., Gälli, R., Stark, W.J., 2008. Removal of oxide nanoparticles in a model wastewater treatment plant: influence of agglomeration and surfactants on clearing efficiency. Environ. Sci. Technol. 42 (15), 5828–5833.

Lin, D., Tian, X., Wu, F., Xing, B., 2010. Fate and transport of engineered nanomaterials in the environment. J. Environ. Qual. 39 (6), 1896–1908.

Liu, J., Hurt, R.H., 2010. Ion release kinetics and particle persistence in aqueous nano-silver colloids. Environ. Sci. Technol. 44, 2169–2175.

Ma, H., Brennan, A., Diamond, S.A., 2012. Phototoxicity of TiO_2 nanoparticles under solar radiation to two aquatic species: daphnia magna and Japanese medaka. Environ. Toxicol. Chem. 31 (7), 1621–1629.

Manigrasso, M., Vernale, C., Avino, P., 2017. Traffic aerosol lobar doses deposited in the human respiratory system. Environ. Sci. Pollut. Res. 24 (16), 1–8.

Maynard, A.D., 2006. Nanotechnology: assessing the risks. Nanotoday 1 (2), 22–33.

Maynard, A.D., Kuempel, E.D., 2005. Airborne nanostructured particles and occupational health. J. Nanopart. Res. 7 (6), 587–614.

McCracken, C., Dutta, P.K., Waldman, W.J., 2016. Critical assessment of toxicological effects of ingested nanoparticles. Environ. Sci.: Nano. 3 (2), 256–282.

Merrifield, D.L., Shaw, B.J., Harper, G.M., Saoud, I.P., Davies, S.J., Handy, R.D., et al., 2013. Ingestion of metal-nanoparticle contaminated food disrupts endogenous microbiota in zebrafish (*Danio rerio*). Environ. Pollut. 174 (5), 157–163.

Mitrano, D.M., Motellier, S., Clavaguera, S., Nowack, B., 2015. Review of nanomaterial aging and transformations through the life cycle of nano-enhanced products. Environ. Int. 77, 132–147.

Monteiller, C., Tran, L., MacNee, W., Faux, S., Jones, A., Miller, B., et al., 2007. The pro-inflammatory effects of low-toxicity low-solubility particles, nanoparticles and fine particles, on epithelial cells in vitro: the role of surface area. Occup. Environ. Med. 64 (9), 609–615.

Munafò, P., Goffredo, G.B., Quagliarini, E., 2015. TiO_2-based nanocoatings for preserving architectural stone surfaces: An overview. Construct Build Mater. 84, 201–218.

Mwaanga, P., Mbulwe, S., Shumbula, P., Nyirenda, J., 2017. Investigating the toxicity of Cu, CuO and ZnO nanoparticles on earthworms in urban soils. J. Pollut. Effects Cont. 5 (3), 1–9.

Nel, A., Xia, T., Mädler, L., Li, N., 2006. Toxic potential of materials at the nanolevel. Science 311 (5761), 622–627.

Nohynek, G.J., Dufour, E.K., 2012. Nano-sized cosmetic formulations or solid nanoparticles in sunscreens: a risk to human health? Arch. Toxicol. 86 (7), 1063–1075.

Nowack, B., Bucheli, T.D., 2007. Occurrence, behaviour and effects of nanoparticles in the environment. Environ. Pollut. 150 (1), 5–22.

Oberdörster, G., Elder, A., Rinderknecht, A., 2009. Nanoparticles and the brain: cause for concern? J. Nanosci. Nanotechnol. 9 (8), 4996–5007.

Park, Y.-H., Bae, H.C., Jang, Y., Jeong, S.H., Lee, H.N., Ryu, W.-I., et al., 2013. Effect of the size and surface charge of silica nanoparticles on cutaneous toxicity. Mol. Cell Toxicol. 9 (1), 67–74.

Part, F., Berge, N., Baran, P., Stringfellow, A., Sun, W., Bartelt-Hunt, S., et al., 2018. A review of the fate of engineered nanomaterials in municipal solid waste streams. Waste Manage. 75, 427–449.

Peijnenburg, W., Praetorius, A., Scott-Fordsmand, J., Cornelis, G., 2016. Fate assessment of engineered nanoparticles in solids dominated media—current insights and the way forward. Environ. Pollut. 218, 1365–1369.

Peralta-Videa, J.R., Zhao, L., Lopez-Moreno, M.L., de la Rosa, G., Hong, J., Gardia-Torresdey, J.L., 2011. Nanomaterials and the environments: a review for the biennium 2008–2010. J. Hazar. Mater. 186 (1), 1–15.

Phenrat, T., Saleh, N., Sirk, K., Tilton, R.D., Lowry, G.V., 2007. Aggregation and sedimentation of aqueous nanoscale zerovalent iron dispersions. Environ. Sci. Technol. 41 (1), 284–290.

Pietroiusti, A., Magrini, A., 2014. Engineered nanoparticles at the workplace: current knowledge about workers' risk. Occup. Med. 64 (5), 319–330.

Qu, X., Alvarez, P.J.J., Li, Q., 2012. Impact of sunlight and humic acid on the deposition kinetics of aqueous fullerene nanoparticles (nC_{60}). Environ. Sci. Technol. 46 (24), 13455–13462.

Ragetti, M.S., Corradi, E., Braun-Fahrländer, C., Schindler, C., de Nazelle, A., Jerrett, M., et al., 2013. Commuter exposure to ultrafine particles in different urban locations, transportation modes and routes. Atmos. Environ. 77 (3), 376–384.

Reidy, B., Haase, A., Luch, A., Dawson, K.A., Lynch, I., 2013. Mechanisms of silver nanoparticle release, transformation and toxicity: a critical review of current knowledge and recommendations for future studies and applications. Materials 6 (6), 2295–2350.

Robertson, T.A., Sanchez, W.Y., Roberts, M.S., 2010. Are commercially available nanoparticles safe when applied to the skin? J. Biomed. Nanotechnol. 6 (5), 452–468.

Roco, M.C., 2011. The long view of nanotechnology development: the national nanotechnology initiative at 10 years. J. Nanopart. Res. 13 (2), 427–445.

Sadrieh, N., Wokovich, A.M., Gopee, N.V., Zheng, J., Haines, D., Parmiter, D., et al., 2010. Lack of significant dermal penetration of titanium dioxide from sunscreen formulations containing nano- and submicron-size TiO_2 particles. Toxicol. Sci. 115 (1), 156–166.

Saleh, N., Kim, H.J., Phenrat, T., Matyjaszewski, K., Tilton, R.D., Lowry, G.V., 2008. Ionic strength and composition affect the mobility of surface-modified Fe^0 nanoparticles in water-saturated sand columns. Environ. Sci. Technol. 42 (9), 3349–3355.

Sayes, C.M., Fortner, J.D., Guo, W., Lyon, D., Boyd, A.M., Ausman, K.D., et al., 2004. The differential cytotoxicity of water-soluble fullerenes. Nano Lett. 4 (10), 1881–1887.

Schulte, P., Geraci, C., Zumwalde, R., Hoover, M., Kuemple, E., 2008. Occupational risk management of engineered nanoparticles. J. Occup. Environ. Hyg. 5 (4), 239–249.

Sen, D., Wolfson, H., Dilworth, M., 2002. Lead exposure in scaffolders during refurbishment construction activity—an observational study. Occup. Med. 52 (1), 49–54.

Staroňová, K., Nielsen, J.B., Roursgaard, M., Knudsen, L.E., 2012. Transport of SiO_2 nanoparticles through human skin. Basic Clin. Pharmacol. Toxicol. 111 (2), 142–144.

Stone, V., Nowack, B., Baun, A., van den Brink, N., von der Kammer, F., Dusinska, M., et al., 2010. Nanomaterials for environmental studies: classification, reference material issues, and strategies for physico-chemical characterisation. Sci. Total Environ. 408 (7), 1745–1754.

Suppan, S., 2013. Nanomaterials in Soil: Our Future Food Chain? Institute for Agriculture and Trade Policy.

Syafiuddin, A., Salmiati, S., Hadibarata, T., Kueh, A.B.H., Salim, M.R., Zaini, M.A.A., 2018. Silver nanoparticles in the water environment in Malaysia: inspection, characterization, removal, modeling, and future perspective. Sci. Rep. 8 (1), 986.

Theng, B.K.G., 2012. Formation and Properties of Clay-Polymer Complexes, second ed. Elsevier, Amsterdam.

Theng, B.K.G., Yuan, G.D., 2008. Nanoparticles in the soil environment. Elements 4 (6), 395–399.

Tiede, K., Boxall, A.B.A., Tear, S.P., Lewis, J., David, H., Hassellöv, M., 2008. Detection and characterization of engineered nanoparticles in food and the environment. Food Addit. Contam. 25 (7), 795–821.

Tran, C.L., Buchanan, D., Cullen, R.T., Searl, A., Jones, A.D., Donaldson, K., 2000. Inhalation of poorly soluble particles. II. Influence of particle surface area on inflammation and clearance. Inhal. Toxicol. 12 (12), 1113–1126.

UN Department of Economic and Social Affairs, 2018. 68% of the World Population Projected to Live in Urban Areas by 2050. <https://www.un.org/development/desa/en/news.html> (accessed 16.05.18.).

Vance, M.E., Kuiken, T., Vejerano, E.P., McGinnis, S.P., Hochella Jr., M.F., et al., 2015. Nanotechnology in the real world: redeveloping the nanomaterial consumer products inventory. Beilstein J. Nanotechnol. 6, 1769–1780.

Voelker, D., Schlich, K., Hohndorf, L., Koch, W., Kuehnen, U., Polleichter, C., et al., 2015. Approach on environmental risk assessment of nanosilver released from textiles. Environ. Res. 140, 661–672.

Wang, Y.G., Li, Y.S., Pennell, K.D., 2008. Influence of electrolyte species and concentration on the aggregation and transport of fullerene nanoparticles in quartz sands. Environ. Toxicol. Chem. 27 (9), 1860–1867.

Watkinson, A.C., Bunge, A.L., Hadgraft, J., Lane, M.E., 2013. Nanoparticles do not penetrate human skin—a theoretical perspective. Pharm. Res. 30 (8), 1943–1946.

Whiteley, C.M., Valle, M.D., Joes, K.C., Sweetman, A.J., 2013. Challenges in assessing release, exposure and fate of silver nanoparticles within the UK environment. Environ. Sci.-Proc. Imp. 15 (11), 2050–2058.

Woskie, S., 2010. Workplace practices for engineered nanomaterial manufacturers. WIREs Nanomed. Nanobiotechnol. 2 (6), 685–692.

Xia, T., Kovochich, M., Brant, J., Hotze, M., Sempf, J., Oberley, T., et al., 2006. Comparison of the abilities of ambient and manufactured nanoparticles to induce cellular toxicity according to an oxidative stress paradigm. Nano Lett. 6 (8), 1794–1807.

Yokel, R.A., MacPhail, R.C., 2011. Engineered nanomaterials: exposures, hazards, and risk prevention. J. Occup. Med. Toxicol. 6 (1), 7.

Zhang, W., Xiao, B., Fang, T., 2018. Chemical transformation of silver nanoparticles in aquatic environments: mechanism, morphology and toxicity. Chemosphere 191, 324–334.

CHAPTER

Presence of nanomaterials on consumer products: food, cosmetics, and drugs*

6

Ana M. Rincon
European Food Safety Authority, Parma, Italy

6.1 Introduction

Materials in the nanoscale are not something new; food is composed of proteins, polysaccharides, and lipids that can exist at the nanoscale or can be organized in functional nanostructures. Humans have been exposed for years to nanoscale particles such as dust, ash, and fine clays through the air by inhalation and from ingestion of water or food.

The application of nanotechnology may result in products with chemical or physical properties or biological effects that differ from those of conventionally manufactured products; therefore the case-by-case evaluation of the safety or effectiveness of nanomaterials or products that involve the application of nanotechnology should consider the unique properties and behaviors that they may exhibit.

This chapter provides information on the substances defined as nanomaterials or engineered nanomaterials (ENMs) that have been officially registered for their use in food, cosmetics, or drugs, as well as information on the requirements for their commercialization, which are different depending on the legal framework of the country. The term "nanomaterial" is more commonly used than "engineered nanomaterials" in the area of food, cosmetics, and drugs.

6.2 European legislation

6.2.1 Food additives

In the European Union (EU), there is no legal definition of "nanomaterial" or "ENM" applicable to food additives as such. The definition of nanomaterial as "a natural, incidental or manufactured material containing particles, in an unbound state or as an aggregate or as an agglomerate and where, for 50% or more of the

* The present chapter is published under the sole responsibility of the author and may not be considered as an EFSA scientific output. The positions and opinions presented in this chapter are those of the author alone and are not intended to represent any official position or scientific works of the European Food Safety Authority (EFSA).

Exposure to Engineered Nanomaterials in the Environment. DOI: https://doi.org/10.1016/B978-0-12-814835-8.00006-6

particles in the number size distribution, one or more external dimensions is in the size range 1–100 nm," according to the Commission Recommendation 2011/696/EC, is not binding because "a recommendation allows institutions to make their view known and to suggest a line of action without imposing any legal obligation on those to whom it is addressed" [EC (European Commission), 2011a].

Regulation (EU) 1169/2011 on the provision of food information to consumer establishes the general principles, requirements, and responsibilities governing food information, and in particular food labeling [EC (European Commission), 2011b]. It lays down the means to guarantee the right of consumers to information and procedures for the provision of food information, and it applies to food business operators at all stages of the food chain and to all foods intended for the final consumer.

The safety of substances to be authorized as food additives as well as the reevaluation of all food additives permitted in the EU before January 20, 2009, is carried out by the European Food Safety Authority (EFSA).

All food additives authorized in the EU are listed in Part B of Annex II of Regulation [EC (European Commission), 2008] no. 1333/2008 on food additives and specific purity criteria have been defined in the Commission Regulation (EU) no. 231/2012 [EC (European Commission), 2012] for them. Currently, no parameters to specify the size of the particles for the food additives authorized in the EU are included as part of their technical specifications.

A program for the reevaluation of food additives that were already authorized in the EU before January 20, 2009 has been set up under Regulation (EU) no. 257/2010 [EC (European Commission), 2010]. Since then, the EFSA Panel on Food Additives and Nutrient Sources added to Food has reevaluated 176 food additives (September 2018), and for some of them the need for further characterization of the fraction of the food additive materials in the nanorange has been identified.

For the risk assessment of ENMs or materials containing a fraction in the nanoscale, the *Guidance on the human and animal risk assessment of the application of nanoscience and nanotechnologies in agri/food/feed* [EFSA (European Food Safety Authority), 2018d] should be followed.

At the time of writing this, no application dossier for a substance, to be authorized as a new food additive, considered ENM has been submitted for its assessment. This can be monitored from the EFSA Register of Questions (EFSA Register of Questions, online).

6.2.2 Novel food

Regulation (EU) 2015/2283 [EC (European Commission), 2015] on novel foods (applicable since January 1, 2018) requires that "Novel foods are subject to the general labeling requirements laid down in Regulation (EU) no. 1169/2011 and other relevant labeling requirements in Union food law." It further establishes, in its Article 33, amendments to the Regulation (EU) 1169/2011 [EC (European Commission), 2011b] and in particular to the definition of "ENMs" as "any intentionally produced material that has one or more dimensions of the order of

100 nm or less or that is composed of discrete functional parts, either internally or at the surface, many of which have one or more external dimensions of the order of 100 nm or less, including structures, agglomerates, or aggregates, which may have a size above the order of 100 nm but retain properties that are characteristic of the nanoscale." This regulation also establishes that food consisting of ENMs should be considered as a novel food.

Furthermore, Regulation (EU) 2015/2283 [EC (European Commission), 2015] laid down that the European Commission (EC) shall adjust and adapt the definition of ENM to technical and scientific progress or to definitions agreed at international level. Within the EC, there is an ongoing work, for the revision of the current definition of nanomaterial in Recommendation 2011/696/EC [EC (European Commission), 2011a].

According to Regulation (EU) 1169/2011 [EC (European Commission), 2011b], the label as "nano" refers to the ingredients present in the form of ENMs. Thus for labeling purposes, only ingredients intentionally engineered and manufactured in the nanoform have to be labeled as "nano." To put it inversely, if an ingredient is found to be nano but it was not intentionally engineered as such—for example, the bulk form is proposed to be added to the product and because of processing or due to the matrix some or most of the ingredients are in nanoform—the labeling requirements of Regulation 1169/2011 would not apply to this ingredient.

6.2.3 Nutrient sources

Annex II of Directive 2002/46/EC [EC (European Commission), 2002] lists the vitamin and mineral substances that may be used in the manufacture of food supplements. No specifications for those substances have been established, and at the time of the assessment of the majority of those substances, in 2009, information on the particle size of the materials was neither considered relevant nor requested.

As can be seen from the EFSA Register of Questions, only an application dossier on silver hydrosol has been submitted twice (2008, 2016) for the assessment of the substance prior to authorization, and in both cases the conclusion of the assessment was that the data submitted was insufficient to conclude on the bioavailability of silver from silver hydrosol and on the safety of the substance [EFSA (European Food Safety Authority), 2008, 2018b]; therefore this product is not authorized to be used in the manufacturing of food supplements in the EU.

Nanotech products as food supplements are advertised in Internet; however, their safety as nanomaterials for that purpose has not been assessed in the EU.

6.2.4 Food contact materials

Food contact materials (FCMs) are either intended to be brought into contact with food, are already in contact with food, or can reasonably be brought into contact with food or transfer their constituents to the food under normal or foreseeable use (https://ec.europa.eu/food/safety/chemical_safety/food_contact_materials_en).

Regulation (EC) no. 1935/2004 [EC (European Commission), 2004] provides the general principles of safety and inertness for all FCMs in the EU. Certain FCMs—ceramic materials, regenerated cellulose film, plastics (including recycled plastic), as well as active and intelligent materials—are covered by specific legislation. The safety of FCMs is evaluated by EFSA.

Only the substances included in the Union list of authorized substances, set out in Annex I to Commission Regulation (EU) no. 10/2011 [EC (European Commission), 2011c], may be intentionally used in the manufacture of plastic layers in plastic materials and articles; as indicated in Article 9, substances in nanoform shall only be used if explicitly authorized and mentioned in the specifications in Annex I.

6.2.5 Cosmetic products

In the EU, "nanomaterial" in cosmetic products refers to an insoluble or biopersistent and intentionally manufactured material and must be labeled in the list of ingredients with the word "nano" in brackets following the name of the substance (http://ec.europa.eu/growth/sectors/cosmetics/products/nanomaterials_en).

According to Art. 13 (1) of Regulation (EC) no. 1223/2009 [EC (European Commission), 2009], responsible persons (i.e., manufacturers, importers, or third persons appointed by them) are required to register cosmetic products on the Cosmetic Products Notification Portal (CPNP). The notification must specify whether the product contains nanomaterials, with their identification and the foreseeable exposure conditions.

In addition, cosmetic products containing nanomaterials other than colorants, preservatives, and UV-filters, and not otherwise restricted by Regulation (EC) no. 1223/2009 [EC (European Commission), 2009], are subject to an additional procedure. They require a specific notification on the CPNP 6 months before placement on the market [Art. 16 (3)]. If the EC has concerns regarding the safety of a nanomaterial, it may request the Scientific Committee on Consumer Safety to perform a risk assessment [e.g., for titanium dioxide (nanoform), styrene/acrylates copolymer (nano), sodium styrene/acrylates copolymer (nano), and colloidal silver (nano)] (https://ec.europa.eu/health/scientific_committees/consumer_safety/opinions_en).

In the EU, certain colorants, preservatives, and UV-filters, including those that are nanomaterials, used in cosmetic products must be authorized by the EC prior to their use.

According to Article 16 (10a) of Regulation (EC) no. 1223/2009 [EC (European Commission), 2009], the EC has to publish a catalog of all nanomaterials used in cosmetic products placed on the market. Nanomaterials used as UV-filters, colorants, and preservatives must be specified in a different section. The catalog should indicate the categories of cosmetic products and the foreseeable exposure conditions. Up to now the Commission has authorized three UV-filters as nanomaterials: titanium dioxide, zinc oxide, and tris−biphenyl triazine. It has also allowed carbon black (nano) for use as a colorant in cosmetic products.

6.2.6 Medicinal products

According to the European Technology Platform, "Nanomedicine is the application of nanotechnology to achieve innovation in healthcare. It uses the properties developed by a material at its nanometric scale 10^{-9} m which often differ in terms of physics, chemistry or biology from the same material at a bigger scale" (https://www.etp-nanomedicine.eu/public/about-nanomedicine/what-is-nanomedicine).

In Europe the responsible agency for the regulation of nanomedicine is the European Medicine Agency (EMA) as well as national regulatory agencies of each member state. EMA is responsible for the scientific evaluation of centralized marketing authorization applications. The centralized marketing authorization, once granted by the EC, is valid in all EU Member States, Iceland, Norway, and Liechtenstein.

In the last few years, nanomedicines have been incorporated in routine clinical treatments. It is important for drug regulators to ensure that the use of nanomedicines is safe and timely for the benefit of public health.

SCENIHR (Scientific Committee on Emerging and Newly Identified Health Risks) (2015) published Guidance on the Determination of Potential Health Effects of Nanomaterials Used in Medical Devices. The guidance provides information to assist with the safety evaluation and risk assessment on the use of nanomaterials in medical devices and highlights the need for specific considerations related to nanomaterials, in view of the possible different properties, interactions, or effects that may differ from bulk form of the same material. For example, information on different parameters for characterization and identification of the nanomaterial intended for use in medical devices have to be provided together with data on methods for the characterization, toxicokinetics testing, and toxicological evaluation. A risk assessment on the nanomaterial to be used in medical devices is needed before its authorization. The potential risk from the use of nanomaterials in medical devices is associated with the possibility of release of free nanoparticles from the device and the duration of the exposure to those nanoparticles.

In addition, the EMA scientific guidelines on nanomedicines help medicine developers to prepare marketing authorization applications [EMA (European Medicine Agency), online1]. The Committee for Medicinal Products for Human Use prepares a document to assist in the generation of relevant data to support a marketing authorization for a specific product.

6.3 Legislation outside Europe

6.3.1 United States of America

In the United States the Food and Drug Administration (FDA) regulates foods (including food additives and dietary supplements), cosmetics, drugs, medical devices, veterinary products, biologics (including vaccines, allergenics, etc.), electronic products that give off radiation and tobacco products. Some of these products may utilize nanotechnology or contain nanomaterials.

The FDA does not have a legal definition for nanotechnology or nanomaterial. However, when referring to nanotechnology, they usually mean the manipulation of materials of extremely small size, usually at dimensions between 1 and 100 nm (https://www.fda.gov/cosmetics/scienceresearch/nanotech/default.htm).

Regulated products, such as dietary supplements (except certain new dietary ingredients), cosmetics (except color additives), and food (except food additives and color additives), are not subject to mandatory premarket review. In these cases the FDA relies on publicly available information, voluntarily submitted information, adverse event reporting (where applicable), and on postmarket surveillance activities, to provide oversight. Where nanotechnology applications are involved, the FDA encourages manufacturers to consult with the agency before placing their products on the market, and this will allow designing any necessary postmarketing safety oversight.

New drugs, new animal drugs, food additives, color additives, certain human devices, and certain new dietary ingredients in dietary supplements are subject to premarket review requirements. This implies that applicants have to submit data to answer questions related to the safety, effectiveness (where applicable), or regulatory status of the product. Individual premarket review procedures include attention to whether the use of nanomaterials suggests the need for additional data on safety or effectiveness.

The Guidance for Industry: Considering Whether an FDA-Regulated Product Involves the Application of Nanotechnology describes the FDA's current thinking on determining whether FDA-regulated products involve nanotechnology (FDA, 2014a). This means whether a material or end product is engineered to have at least one external dimension, or an internal or surface structure, in the nanoscale range (approximately 1–100 nm); or whether a material or end product is engineered to exhibit properties or phenomena, including physical or chemical properties or biological effects, that are attributable to its dimension(s), even if these dimensions fall outside the nanoscale range, up to 1 μm (1000 nm).

The Guidance for Industry: Assessing the Effects of Significant Manufacturing Process Changes, Including Emerging Technologies, on the Safety and Regulatory Status of Food Ingredients and Food Contact Substances, Including Food Ingredients that are Color Additives describes the factors to be considered when determining whether a significant change in manufacturing process for a food substance already in the market affects its identity, safety, and/or regulatory status and warrants a regulatory submission to FDA (2014b). Intentional alterations of the particle size distribution on the nanometer scale of an already-regulated material can sometimes be a significant manufacturing change that would require reassessment by the FDA.

No official list of regulated and nonregulated product considered as nanomaterials is available in the United States.

6.3.2 Canada

Health Canada regulates drugs, medical device, cosmetics, food additives, and natural health products among others (https://www.canada.ca/en/health-canada/corporate/mandate/regulatory-role/what-health-canada-regulates.html).

Currently, there are no regulations specific to nanotechnology-based health and food products. Health Canada relies on authorities within existing legislative and regulatory frameworks, which require the assessment of potential risks and benefits of products to the health and safety of Canadians before they can be authorized for sale (https://www.canada.ca/en/health-canada/services/drugs-health-products/nanotechnology-based-health-products-food.html).

Health Canada has adopted a working definition for nanomaterials that is described in the Policy Statement on Health Canada's Working Definition for Nanomaterial (Health Canada, online1), and it is intended to be used as a tool to help in gathering safety information about nanomaterials. Any manufactured substance or product and any component material, ingredient, device, or structure is considered as a nanomaterial if it is at or within the nanoscale in at least one external dimension or has internal or surface structure at the nanoscale, or it is smaller or larger than the nanoscale in all dimensions and exhibits one or more nanoscale properties/phenomena.

Health Canada together with Environment Canada conducts assessments of the potential environmental and human health risks arising from environmental exposure to new substances, including nanomaterials. Nanotechnology applications, used in food additives, novel foods, or in FCMs (e.g., packaging materials), would be regulated under the Food and Drug Regulations (Health Canada, online2). These applications would be subject to the same regulations that currently apply to conventional food products. Health Canada examines and evaluates the potential health risks of nanotechnology-based food products on a case-by-case basis.

6.3.3 Australia and New Zealand

In Australia, there are different bodies regulating nanotechnology and the use of nanomaterials in commercial products. Each body works within a different legislative framework relating to their own specific field, such as medicine, food, pesticides, veterinary medicine, cosmetics. The NICNAS (National Industrial Chemicals Notification and Assessment Scheme) (online1) is responsible for the regulation of industrial nanomaterials used in products such as paints, dyes, inks, plastics, cosmetics, consumer goods, and surface coatings. The FSANZ (Food Standards Australia New Zealand) (online) regulates nanotechnologies in foods, food packaging, and FCMs. The Therapeutic Goods Administration (TGA) manages nanoparticles in therapeutic goods and medical devices. In addition, the

Australian Competition and Consumer Commission regulates all consumer products containing nanomaterials that do not fall under other regulatory jurisdictions.

In Australia the NICNAS (National Industrial Chemicals Notification and Assessment Scheme) (online2) has developed a working definition for an "industrial nanomaterial" for regulatory purposes "... industrial materials intentionally produced, manufactured or engineered to have unique properties or specific composition at the nanoscale, that is a size range typically between 1 and 100 nm, and is either a nano-object (i.e. that is confined in one, two, or three dimensions at the nanoscale) or is nanostructured (i.e. having an internal or surface structure at the nanoscale)."

The FSANZ has adopted a range of strategies to manage any potential risks associated with nanotechnologies in foods, with the aim of ensuring that public health and safety is protected. The Application Handbook was amended in 2008 to ensure that any application to approve the use of nanotechnology in food provides appropriate information for FSANZ to conduct a thorough risk assessment [FSANZ (Food Standards Australia New Zealand), 2016a]. The FSANZ has set up a Scientific Nanotechnology Advisory Group (SNAG) comprising experts in the fields of nanosafety, pharmacology, nanofood technology, toxicology, and nanometrology. The SNAG will advise on the development of guidance for a range of stakeholders, future uses of nanotechnology in food and food packaging, and national/international legislation and policy.

The FSANZ (Food Standards Australia New Zealand) (2016b) published a report on the use of nanotechnology in food additives and packaging focused on health risks associated with oral ingestion of titanium dioxide, silicon dioxide, and silver in food. Based on the weight of evidence on nanoscale titanium dioxide, silicon dioxide, and silver, a significant health risk for these food grade materials was not considered.

Cosmetics fall under the assessment of the NICNAS. A declaration confirming whether or not the notified substance is in the nanoform is required by all introducers of new chemicals [NICNAS (National Industrial Chemicals Notification and Assessment Scheme), online3]. Regarding existing chemicals, NICNAS can initiate reviews of them (including their nanoforms) when health or environmental concerns are identified.

The TGA is a statutory body responsible for the regulation on nanomedicines (including sunscreens). Nanomedicines in the form of metal oxides (e.g., zinc oxide), liposomes (e.g., amphotericin B), protein–drug conjugates (e.g., albumin-bound paclitaxel), polymeric nanoparticles (e.g., sevelamer), and emulsions (e.g., cyclosporine) have been registered in Australia [TGA (Therapeutic Goods Administration), 2016a]. A scientific review report on the safety of titanium dioxide and zinc dioxide nanoparticles used in sunscreens was published in 2016 and concluded that neither of them is likely to cause harm when used as ingredients in sunscreens and when sunscreens are used as directed [TGA (Therapeutic Goods Administration), 2016b].

6.4 Use of nanomaterials in the European Union

There is a public perception that nanomaterials may pose a risk to consumers and this is fueled for the lack of knowledge of which nanomaterials are actually on the market.

Several initiatives have been unsuccessfully attempted to obtain information on products containing nanomaterials on the market via voluntary notifications schemes; there were calls by some EU Member States and nongovernmental organizations. In this respect, the EC has published an impact assessment on a Delegation Agreement with the European Chemicals Agency on the European Union Observatory for Nanomaterials and the European Union Chemical Legislation Finder in the framework of the COSME program (EC, 2017). The product databases only cover a very small part of the real market, and they use search functions to the term "nano" but do not verify whether the product indeed contains nanomaterials.

6.4.1 Food additives

According to the specifications for food additive authorized in the EU {Commission Regulation (EU) no. 231/2012 [EC (European Commission), 2012]}, none of them is described as a nanomaterial. During the reevaluation of food additives permitted in the EU before January 20, 2009 {Regulation (EU) no. 257/2010 [EC (European Commission), 2010]}, the uncertainty about the percentage of the material in the nanoform when used as a food additive has been identified in some cases {e.g., iron oxide and hydroxides (E 172) [EFSA (European Food Safety Authority), 2015], titanium dioxide (E 171) [EFSA (European Food Safety Authority), 2016a], silver (E 174) [EFSA (European Food Safety Authority), 2016b], gold (E 175) [EFSA (European Food Safety Authority), 2016c], silicon dioxide (E 551) [EFSA (European Food Safety Authority), 2018a], or silicates (E 552–553) [EFSA (European Food Safety Authority), 2018e]}.

The EC has established a follow-up approach for those food additives already evaluated by EFSA and for which some concerns have been identified. This includes the request for additional data to business operators in order to better characterize the identity of the food additive regarding the presence or the absence of nanoparticles of the materials [EC (European Commission), online].

No specific nanomaterial proposed to be used as a food additive in the EU has been evaluated (EFSA Register of Questions, online).

6.4.2 Novel food

From the Commission Implementing Regulation (EU) 2017/2470 (EC, 2017) on establishing the Union list of novel foods in accordance with Regulation (EU)

2015/2283 [EC (European Commission), 2015] of the European Parliament and of the Council on Novel Foods, it can be seen that none of the substances described falls under the definition of nanomaterial or ENM. Only for trans-resveratrol the particle size is described as 100% less than 62.23 μm without any limitation for the minimum size of the particle.

6.4.3 Food supplements

Annex II of Directive 2002/46/EC lists the minerals and vitamins that can be used for the manufacturing of food supplements. No specifications for those minerals are established in the legislation and, therefore apart from the name of the chemical, too much information is not available.

Most of the sources of nutrients included in Annex II of Directive 2002/46/EC were evaluated at a time when the knowledge and the concerns of nanomaterials were different from the current ones.

Calcium silicate and silicic acid were evaluated in 2009 [EFSA (European Food Safety Authority), 2009], and both are described in the corresponding assessment in the range of nanoforms. However, this is not clearly specified in Annex II of Directive 2002/46/EC [EC (European Commission), 2002], neither in the label of the products containing silicic acid as a source of silicon.

6.4.4 Food contact materials

From Annex I to Commission Regulation (EU) no. 10/2011 [EC (European Commission), 2011c] (latest consolidated version January 18, 2018), only seven substances are listed as being present in their nanoform: titanium nitride, zinc oxide, and zinc oxide coated with [3-(methacryloxy)propyl] trimethoxysilane; and four different polymers [(butadiene, ethyl acrylate, methyl methacrylate, and styrene) copolymer cross-linked with divinylbenzene (butadiene, ethyl acrylate, methyl methacrylate, and styrene) copolymer not cross-linked (methacrylic acid, ethyl acrylate, *n*-butyl acrylate, methyl methacrylate, and butadiene) copolymer, and (butadiene, ethyl acrylate, methyl methacrylate, and styrene) copolymer cross-linked with 1,3-butanediol dimethacrylate].

From the EFSA Register of Questions (online) (accessed on 25.09.18), it can be observed that a risk assessment on selenium nanoparticles (as an active substance) has been published recently [EFSA (European Food Safety Authority), 2018c]; the evaluation of "nano-hexadecyltrimethylammonium bromide modified montmorillonite organoclay for use as additive in plastics" has been ongoing since 2013, and in 2018 a new evaluation on the safety of silver as nanomaterial for its use as additive in plastics has been received.

6.4.5 Cosmetics

According to Article 16 (10a) Regulation (EC) no. 1223/2009 [EC (European Commission), 2009], the EC has to publish a catalog of all nanomaterials used in cosmetic products placed on the market. A catalog of nanomaterial used in cosmetic products is available online http://ec.europa.eu/docsroom/documents/24521.

Nanomaterials in the catalog are classified as colorants, UV-filters, preservatives, and other uses. The catalog should indicate the categories of cosmetic products and the foreseeable exposure conditions.

Currently, colorants [Acid Yellow 23 (tartrazine), barium sulfate, carbon black, CI 77288 (chromium oxide), CI 77491 (red iron oxide), CI 77499 (red iron oxide), CI 77510 (ferric ferrocyanide), CI 77891 (titanium dioxide), copper, gold, silvera, and Pigment Red 57 (litholrubine BK)], UV-filters (bis-ethylhexyloxyphenol methoxyphenyl triazine, ethylhexyl methoxycinnamate, methylene bis-benzotriazolyl tetramethylbutylphenol, titanium dioxide, tris–biphenyl triazine, and zinc oxide), and substances with other functions than colorant, preservative, and UV-filter (alumina, cellulose, colloidal copper, colloidal gold, colloidal platinum, colloidal silver, fullerenes, gold thioethylamino hyaluronic acid, hydrated silica, hydroxyapatite, lithium magnesium sodium silicate, platinum, platinum powder, retinol, sapphire powder, silica, silica dimethicone silylate, silica dimethyl silylate, silica silylate, sodium magnesium fluorosilicate, sodium magnesium silicate, sodium propoxyhydroxypropyl thiosulfate silica, styrene/acrylates copolymer, tin oxide, and tocopherol acetate) are included in the catalog of nanomaterial [version 1 (31.12.16) (accessed on 11.07.18)]. These represent less than 1% of cosmetic products notified in the CPNP. The catalog has only an informative value, and it is not a list of authorized nanomaterials to be used in cosmetics in the EU.

Up to now, the Commission has authorized three UV-filters consisting of nanomaterials: titanium dioxide, zinc oxide, and tris–biphenyl triazine. It has also allowed carbon black (nano) for use as a colorant in cosmetic products (http://ec.europa.eu/growth/sectors/cosmetics/products/nanomaterials_en).

6.4.6 Medicines

From the NETP (Nanomedicine European Technology Platform) (online1), an overview of the development of nanotechnology to be applied in medicine is available: nanomedicine has been proposed to treat cancer, atherosclerosis, diabetes, Alzheimer's disease, arthritis, and infectious diseases. Nanomedicine has also applications in tools and formulations in the area of ophthalmology, to combat antimicrobial resistance and for tissue engineering. However, there is not any official list that compiles all the authorized uses of nanomedicines.

Some examples of nanomedicines from the NETP are the magnetic properties of iron oxide nanoparticles that make them suitable imaging agents with MRI scans; nanoparticles of iron oxide thanks to their small size and concentration in the tumor allow a very high resolution and an accurate mapping of lesions. Therefore nanoparticles of iron oxide help doctors in planning the removal of a tumor [NETP (Nanomedicine European Technology Platform), online2]. Another example in the area of cancer is gold nanorods that can carry chemotherapy drugs and locally excited in the tumor by infrared light. The induced heat stimulates the release of the encapsulated drug and, in addition, helps destroy the cancer cells, resulting in a combined effect of enhanced delivery, and intrinsic therapy.

Iron oxide nanoparticles with specific coating can bind to specific tumors and being used as cancer biomarkers due to their magnetic properties. Thanks to the size of the nanoparticles and the concentration that they can reach in the tumors, a very high resolution, and an accurate mapping of lesions would be possible (Leng et al., 2018; Thoidingjam and Tiku, 2017). The injection of nanoparticles in the tumor and their activation to produce heat and destroy cancer cells locally by magnetic fields; X-ray or light is another area under development. Gold nanorods are good candidates for being used in chemotherapy drugs and locally excited in the tumor by infrared light (Liao et al., 2015; An et al., 2017).

As previously mentioned (Section 6.3.6) a safety evaluation of nanomaterials to be used in medical devices (wound care materials, uncured dental and bone fillers, devices applied in the respiratory tract, etc.) is needed before their authorization [SCENIHR (Scientific Committee on Emerging and Newly Identified Health Risks), 2015]. From the data requirement guidelines available at the EMA (European Medicine Agency) (online2), applications on injected iron-based nanocolloidal products and injected liposomal products are in progress. Block-copolymer-micelle and surface coating nanomedicines are other areas being evaluated for their application.

6.5 Existing inventories

There are a number of online inventories containing information about nanomaterials and their use.

The Consumer Product Inventory (http://www.nanotechproject.org/cpi) aims to create a "living" international inventory for the exchange of accurate information on nano-enabled consumer products. However, the latest update was on October 3, 2011 (accessed on 25.04.18). The inventory at that time included 1317 products and states that Europe was considered as the second greatest contributor, after the United States, as "manufacturer-identified nanotechnology-enabled consumer products."

In the EU, some Member States have their own databases. The Nanobuktdatenbank (https://www.bund.net/chemie/nanotechnologie/) is maintained by the Bund Fur Umwelt und Naturschutz (BUND), a nongovernmental organization for environmental and nature protection. The database contains about 200 "nanoproducts" (accessed on 26.04.18), 50 of them belonging to the category "Lebensmittel" (food).

Belgium has a nanoregister (https://www.health.belgium.be/en/environment/chemical-substances/nanomaterials/register) where companies have to register the nanomaterials they place on the market, but the information on that register is not publicly available.

France has introduced a notification system for substances in nanoform, including such substances in mixtures, and in articles, if intentionally released, based on decree 2012-232 on Annual Declaration that entered into force on May 1, 2013 (https://www.r-nano.fr/?locale = en). The French register of nanomaterials is not publicly available.

The Danish nanodatabase (http://nanodb.dk/) contains 3036 products (accessed on 25.09.18), in which the products containing nanomaterials are classified according to eight different categories (e.g., food and beverage). It is possible to search by substance and category; however, it is unknown which approach is used to build such a database because none of the ingredients listed in the label of food products inserted in the database are indicated as nanomaterial.

The Australian Inventory of Chemical Substances (https://www.nicnas.gov.au/chemical-inventory-AICS) is a database of chemicals available for industrial use in Australia. A full list of substances as nanomaterial included in that database is not accessible as such.

6.6 Conclusion

Nanotechnology is an emerging technology that can be used in a broad range of products. The peer-reviewed literature gives us information on the scientific areas under development, but their application on market products is not guaranteed. Innovative products can have advantages, but their safety has to be assessed to demonstrate that their use does not produce adverse effects on human health and/or on the environment. Till today, the evidence that nanotechnologies are being used in the food industry on a wide scale is very limited, although a lot of research is being published on potential future applications. A few products defined as nanomaterials are used in cosmetic products and despite the lack of an official list of use of nanomedicines, it seems that this is an area on progress of innovative products to the market.

There is not any publicly available updated catalog of products containing nanomaterials on the market, and this undermines the trust of the consumers on their safety.

Regulators are continually revising new data, monitoring activities, and challenges emerged related to their activities and legislation framework to ensure that the use of products on the market is safe.

Key points

- Nanotechnology is an emerging technology.
- The terms "nanomaterials" and "engineered nanomaterials" are generally used interchangeably despite having different connotations.
- The peer-reviewed literature gives us information on the scientific areas under development, but their application on market products is not guaranteed.

- There is no official database of nanomaterials and/or engineered nanomaterials used in the area of food, cosmetics, and medicines, being the latest an area on progress for innovative products on the market.
- The safety of the products, including nanomaterials and/or engineered nanomaterials, on the market is a priority for regulators and policymakers.

References

An, L., Wang, Y., Tian, Q., Yang, S., 2017. Small gold nanorods: recent advances in synthesis, biological imaging and cancer therapy. Materials 10, 1372. Available from: https://doi.org/10.3390/ma10121372.

EC (European Commission), 2002. Directive 2002/46/EC of the European Parliament and of the Council of 10 June 2002 on the approximation of the laws of the member states relating to food supplements. Off. J. Eur. Communities 183, 51–57.

EC (European Commission), 2004. Regulation (EC) no 1935/2004 of the European Parliament and of the Council of 27 October 2004 on materials and articles intended to come into contact with food and repealing Directives 80/590/EEC and 89/109/EEC. Off. J. Eur. Communities 338, 4–17.

EC (European Commission), 2008. Regulation (EC) no 1333/2008 of the European Parliament and of the Council of 16 December 2008 on food additives. Off. J. Eur. Communities 354, 16–33.

EC (European Commission), 2009. Regulation (EC) no 1223/2009 of the European Parliament and of the Council of 30 November 2009 on cosmetic products. Off. J. Eur. Communities 342, 59–209.

EC (European Commission), 2010. Commission Regulation (EU) no 257/2010 of 25 March 2010 setting up a programme for the re-evaluation of approved food additives in accordance with Regulation (EC) no 1333/2008 of the European Parliament and of the Council on Food Additives. Off. J. Eur. Communities 80, 19–27.

EC (European Commission), 2011a. Commission recommendation of 18 October 2011 on the definition of nanomaterial. Off. J. Eur. Communities 275, 38–40.

EC (European Commission), 2011b. Regulation (EU) no 1169/2011 of the European Parliament and of the Council of 25 October 2011 on the provision of food information to consumers, amending regulations (EC) no 1924/2006 and (EC) no 1925/2006 of the European Parliament and of the Council, and repealing Commission Directive 87/250/EEC, Council Directive 90/496/EEC, Commission Directive 1999/10/EC, Directive 2000/13/EC of the European Parliament and of the Council, Commission Directives 2002/67/EC and 2008/5/EC and Commission Regulation (EC) no 608/2004. Off. J. Eur. Communities 304, 18–63.

EC (European Commission), 2011c. Commission regulation (EU) no 10/2011 of 14 January 2011 on plastic materials and articles intended to come into contact with food. Off. J. Eur. Communities 12, 1–89.

EC (European Commission), 2012. Commission Regulation (EU) no 231/2012 of 9 March 2012 laying down specifications for food additives listed in annexes II and III to regulation (EC) no 1333/2008 of the European Parliament and of the council. Off. J. Eur. Communities 83, 1–295.

EC (European Commission), 2015. Regulation (EU) 2015/2283 of the European Parliament and of the council of 25 November 2015 on novel foods, amending regulation (EU) no 1169/2011 of the European Parliament and of the council and repealing regulation (EC) no 258/97 of the European Parliament and of the council and commission regulation (EC) no 1852/2001. Off. J. Eur. Communities 327, 1–22.

EC (European Commission), online. Available from: <https://ec.europa.eu/food/safety/food_improvement_agents/additives/re-evaluation_en>.

EC (European Commission), 2017. Commission Staff Working Document. Impact Assessment. Commissio Implementing Decision on a Delegation Agreement with the European Chemicals Agency on the European Union Observatory for Nanomaterials and the European Union Chemical Legislation Finder in the framework of the COSME programme. Brussels, 2017. SWD(2017) 138 final, 1–80.

EFSA (European Food Safety Authority), 2008. Scientific statement of the panel on food additives and nutrient sources added to food on a request from the commission on silver hydrosol as a source of silver added for nutritional purposes to food supplements. EFSA J. 884, 1–3.

EFSA (European Food Safety Authority), 2009. Scientific opinion of the EFSA Panel on Food Additives and Nutrient Sources added to Food. Calcium silicate and silicon dioxide/silicic acid gel added for nutritional purposes to food supplements. EFSA J. 1132, 1–24.

EFSA (European Food Safety Authority), 2015. EFSA Panel on Food Additives and Nutrient Sources added to Food. Scientific opinion on the re-evaluation of iron oxides and hydroxides (E 172) as food additives. EFSA J. 13 (12), 4317. 57 pp.

EFSA (European Food Safety Authority), 2016a. EFSA Panel on Food Additives and Nutrient Sources added to Food. Scientific opinion on the re-evaluation of titanium dioxide (E 171) as a food additive. EFSA J. 14 (9), 4545. 83 pp.

EFSA (European Food Safety Authority), 2016b. EFSA Panel on Food Additives and Nutrient Sources added to Food. Scientific opinion on the re-evaluation of silver (E 174) as food additive. EFSA J. 14 (1), 4364. 64 pp.

EFSA (European Food Safety Authority), 2016c. EFSA Panel on Food Additives and Nutrient Sources added to Food. Scientific opinion on the re-evaluation of gold (E 175) as a food additive. EFSA J. 14 (1), 4362. 43 pp.

EFSA (European Food Safety Authority), 2018a. EFSA Panel on Food Additives and Nutrient Sources added to Food, Younes M, Aggett P, Aguilar F, Crebelli R, Dusemund B, Filipič M, Frutos MJ, Galtier P, Gott D, Gundert-Remy U, Kuhnle GG, Leblanc J-C, Lillegaard IT, Moldeus P, Mortensen A, Oskarsson A, Stankovic I, Waalkens-Berendsen I, Woutersen RA, Wright M, Boon P, Chrysafidis D, Gürtler R, Mosesso P, Parent-Massin D, Tobback P, Kovalkovicova N, Rincon AM, Tard A and Lambré C, 2018. Scientific opinion on the re-evaluation of silicon dioxide (E 551) as a food additive. EFSA J. 16 (1), 5088. 60 pp.

EFSA (European Food Safety Authority), 2018b. EFSA Panel on Food Additives and Nutrient Sources added to Food, Younes M, Aggett P, Aguilar F, Crebelli R, Dusemund B, Filipič M, Frutos MJ, Galtier P, Gott D, Gundert-Remy U, Kuhnle GG, Lambré C, Leblanc J-C, Lillegaard IT, Moldeus P, Mortensen A, Oskarsson A, Stankovic I, Waalkens-Berendsen I, Woutersen RA, Wright M, Tobback P, Rincon AM, Smeraldi C and Gott D, 2018. Scientific opinion on the safety and bioavailability of silver hydrosol as a source of silver added for nutritional purposes to food supplements. EFSA J. 14 (3), 5237. 13 pp.

EFSA (European Food Safety Authority), 2018c. Panel on Food Contact Materials, Enzymes, Flavourings and Processing Aids, Silano V, Bolognesi C, Chipman K, Cravedi J-P, Engel K-H, Fowler P, Franz R, Grob K, G Gürtler R, Husøy T, Kärenlampi S, Mennes W,

Milana MR, Pfaff K, Riviere G, Srinivasan J, Tavares Poças MF, Tlustos C, Wölfle D, Zorn H, Kolf-Clauw M, Lampi E, Svensson K, Lioupis A and Castle L, 2018. Scientific Opinion on the safety assessment of the active substance selenium nanoparticles, for use in active food contact materials. EFSA J. 16 (1), 5115. 7 pp.

EFSA (European Food Safety Authority), 2018d. EFSA Scientific Committee, Hardy A, Benford D, Halldorsson T, Jeger MJ, Knutsen HK, More S, Naegeli H, Noteborn H, Ockleford C, Ricci A, Rychen G, Schlatter JR, Silano V, Solecki R, Turck D, Younes M, Chaudhry Q, Cubadda F, Gott D, Oomen A, Weigel S, Karamitrou M, Schoonjans R and Mortensen A, 2018. Guidance on risk assessment of the application of nanoscience and nanotechnologies in the food and feed chain: Part 1, Human and animal health. EFSA J. 16 (7), 5327. 95 pp.

EFSA (European Food Safety Authority), 2018e. EFSA Panel on Food Additives and Nutrient Sources added to Food, Younes M, Aggett P, Aguilar F, Crebelli R, Dusemund B, Filipič M, Frutos MJ, Galtier P, Gott D, Gundert-Remy U, Kuhnle GG, Leblanc J-C, Lillegaard IT, Moldeus P, Mortensen A, Oskarsson A, Stankovic I, Waalkens-Berendsen I, Woutersen RA, Wright M, Boon P, Gürtler R, Mosesso P, Parent-Massin D, Tobback P, Chrysafidis D, Rincon AM, Tard A and Lambré C, 2018e. Scientific opinion on the re-evaluation of calcium silicate (E 552), magnesium silicate (E 553a(i)), magnesium trisilicate (E 553a(ii)) and talc (E 553b) as food additives. EFSA Journal 2018;16(8):5375, 50 pp.

EFSA Register of Questions, online. Available at: http://registerofquestions.efsa.europa.eu/roqFrontend/ListOfQuestionsNoLogin?0&panel=ALL>.

EMA (European Medicine Agency), online1. Available from: <http://www.ema.europa.eu/ema/index.jsp?curl = pages/regulation/general/general_content_000564.jsp&mid = WC0b01ac05806403e0>.

EMA (European Medicine Agency), online2. Available from: <http://www.ema.europa.eu/ema/index.jsp?curl = pages/regulation/general/general_content_000564.jsp&mid = WC0b01ac05806403e0>.

FDA (Food and Drug Administration), 2014a. Guidance for Industry. Considering Whether an FDA-Regulated Product Involves the Application of Nanotechnology. U.S. Department of Health and Human Services. Food and Drug Administration. Office of the Commissioner. June 2014, 1–14.

FDA (Food and Drug Administration), 2014b. The Guidance for Industry: Assessing the Effects of Significant Manufacturing Process Changes, Including Emerging Technologies, on the Safety and Regulatory Status of Food Ingredients and Food Contact Substances, Including Food Ingredients that are Color Additives. U.S. Department of Health and Human Services. Food and Drug Administration. Center for Food Safety and Applied Nutrition. June 2014, 1–29.

FSANZ (Food Standards Australia New Zealand), 2016a. Application Handbook. 2016. Available online from: <http://www.foodstandards.gov.au/code/changes/Documents/Application%20Handbook%20as%20at%201%20March%202016.pdf>.

FSANZ (Food Standards Australia New Zealand), 2016b. Potential Health Risks Associated with Nanotechnologies in Existing Food Additives. Available online from: <https://www.foodstandards.gov.au/publications/Documents/Safety%20of%20nanotechnology%20in%20food.pdf>.

FSANZ (Food Standards Australia New Zealand), online. Available from: <http://www.foodstandards.gov.au/consumer/foodtech/nanotech/Pages/default.aspx>.

Health Canada, online1. Available from: <https://www.canada.ca/en/health-canada/services/science-research/reports-publications/nanomaterial/policy-statement-health-canada-working-definition.html>.

Health Canada, online2. Available from: <https://www.canada.ca/en/health-canada/services/science-research/emerging-technology/nanotechnology/food-food-packaging.html>.

Leng, F., Liu, F., Yang, Y., Wu, Y., Tian, W., 2018. Strategies on nanodiagnostics and nanotherapies of the three common cancers. Nanomaterials 8, 202. Available from: https://doi.org/10.3390/nano8040202.

Liao, J., Li, W.T., Peng, J.R., Yang, Q., Li, H., Wei, Y.Q., et al., 2015. Combined cancer photothermal-chemoterapy based on doxorubicin/gold nanorod-loaded pymersomes. Theranostics 5 (4), 345−356.

NETP (Nanomedicine European Technology Platform), online1. Available from: <https://www.etp-nanomedicine.eu/public/about-nanomedicine/nanomedicine-applications>.

NETP (Nanomedicine European Technology Platform), online2. Available from: <https://www.etp-nanomedicine.eu/public/about-nanomedicine/nanomedicine-applications/nanomedicine-in-cancer>.

NICNAS (National Industrial Chemicals Notification and Assessment Scheme), online1. Available from: <https://www.nicnas.gov.au/chemical-information/Topics-of-interest2/subjects/nanomaterials-nanotechnology>.

NICNAS (National Industrial Chemicals Notification and Assessment Scheme), online2. Available from: <https://www.nicnas.gov.au/notify-your-chemical/data-requirements-for-new-chemical-notifications/data-requirements-for-notification-of-new-industrial-nanomaterials/nicnas-working-definition-of-industrial-nanomaterial>.

NICNAS (National Industrial Chemicals Notification and Assessment Scheme), online3. Available from: <https://www.nicnas.gov.au/chemical-information/Topics-of-interest2/subjects/nanomaterials-nanotechnology/nicnas-regulatory-activities-in-nanomaterials>.

SCENIHR (Scientific Committee on Emerging and Newly Identified Health Risks), 2015. Opinion on the determination of potential health effect of nanomaterials used in medical devices. Available online from: <http://ec.europa.eu/health/scientific_committees/emerging/docs/scenihr_o_045.pdf>.

Thoidingjam, S., Tiku, A.B., 2017. New developments in breast cancer therapy: role of iron oxide nanoparticles. Adv. Nat. Sci.: Nanosci. Nanotechnol. 8, 023002 (9 pp).

TGA (Therapeutic Goods Administration), 2016b. Literature review on the safety of titanium dioxide and zinc oxide nanoparticles in sunscreens, Version 1.1. Available from: <https://www.tga.gov.au/sites/default/files/nanoparticles-sunscreens-review-_2016_1.pdf>.

TGA (Therapeutic Goods Administration). 2016a. Regulation of nanomedicines by the Therapeutic Goods Administration. October 2016. Available at: https://www.tga.gov.au/sites/default/files/tga-presentation-nanoparticle-therapeutics-2016-20-october-2016.pdf.

SECTION

III

Advances in Engineered Nanomaterials' Application to Biology and Medicine, From Research to Practice

CHAPTER

Innovation in procedures for human and ecological health risk assessment of engineered nanomaterials

7

Arturo A. Keller and Nicol Parker

Bren School of Environmental Science & Management, University of California, Santa Barbara, CA, United States

7.1 Introduction

The goal of human and ecological health risk assessment is to determine whether the presence of a chemical in a particular environment will pose a health risk to potential receptors, and it is the basis for developing risk management actions to reduce the health risks below an acceptable threshold (NRC, 1983; Bridges, 2003). Risk management actions may involve reducing exposure via institutional controls, installing treatment systems to reduce emissions, or remediating a contaminated site to acceptable levels; actions may also involve substituting a chemical in a formulated product for another that poses lower health risks to humans or ecological receptors, with a similar functionality.

Human and ecological health risk assessment requires a determination of the likelihood of exposure to a contaminant at a given concentration level for a certain amount of time (i.e., the dose) and the expected health effects at the given dose (Suter, 2007). The likelihood of exposure can be determined either via actual measurements of the concentrations of the contaminant in the environmental compartment of interest (e.g., indoor air, outdoor air, freshwater, groundwater, soils, and food), or via predictive modeling, particularly when observed concentrations are sparse or not available. Predictive exposure modeling requires a solid understanding of the emissions of the contaminant(s) of interest from different sources, the transport of the chemicals within different media (e.g., air, water, and soils) and across different media, as well as the fate of the chemicals in the different media (e.g., transformations, sorption, or attachment to solid phases). Deleterious health effects (i.e., the hazard) are generally determined via controlled exposure studies using the target organisms or more likely the organisms' representative of different taxa. Health effects are then extrapolated to other organisms using a number of procedures. Health effects may be acute (e.g., high doses leading to almost immediate mortality or evident serious health effects) or chronic

Exposure to Engineered Nanomaterials in the Environment. DOI: https://doi.org/10.1016/B978-0-12-814835-8.00007-8

(e.g., lower doses leading to reduction in life span, reproduction ability, endocrine disruption, difficulty in breathing, cancer, and mutagenicity).

7.2 Challenges in conducting risk assessment for engineered nanomaterials

While determining that these risks may present difficulties even for well-studied chemicals, the challenges are even more daunting for emerging contaminants, and particularly for nanomaterials (Savolainen et al., 2010a; Kuempel et al., 2012; Becker, 2013). The scientific knowledge needed to understand their fate and transport in different media is still evolving, but it is clear that the processes that govern the behavior of nanomaterials are generally quite different from those of organic chemicals, metals, and other inorganic ions, and even to some extent larger scale colloids (Lowry and Casman, 2009; Garner and Keller, 2014; Cornelis et al., 2014; Collin et al., 2014; Peijnenburg et al., 2015; Lin et al., 2010; Batley et al., 2013). Thus traditional fate and transport paradigms and the related models cannot be used to predict the behavior of nanomaterials.

Engineered nanomaterials (ENMs) possess unique properties that distinguish them from other chemicals, even from the atoms or molecules they are composed of, or their larger scale (often denominated "bulk") materials. ENMs are aggregates of molecules that form particles or materials with at least one dimension smaller than 100 nm. ENMs are specifically designed ENMs, typically with a well-defined composition, to achieve a specific functionality. The huge surface area to volume ratio of ENMs means that most atoms or molecules in the material are on the surfaces that interact with the medium (e.g., air, water, a specific biological fluid, or tissue) and the other molecules that compose the medium (e.g., ions, biomolecules and other organic molecules, and cell membranes). Thus relative to bulk materials, ENMs present a much more reactive surface. In addition the surface energy of ENMs leads to interactions to minimize the surface area, leading to additional aggregation of particles until an equilibrium is established, leading in general to aggregate dimensions greater than 100 nm. When ENMs are synthesized, either via natural processes or industrially, the concentration of ENMs may be large enough to lead to many interactions between similar ENMs, leading to homoaggregation, unless a stabilizing agent is present or added during the synthesis of ENMs (Kulkarni and Muddapur, 2014; Groeneveld and de Mello Donegá, 2014; Mäkelä et al., 2017; Duan et al., 2015; Zhao et al., 2011). However, when ENMs are released into environmental media, the concentration of other natural particles (e.g., particles suspended in aqueous media, aerosols) may be in the order of magnitude greater than the concentration of the original ENMs, leading to heteroaggregation (Wang et al., 2015a,b; Shen et al., 2014; Praetorius et al., 2014; Quik et al., 2014; Luo et al., 2017; Abdel-Fattah et al., 2013; Zhou et al., 2012). Homo- and heteroaggregation are not relevant for molecules, and although these processes are relevant for colloids, they are magnified for ENMs due to their size. When nanomaterials are heteroaggregate, the size of the aggregate is generally large enough to lead to an accelerated deposition, either as aerosols depositing on soil surfaces, or suspended particles depositing on the sediment bed in aqueous environments. ENMs and their homoaggregates may also deposit directly, albeit at a slower rate given their smaller size. Resuspension of ENMs, homo-, and heteroaggregates is possible, via wind or water currents, or bioturbation and other biological processes. In addition the deposited materials can be transported downward into soils or sediments, resulting in deeper burial. Therefore besides chemical transformations common to most chemicals, physical transformations such as homo- and heteroaggregation need to be considered when evaluating the health and environmental risks posed by ENMs.

In addition the reactive nature of many nanomaterials, particularly certain ENMs (e.g., Ag(0), Fe(0), Cu(0), CuO, and ZnO) can lead to interactions with oxidizing agents, sulfide, and other ions present in the aqueous medium, leading to the transformation of the ENM surface and in many cases to the release of metal ions (e.g., Ag^{+}, Cu^{2+}, and Zn^{2+}), which may have additional reactions to form oxides, hydroxides, sulfides, and other complexes, including metalorganics (Phenrat et al., 2008; Lowry and Casman, 2009; Keller et al., 2010; Thio et al., 2011). While these processes also occur for bulk materials, the large surface area of ENMs accelerates the pace of transformation, resulting in a much faster oxidation/sulfidation of the ENM surface leading to release of the metal ions.

ENMs may also interact with biomolecules and biological surfaces present in the medium, for example, with proteins, leading to the formation of protein coronas, or with terrestrial plant root exudates, which may accelerate the pace of transformation of the ENMs. The interaction of ENMs with different cell membranes has been documented in a large number of studies (Cornelis et al., 2014; Collin et al., 2014; Melegari et al., 2013; Wang et al., 2011; Hanna et al., 2013; Keller et al., 2018). The uptake of ENMs by organisms at many ecological levels has been documented, from bacteria and protozoa, to mussels, plants, and other complex organisms (Bouldin et al., 2008; Ma et al., 2010; Miao et al., 2010; Fortin and Campbell, 2000). Trophic transfer of ENMs has also been demonstrated. While some of these processes have correspondence with those of organic chemicals, the underlying mechanisms are generally quite different and require a distinct mathematical representation in environmental fate and transport models.

Predicting the concentrations of ENMs thus requires novel experimental and modeling approaches. An important challenge for ENM risk assessment has been the difficulties in analyzing environmental samples (air, water, soil, or biotissue) to determine ENM concentrations. This is feasible for indoor air and workplace surfaces in controlled occupational environments and laboratory settings, but it has proven to be a formidable challenge for outdoor locations. While it is almost certain that ENMs will be detected in the sample, since they are ubiquitous in the environment from natural processes, determining whether they are introduced ENMs has been nearly impossible in most cases. Analytical instruments and methods exist for detecting ENMs, determining their size and size distribution, concentration, and shape. However, until recently the only approaches available for determining the elemental composition of metal-based ENMs were to completely dissolve them and then use atomic absorption (AA) or inductively coupled plasma (ICP) to determine the elements present, but linking size, concentration, and composition were not direct but rather inferred. Recently the advent of commercial single-particle ICP-MS offers a new tool that can be used to determine the composition of the actual ENMs without having to dissolve them and therefore one can determine size, concentration, and composition at once. This has allowed the detection of ENMs in water, food, and biological tissues, when a known ENM is present (Hadioui et al., 2015; Mitrano et al., 2012a,b; Verleysen et al., 2015; Keller et al., 2018; Fréchette-Viens et al., 2017; Schwertfeger et al., 2016; Laborda et al., 2016). However, differentiating between natural ENMs and ENMs is still a challenge for most metal-based ENMs, since the background concentration of many natural ENMs is high enough to compete with those of the introduced ENMs. For example, for TiO_2, there are sufficient natural ENMs that even clever isotopic or related elemental composition fingerprinting has not been conclusive enough (Praetorius et al., 2017). For other ENMs such as those based on silica, iron oxides, and other common elements, it will require novel approaches or instruments to tease out the different sources of ENMs in a particular sample.

To date, accurate determination of the concentration of carbon-based ENMs still presents many challenges. One approach is to collect and count them via electron microscopy (Dahm et al., 2015), but in addition to being resource-intensive and time-consuming, it can lead to some unintended bias, depending on how the sample was collected and analyzed. It can be improved by determining the elemental carbon in the sample (Dahm et al., 2015), provided no other carbon sources are present in the sample. Spectroscopic and near-infrared fluorescence can also be used to determine the concentrations, albeit limited to higher concentrations (Petersen et al., 2016). Thermal approaches such as thermogravimetric analysis and combustion can also be employed (Petersen et al., 2016). Other indirect approaches are based on measuring the concentration of metal impurities, from their synthesis, using ICP or AA (Hanna et al., 2014). This assumes that the concentration of impurities is fairly homogeneous, but in fact this may not be the case. This is an area where significant research is needed to develop rapid analytical methods that can detect ENMs at the low concentrations expected in environmental media.

Determining health effects of ENMs on different ecological receptors has considerably progressed in the past decade. In many cases, the dominant health effect is due to released metal ions. Thus ENMs such as Ag, Fe, Cu, CuO, $Cu(OH)_2$, Cu_2O, and ZnO have been clearly shown to cause effect due to the released metal ion. While some studies have determined a slightly lower toxicity threshold for the ENM (e.g., nano-CuO) than for the equivalent dissolved metal ion (e.g., Cu^{2+}), by and large the effect seems to be due to the metal. When ENMs are taken up by different organisms, they can dissolve internally, resulting in a much higher impact than when the equivalent metal ions are present in the surrounding waters, since some organisms have mechanisms to exclude the uptake of different metal ions when present at high concentrations. Some ENMs are also directly involved in the generation of reactive oxygen species (ROS) (Bodaghi et al., 2013; Zhao et al., 2017; Kalatehjari et al., 2015; Wu et al., 2015; Adeleye et al., 2016; Wang et al., 2013), for example, TiO_2, and impose health effects due to the oxidation related to ROS. Other ENMs can also readily participate in redox reactions (e.g., nano-CeO_2) (Xia et al., 2008; García et al., 2011), which can serve to provide or remove electrons to biomolecules or biological surfaces, leading to damaging changes. These health effects are discussed in more detail in other chapters.

7.3 Innovative approaches in engineered nanomaterial health risk assessment

7.3.1 Estimating predicted environmental concentrations

As mentioned above, there have been significant advances in measuring the concentrations of several ENMs, and to date there are no standard methods. Therefore novel modeling approaches are being developed to determine predicted environmental concentrations (PECs), which can be used for risk assessment. As a first approximation, material flow analyses (MFAs) have been used to better understand the flow of ENMs from their synthesis to the incorporation into nano-enabled products, their use in society, and their eventual methods of disposal and recycling (Sun et al., 2014; Gottschalk and Nowack, 2011; Gottschalk et al., 2009; Mueller et al., 2008; Keller et al., 2013; Lazareva and Keller, 2014; Keller and Lazareva, 2014). Based on the MFA, one can establish a range of amounts (mass) of ENMs that may enter the environment via release to the atmosphere, waterbodies, or soil and other urban surfaces during the life cycle of an ENM. Most recently, the time-dependent nature of ENM use has been taken into consideration in dynamic MFA models (Sun et al., 2017; Bornhöft et al., 2016; Song et al., 2017). MFAs rely on a number of estimates, measurements, and assumptions that are fraught with uncertainty. To begin with, there is no official database that provides the actual amount of ENMs produced globally or even in any single country. As is the case with most chemicals, production and sales information is considered highly valued proprietary information by the manufacturers, even though it is difficult to understand why historical information is not more forthcoming, given that from a marketing and strategic perspective the highest value is on current and future estimates.

Case study 1 Dynamic material flow analysis (MFA) for engineered nanomaterials (ENMs) in the United States.

ENMs have been in use for decades, sold as ultrafine powders for a number of applications. However, production estimates for nanoscale materials are only available for the past two decades. Nevertheless, there has been an increasing accumulation of ENMs in various product categories over this time. Fig. 7.1 presents an estimate of the ENM material flows in the United States by 2020 for four applications that have the most significant release to the environment. For the coating and paints sector, there is a significant stock that builds up every year, while for the other applications there is little or no accumulation in inventory. The majority of the release is to landfills and soils, but an increasing fraction is released to air and surface water. This will increase the concentration present in these media. Methods used to estimate MFA are discussed in more detail in Song et al. (2017).

There are market studies and other estimates of the fraction of the production that is used in different applications; however, it is typically available at a highly aggregated level (e.g., automotive, construction, medical, food, and agriculture). This information is extremely useful, but the level of aggregation leads to potentially large errors in estimating the release of ENMs to different environmental media based on the broad application. For example, ENMs in some medical applications may be used for diagnosis, with almost 100% of the ENMs being excreted and thus routed to a wastewater treatment plant (WWTP), or in a medical patch for disinfection, almost 100% of the ENMs will end up in a landfill, or applied dermally, where a fraction of the ENMs will be washed away during showering with a fraction going to a WWTP, and another fraction to a waterbody via direct contact. To complicate this, other ENMs may be in medical equipment, within

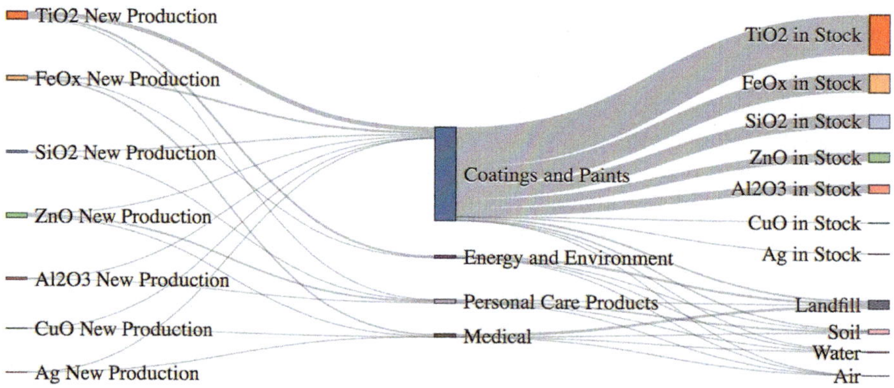

FIGURE 7.1

The US material flows by 2020 for ENMs in selected applications. The authors would like to thank Runsheng Song for his assistance in the preparation of this figure. *ENM*, Engineered nanomaterial.

the electronics or optical detectors, and thus have no mechanism of release to the environment; they may either be eventually landfilled or recycled, depending on their value, local practices, and regulations.

For some applications, such as ENMs in food and drinks (Dudkiewicz et al., 2011; Blasco and Picó, 2011; Reed et al., 2014), or personal care products (Keller et al., 2014), there is sufficient information to establish the major flows to WWTPs and to the environment. Release or emission factors of ENMs in food or personal care products are nearly 100% to WWTP; thus the uncertainty is relatively small. For paints and coatings (Song et al., 2017), another important ENM application with high likelihood of release to the environment, there are a small number of studies documenting release estimates, from which one must perform significant extrapolation, particularly with regard to the time-dependent nature of the release. Since the few experimental studies available have been relatively of short term compared to the decades of expected actual use, the pattern of release over the life of a paint or coating is not well understood. In any case, consideration of release over multiple years introduces a more realistic scenario for many ENM applications.

Once the ENMs enter the waste streams, they are routed to WWTPs, waste incineration plants (WIPs), or solid-waste disposal (i.e., landfills); a small fraction may return to ENM-enabled products in a recycling loop, if the component with ENMs is itself recycled. Few ENM applications have a significant recycling loop. There are a growing number of studies on the fate of ENMs in WWTPs (Westerhoff et al., 2011; Kiser et al., 2009; Limbach et al., 2008; Kaegi et al., 2011; Hou et al., 2012; Azodi et al., 2016; Hendren et al., 2013; Brar et al., 2010; Kunhikrishnan et al., 2015). By and large the evidence indicates that most ENMs present in the incoming stream will either aggregate substantially and settle out in the sludge (biosolids) or be transformed in the very reactive environment of WWTPs, passing first through anaerobic processes that can sulfidate some ENMs, then through aeration processes that promote oxidation of the ENM surfaces. At the end of the WWTP process the fraction of ENMs remaining in the treated water effluent is predicted to be small (0%–10%), and in many cases heavily transformed (i.e., dissolved, oxidized, and sulfidated). The majority of ENMs in raw wastewater end up in WWTP sludge and are disposed of via application to agricultural soils, incineration, or landfill. In the United States, about 50% of biosolids are disposed to agricultural soils (USEPA, 2018). Due to the substantial fraction applied to soils, there is high potential for release to biosolids soils, albeit many ENMs will be transformed.

ENMs can also enter the environment (air) via WIPs from the incineration of biosolids or other waste streams with ENM-enabled products. Information on typical paths for waste disposal is available by country, and in some cases even at the urban area level. Some recent studies (Vejerano et al., 2014; Walser et al., 2012) have quantified the release of ENMs via the gas phase during incineration; however, most ENMs end up in the ashes or slag of the WIP, except for carbonaceous ENMs that are generally transformed to CO_2. ENMs disposed to landfills in the United States are generally assumed to remain contained; however, for other countries, different assumptions and release rates may have the need for determination. Estimates of material flows through waste disposal routes serve to inform MFAs with regard to the release of ENMs to waste and the environment. MFAs cannot account for transport and fate processes between environmental media (e.g., sedimentation, advection, and transformation), thereby requiring more sophisticated models for the development of PECs.

A number of models have been recently developed to model the behavior of ENMs after their release to air, water, or soil. Two distinct modeling approaches have been employed: multimedia models that discretize the environment into different compartments that are linked by mass-transfer processes to convey ENMs from one compartment to another (e.g., deposition of aerosols from atmosphere to soils or waterbodies, resuspension of ENMs from the sediment bed to the suspended sediment compartment) and unidirectional transport models for rivers. Examples of multimedia models include SimpleBox4Nano (Meesters et al., 2014), MENDNANO (Liu and Cohen, 2012), and nanoFate (Garner et al., 2017), which have different levels of detail and discretization of the environmental compartments. Some of these models have the ability to consider time-dependent variations in the release of ENMs (e.g., due to seasonality or other factors) as well as the effect of time-dependent meteorological processes, which play an important role in transferring ENMs from one environmental compartment to another (e.g., washout of aerosols from the atmosphere, mobilization of suspended sediments from freshwater to the marine compartment). Consideration of the temporal factors is important to estimate peaks in PECs that may occur with time, which may lead to higher health risk. A key distinction of nanoFate is the ability to track the dissolved metal ions to estimate PECs for both ENMs and metal ions in every compartment which are important when considering the overall risk of the use of ENMs.

The second class of models considers the unidirectional transport of ENMs in rivers. Examples of these models include those by Praetorius et al. (2012), Markus et al. (2016), NanoDUFLOW (Quik et al., 2015; de Klein et al., 2016), and Dale et al. (2015). The Praetorius and Markus models are based on a very detailed hydrologic model of the Rhine River. ENMs are released from a point source (WWTP), and the concentration of downstream is predicted for this system. NanoDUFLOW has been implemented for the Dommel River in the Netherlands and also provides ENM concentration at a distance from a point source. The Dale model includes the application of ENMs in agriculture and models the contribution of point and nonpoint sources in a Virginia watershed. These models have a much more detailed spatial representation of the hydrologic and sediment transport processes within a river, although their representation of atmospheric and soil processes are generally more limited relative to the multimedia models. The main finding from these models is that the highest water column PECs are expected near the points of release, and that sedimentation and reactions within the sediments are important when considering PECs of ENMs released to the compartment.

Case study 2 Predicted Environmental Concentrations for a Watershed.

The environmental distribution for engineered nano-TiO_2 was evaluated in the watershed of the greater Salem, Oregon region using nanoFate. This watershed was chosen as it is the representative of regions of the United States where there is high-density urban land embedded within a largely agricultural region (Fig. 7.2). Emission of nano-TiO_2 considered land use fractions (1% freshwater, 59% undeveloped, 8% urban, and 32% agricultural), population density per land use, nano-TiO_2 product types, demand, the influence of waste disposal practices, and release to specific environmental compartments (e.g., air, soil, and water) (Garner et al., 2017; Keller and Lazareva, 2014; OECD, 2018; SEDAC, 2018; Future Markets, Inc., 2012).

Predicted environmental concentrations (PECs) in different compartments vary by orders of magnitude, air, and freshwater. PECs have significant variability, and the highest PECs are in agricultural soil to which treated sewage sludge (biosolids) is applied. The high variability between concentrations in the respective compartments is largely attributable to differences in nano-TiO_2 loads released directly to the compartment. The variability in freshwater PECs (Fig. 7.3) is largely attributable to runoff and wet deposition during precipitation events. For this region, PECs are considerably lower than toxicity thresholds for a number of sensitive species, indicating no current ecological health risks.

7.3.2 Estimating health thresholds

Since establishing ENM hazard based on actual exposure of humans is not feasible, there has been a need to consider other approaches, either using surrogate organisms (e.g., rodents) that may have relatively similar modes of exposure and damage, or by using novel cell-based assays. The latter, cell-based assays, can inform whether an exposure of a particular type of cell (e.g., liver, kidney, and lung) to a certain concentration of an ENM can lead to cell damage and provides information on the molecular mechanisms of toxicity (Liu et al., 2013a; Liu et al., 2013b; Zhang et al., 2012; Kar et al., 2014; Chau and Yap, 2012; Fourches et al., 2010; Burello and Worth, 2011; Burello, 2017a). The challenges with the first approach (i.e., surrogate organisms) are that body mass is usually substantially different, and the particular organ(s) or mechanisms of toxicity may be quite different in humans than in the model organisms. For the second approach the travel path from the point and mode of exposure (e.g., inhalation, ingestion, and dermal) to the affected cells may play a substantial role in modifying the ENMs, and it is difficult to determine the actual ENM concentration that will occur within an organ based on the external ENM concentration at which the person is exposed to. These two approaches are complementary, and while neither one can directly provide the toxicity thresholds, they serve to provide useful information for establishing mechanisms and the likelihood of a health effect if a given ENM is present at a particular concentration in a specific organ. To be able to predict the potential human health toxicity of ENMs more broadly, there are a growing number of quantitative structure–activity

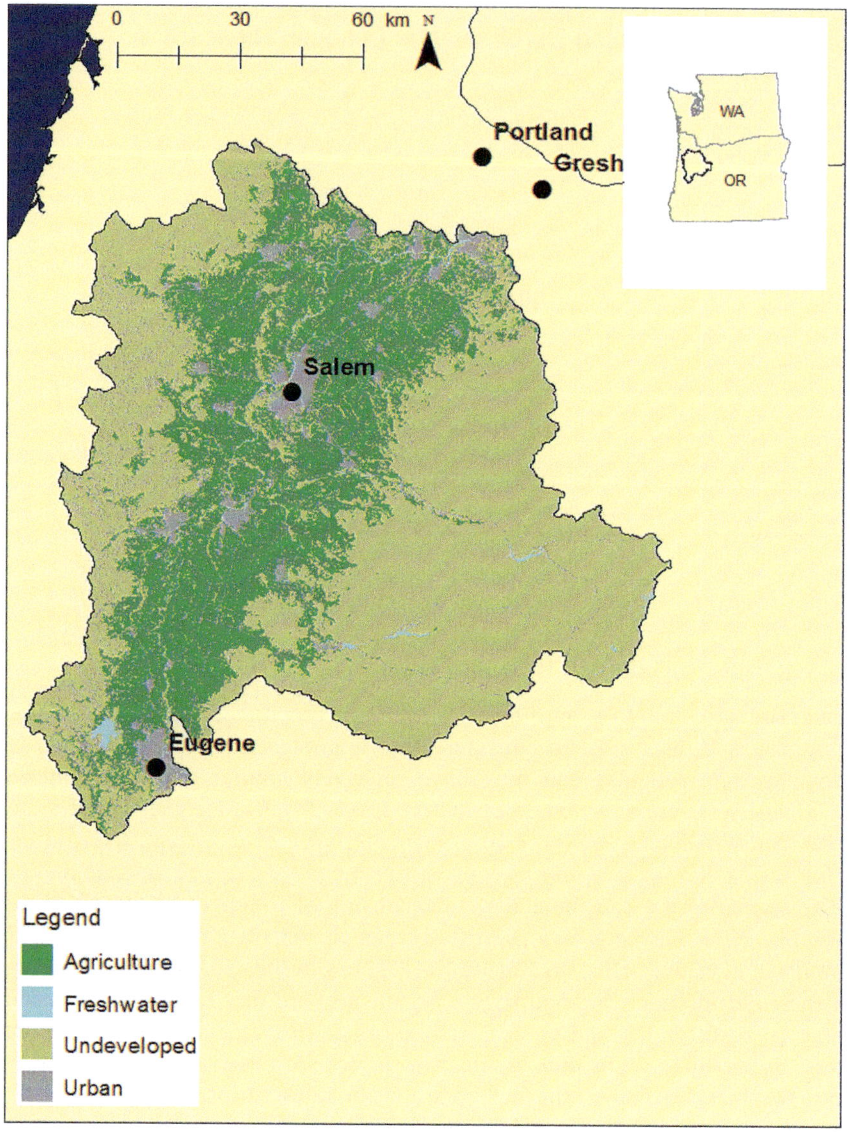

FIGURE 7.2

Watershed for the greater Salem, Oregon region.

relationships (QSARs) for different ENM classes, characteristics (e.g., solubility, reactivity, zeta potential, and band gap), and endpoints [e.g., cellular uptake; ATP, lactate dehydrogenase (LDH), and MTS assays; apoptosis; mitochondrial potential] (Burello, 2017b; Oksel et al., 2015). These QSARs can help in establishing health thresholds that can be used in ENM risk assessment. However,

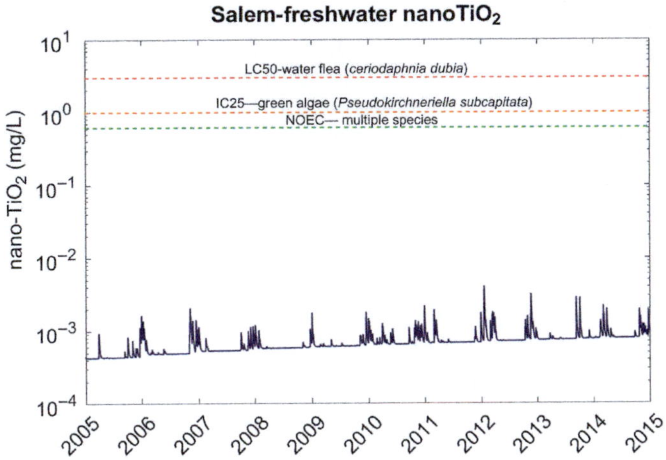

FIGURE 7.3

Risk from nano-TiO_2 for sensitive aquatic organisms.

QSARs estimations are associated with at least the same level of uncertainty of the experimental data used to build them (normally toxicological tests on model organisms) and should be used with caution for predictive estimation of health thresholds.

Over the past decade the number of ecotoxicological studies of different ENMs for a wide range of organisms has increased substantially. Earlier studies generally suffered from incomplete ENM characterization as well as inaccurate representation of the actual exposure dose, as many studies simply assumed that if the organisms are placed in a suspension at a known initial ENM concentration, it would be the representative of the exposure concentration. For many ENMs, this is an invalid assumption due to the aggregation and transformation processes that occur rapidly in many test media. Thereby, establishing the actual dose is nontrivial, and this issue is still relevant in many recent studies. Nevertheless, available ecotoxicological information for ENMs has been used in the construction of species sensitivity distributions (SSDs) for different ENMs (Garner et al., 2015), which can then be used to establish ecotoxicity thresholds such as the hazard concentration at which 5% of the species would experience adverse effects (HC_5). Effect concentrations such as no observed adverse effect levels (NOAELs), lowest observable adverse effects levels, and other effect concentrations or doses can also be used for individual species or for different trophic levels. To date, there is sufficient ecotoxicological information to inform SSDs for freshwater organisms for a dozen ENMs, but there is a need for more studies in marine and terrestrial environments to develop the corresponding SSDs.

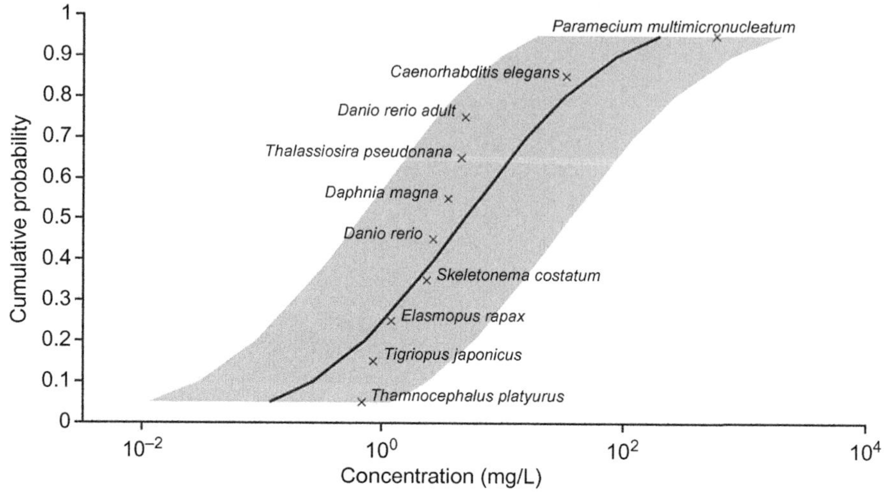

FIGURE 7.4

Species sensitivity distribution for nano-ZnO.

Adapted from Garner, K.L., et al., 2015. Species sensitivity distributions for engineered nanomaterials. Environ. Sci. Technol. 49 (9), 5753–5759.

Case study 3 Species sensitivity distribution (SSD) for nano-ZnO.

Given its toxicity, there have been a substantial number of ecotoxicological studies for nano-ZnO. These studies cover a wide range of organisms [e.g., *Danio rerio* (Yu et al., 2011; Xiong et al., 2011), *Daphnia magna* (Blinova et al., 2010; Heinlaan et al., 2008; Zhu et al., 2009), *Elasmopus rapax* (Wong et al., 2010), *Escherichia coli* (Li et al., 2011), *Leptocheirus plumulosus* (Hanna et al., 2013), *Lytechinus pictus* (Fairbairn et al., 2011), *Pseudokirchneriella subcapitata* (Aruoja et al., 2009), *Saccharomyces cerevisiae* (Kasemets et al., 2009), *Skeletonema costatum* (Wong et al., 2010), *Thalassiosira pseudonana* (Wong et al., 2010), *Thamnocephalus platyurus* (Heinlaan et al., 2008), *Tigriopus japonicus* (Wong et al., 2010), *Vibrio fischeri* (Heinlaan et al., 2008)]. The corresponding SSD for nano-ZnO (Fig. 7.3) indicates that while many freshwater organisms are affected even below or at 1 mg/L, others can tolerate 10 mg/L or even more than 100 mg/L. While there is considerable uncertainty (as denoted by the gray envelope in Fig. 7.4) in the SSD, it is useful to consider this in the risk assessment, to compare against the predicted environmental concentrations.

7.3.3 Examples of health risk assessments

Risk assessments of ENMs can be generally divided into those focused on indoor (mostly occupational) human health risk, and outdoor human and ecological health risks. While there has been considerable progress in our collective understanding of the issues surrounding occupational health risks of ENMs in the past decade, there are still many challenges ahead. Early work identified possible approaches for determining exposure and hazard (Schulte et al., 2008) and identified approaches for controlling exposure (Schulte et al., 2008) and the development of occupational exposure limits (Schulte et al., 2010). A number of studies provide examples of how to determine exposure and translate toxicological information into values that can be used to assess risk (O'Shaughnessy, 2013; Shepard and Brenner, 2013; Fransman et al., 2017; Tsang et al., 2017). Recent reviews point the need to continue developing exposure metrics and approaches to determining the key human health effects of ENMs (Savolainen et al., 2010b) and solving challenges with regard to extrapolating dose–response to establish occupational exposure limits (Leso et al., 2017). Since it is clear that toxicological data will not be available for all possible ENM compositions, sizes, and other characteristics, there have been a number of proposals to group ENMs in different manners (Landvik et al., 2018). However, there is no clear consensus on the best approach to do so. The use of alternative testing strategies is an option to make progress on the assessment of health hazards (Hjorth et al., 2017).

Outdoor ENM risk assessments have had to rely on modeling to estimate PECs. For example, early work by Nowack's group served to develop the first PECs that were compared to available ENM ecotoxicological thresholds to determine whether there was minimal risk, substantial risk, or more information would be needed to make a determination of the health risk. These PECs were developed at the country or continental level, providing an early assessment of the magnitude of ENM risks. PECs were derived using a stochastic, probabilistic modeling for ENM emission, quantities of which were used as input to an MFA (Sun et al., 2014). PECs were estimated for the EU and Switzerland, accounted for environmental compartment sizes and emissions (compartment processes such as deposition, sedimentation, and resuspension were not accounted for). In a follow-up study the PECs were used to evaluate environmental risks to organisms based on probabilistic SSDs of five ENMs, nano-Ag, carbon nanotubes, fullerenes, nano-TiO_2, and nano-ZnO (Coll et al., 2016). The risk was quantified using a risk characterization ratio, which is the quotient of the PEC and probable no effect concentration where if the quotient is greater than 1, it is indicative of risk for adverse effects, and a quotient less than 1 indicates no risk. These PECs and effect concentrations were calculated probabilistically utilizing Monte Carlo simulations to derive probability density curves. SSD curves predicted for freshwater ENM sensitivity in the order of greatest to least, Ag > ZnO > fullerenes > TiO_2 > CNT; in soil, the predicted sensitivity was ZnO > fullerenes > Ag > CNT > TiO_2. Comparison of PECs to probabilistic SSDs predicted either low or no risk for ENMs, with a risk characterization ranking from greatest to least in freshwater of ZnO > Ag > TiO_2 > fullerenes = CNT, and for soil, TiO_2 > Ag = ZnO = CNT = fullerenes.

More recently, Keller's group used the nanoFate model to predict environmental concentrations, using the San Francisco Bay as a case study (Garner et al., 2017). nanoFate can be applied to any region in the world, provided that the local parameters are adjusted, and can be scaled down to study a near-field release, for example, downstream of a WWTP, to better understand local PECs. Four ENMs (i.e., TiO_2, CeO_2, ZnO, and CuO) were considered in the San Francisco Bay case study, covering a range of ENM characteristics, production levels, and applications. The resulting time-dependent PECs were compared to ecotoxicological thresholds for the various ENMs. For freshwater, the HC_5 based on the NOAEL was used. Results of the San Francisco case study predict the HC_5 is being exceeded for nano-TiO_2 within the freshwater column, particularly during episodic inputs due to precipitation events, and that Zn^{2+} concentrations, due to the input of transformed nano-ZnO, may also exceed the HC_5 for Zn^{2+} during these events. PECs of nano-CeO_2 and nano-CuO in freshwater are generally below their corresponding HC_5. For terrestrial compartments (soils) the highest PECs are in agricultural soils treated with WWTP biosolids that contain ENMs. Although at present the soil PECs for nano-TiO_2 and nano-ZnO are below their soil HC_5s, the PECs are continuing to rise as the use of these ENMs continues to increase and may in the coming years exceed the corresponding soil PECs. Toxicity data was more limited for nano-CuO and nano-CeO_2. For nano-CuO, soil PECs are below the half maximal effective concentration of soil microbe growth inhibition, and there was no available information to determine the risk of nano-CeO_2 in soils.

Shortly after Giese et al. (2018), a study considered the past, current, and future use of CeO_2, SiO_2, and Ag ENMs in Germany and used probabilistic MFAs and SSDs for each ENM to determine PECs and risk. This study evaluated the potential accumulation of these ENMs in different environmental compartments since 1980 and up to 2050. As most previous studies, there is a wide range of PECs even for a single year. For example, nano-CeO_2 PECs could range from 1 pg/L to a few hundred ng/L (from 2017 to 2050), indicating that when the uncertainties in the various elements of the estimates are taken into account, there could be several orders of magnitude difference in results. A key finding is that for the ENMs evaluated, their use in Germany generally poses relatively low risk to freshwater column organisms. However, they caution that PECs of ENMs near-point sources may be considerably higher and could be above ecotoxicity thresholds.

7.4 Conclusion

While all health risk assessments involve uncertainties, ENMs pose a number of major challenges. Over the past couple of decades there has been a major leap in our understanding of the issues, and the acquisition of knowledge and data needed to perform these risk assessments. On the exposure assessment side the possibility of being able to collect occupational and environmental samples and use standardized methods of analysis, at least for a growing subset of ENMs, seems within reach. To support empirical risk assessment, especially for ENMs for which monitoring technologies are unlikely to be available in the near future, a number of

ENM fate and transport models have been developed that can serve to provide PECs; however, there is a need to validate these models with monitoring data. These models also require a number of ENM characteristics and process data (e.g., rates of dissolution and transformation in different environmental conditions), and there is a major need to develop predictive tools for these aspects (Burello, 2017b). Undoubtedly, more advanced models and measurement approaches will be developed to address the current limitations.

On the hazard assessment side, there is a significant amount of progress on the evaluation of the ecotoxicity for a growing number of ENMs and organisms, and there are approaches, such as the SSDs, that can serve to provide rough estimates of the threshold levels, which may result in ecological risk. Our understanding of the human health effects of ENMs has also progressed considerably but by and large relies on either extrapolating from animal to human effects, with its corresponding uncertainties, or the use of alternative approaches such as high-throughput studies with cells, with a gap between cellular response and organ or organismal response. While a number of QSARs have been developed to address these gaps, they focus on the effects to cells, due to data availability. These QSARs can only be used as "weight of evidence" rather than directly in risk assessment (Burello, 2017b).

Acknowledgments

This work was partially supported by the National Science Foundation and the US Environmental Protection Agency (USEPA) under NSF-EF0830117. Any opinions, findings, conclusions, or recommendations expressed in this material are those of the authors and do not necessarily reflect the views of the funding agencies.

References

Abdel-Fattah, A., et al., 2013. Dispersion stability and electrokinetic properties of intrinsic plutonium colloids: implications for subsurface transport. Environ. Sci. Technol. Available from: https://doi.org/10.1021/es304729d.

Adeleye, A.S., et al., 2016. Influence of phytoplankton on fate and effects of modified zerovalent iron nanoparticles. Environ. Sci. Technol. 50 (11), 5597–5605. Available from: http://pubs.acs.org/doi/abs/10.1021/acs.est.5b06251 (accessed 30.11.16.).

Aruoja, V., et al., 2009. Toxicity of nanoparticles of CuO, ZnO and TiO_2 to microalgae *Pseudokirchneriella subcapitata*. Sci. Total Environ. 407 (4), 1461–1468. Available from: https://www.sciencedirect.com/science/article/pii/S0048969708010784 (accessed 01.06.18.).

Azodi, M., Sultan, Y., Ghoshal, S., 2016. Dissolution behavior of silver nanoparticles and formation of secondary silver nanoparticles in municipal wastewater by single-particle ICP-MS. Environ. Sci. Technol. 50 (24), 13318–13327. Available from: http://pubs.acs.org/doi/10.1021/acs.est.6b03957 (accessed 19.12.17.).

Batley, G.E., Kirby, J.K., McLaughlin, M.J., 2013. Fate and risks of nanomaterials in aquatic and terrestrial environments. Acc. Chem. Res. 46 (3), 854–862. Available from: http://pubs.acs.org/doi/10.1021/ar2003368 (accessed 03.05.18.).

Becker, S., 2013. Nanotechnology in the marketplace: how the nanotechnology industry views risk. J. Nanopart. Res. 15 (5), 1426. Available from: http://link.springer.com/10.1007/s11051-013-1426-7 (accessed 01.04.14.).

Blasco, C., Picó, Y., 2011. Determining nanomaterials in food. TrAC, Trends Anal. Chem. 30 (1), 84–99. Available from: https://www.sciencedirect.com/science/article/pii/S0165993610002724 (accessed 09.07.18.).

Blinova, I., et al., 2010. Ecotoxicity of nanoparticles of CuO and ZnO in natural water. Environ. Pollut. 158 (1), 41–47. Available from: https://www.sciencedirect.com/science/article/pii/S0269749109004266 (accessed 01.06.18.).

Bodaghi, H., et al., 2013. Evaluation of the photocatalytic antimicrobial effects of a TiO_2 nanocomposite food packaging film by in vitro and in vivo tests. LWT—Food Sci. Technol. 50 (2), 702–706.

Bornhöft, N.A., et al., 2016. A dynamic probabilistic material flow modeling method. Environ. Modell. Softw. 76, 69–80. Available from: https://www.sciencedirect.com/science/article/pii/S1364815215301018 (accessed 18.12.17.).

Bouldin, J.L., et al., 2008. Aqueous toxicity and food chain transfer of quantum dots in freshwater algae and *Ceriodaphnia dubia*. Environ. Toxicol. Chem. 27 (9), 1958. Available from: http://doi.wiley.com/10.1897/07-637.1 (accessed 02.07.18.).

Brar, S.K., et al., 2010. Engineered nanoparticles in wastewater and wastewater sludge—evidence and impacts. Waste Manage. 30, 504–520. Available from: http://www.sciencedirect.com/science/article/pii/S0956053X09004607.

Bridges, J., 2003. Human health and environmental risk assessment: the need for a more harmonised and integrated approach. Chemosphere 52 (9), 1347–1351. Available from: https://www.sciencedirect.com/science/article/pii/S0045653503004697 (accessed 03.05.18.).

Burello, E., 2017a. A mechanistic model for predicting lung inflammogenicity of oxide nanoparticles. Toxicol. Sci. 159 (2), 339–353. Available from: http://academic.oup.com/toxsci/article/159/2/339/3896879/A-Mechanistic-Model-for-Predicting-Lung (accessed 03.05.18.).

Burello, E., 2017b. Review of (Q)SAR models for regulatory assessment of nanomaterials risks. NanoImpact 8, 48–58. Available from: https://www.sciencedirect.com/science/article/pii/S2452074817300538 (accessed 03.05.18.).

Burello, E., Worth, A.P., 2011. QSAR modeling of nanomaterials. Wiley Interdiscip. Rev. Nanomed. Nanobiotechnol. 3 (3), 298–306. Available from: http://doi.wiley.com/10.1002/wnan.137 (accessed 03.05.18.).

Chau, Y.T., Yap, C.W., 2012. Quantitative nanostructure–activity relationship modelling of nanoparticles. RSC Adv. 2 (22), 8489. Available from: http://xlink.rsc.org/?DOI = c2ra21489j (accessed 03.05.18.).

Coll, C., et al., 2016. Probabilistic environmental risk assessment of five nanomaterials (nano-TiO_2, nano-Ag, nano-ZnO, CNT, and fullerenes). Nanotoxicology 10 (4), 436–444. Available from: http://www.tandfonline.com/doi/full/10.3109/17435390.2015.1073812 (accessed 08.09.16.).

Collin, B., et al., 2014. Environmental release, fate and ecotoxicological effects of manufactured ceria nanomaterials. Environ. Sci.: Nano 1 (6), 533–548. Available from: http://pubs.rsc.org/en/Content/ArticleHTML/2014/EN/C4EN00149D (accessed 27.10.14.).

Cornelis, G., et al., 2014. Fate and bioavailability of engineered nanoparticles in soils: a review. Crit. Rev. Environ. Sci. Technol. 44 (24), 2720–2764. Available from: http://www.tandfonline.com/doi/full/10.1080/10643389.2013.829767 (accessed 26.09.16.).

Dahm, M.M., et al., 2015. Carbon nanotube and nanofiber exposure assessments: an analysis of 14 site visits. Ann. Occup. Hyg. 59 (6), 705–723. Available from: https://academic.oup.com/annweh/article-lookup/doi/10.1093/annhyg/mev020 (accessed 01.06.18.).

Dale, A.L., Lowry, G.V., Casman, E.A., 2015. Stream dynamics and chemical transformations control the environmental fate of silver and zinc oxide nanoparticles in a watershed-scale model. Environ. Sci. Technol. 49 (12), 7285–7293. Available from: http://pubs.acs.org/doi/10.1021/acs.est.5b01205 (accessed 18.12.17.).

de Klein, J.J.M., et al., 2016. Towards validation of the nanoDUFLOW nanoparticle fate model for the river Dommel, The Netherlands. Environ. Sci.: Nano 3 (2), 434–441. Available from: http://xlink.rsc.org/?DOI = C5EN00270B (accessed 01.06.18.).

Duan, H., Wang, D., Li, Y., 2015. Green chemistry for nanoparticle synthesis. Chem. Soc. Rev. 44 (16), 5778–5792. Available from: http://xlink.rsc.org/?DOI = C4CS00363B (accessed 03.05.18.).

Dudkiewicz, A., et al., 2011. Characterization of nanomaterials in food by electron microscopy. TrAC, Trends Anal. Chem. 30 (1), 28–43. Available from: https://www.sciencedirect.com/science/article/pii/S0165993610002967 (accessed 09.07.18.).

Fairbairn, E.A., et al., 2011. Metal oxide nanomaterials in seawater: linking physicochemical characteristics with biological response in sea urchin development. J. Hazard. Mater. 192 (3), 1565–1571.

Fortin, C., Campbell, P.G.C., 2000. Silver uptake by the green alga *Chlamydomonas reinhardtii* in relation to chemical speciation: influence of chloride. Environ. Toxicol. Chem. 19 (11), 2769–2778. Available from: http://doi.wiley.com/10.1002/etc.5620191123 (accessed 02.07.18.).

Fourches, D., et al., 2010. Quantitative nanostructure–activity relationship modeling. ACS Nano 4 (10), 5703–5712. Available from: http://pubs.acs.org/doi/10.1021/nn1013484 (accessed 03.05.18.).

Fransman, W., et al., 2017. Comparative human health impact assessment of engineered nanomaterials in the framework of life cycle assessment. Risk Anal. 37 (7), 1358–1374. Available from: http://doi.wiley.com/10.1111/risa.12703 (accessed 01.06.18.).

Fréchette-Viens, L., Hadioui, M., Wilkinson, K.J., 2017. Practical limitations of single particle ICP-MS in the determination of nanoparticle size distributions and dissolution: case of rare earth oxides. Talanta 163, 121–126. Available from: https://www.sciencedirect.com/science/article/pii/S0039914016308475?via%3Dihub (accessed 19.12.17.).

Future Markets, Inc., 2012. The Global Market for Nanomaterials 2002-2016. Production Volumes, Revenues and End Use Markets; p. 371 (http://www.futuremarketsinc.com/. Accessed Jan 29, 2013).

García, A., et al., 2011. Acute toxicity of cerium oxide, titanium oxide and iron oxide nanoparticles using standardized tests. Desalination 269 (1–3), 136–141.

Garner, K.L., et al., 2015. Species sensitivity distributions for engineered nanomaterials. Environ. Sci. Technol. 49 (9), 5753–5759.

Garner, K.L., Keller, A.A., 2014. Emerging patterns for engineered nanomaterials in the environment: a review of fate and toxicity studies. J. Nanopart. Res. 16 (8), 2503. Available from: http://link.springer.com/10.1007/s11051-014-2503-2 (accessed 28.07.14.).

Garner, K.L., Suh, S., Keller, A.A., 2017. Assessing the risk of engineered nanomaterials in the environment: development and application of the nanoFate model. Environ. Sci. Technol. 51 (10), 5541–5551. Available from: http://pubs.acs.org/doi/abs/10.1021/acs.est.6b05279 (accessed 12.05.17.).

Giese, B., et al., 2018. Risks, release and concentrations of engineered nanomaterial in the environment. Sci. Rep. 8 (1), 1565. Available from: http://www.nature.com/articles/s41598-018-19275-4 (accessed 03.05.18.).

Gottschalk, F., et al., 2009. Modeled environmental concentrations of engineered nanomaterials (TiO_2, ZnO, Ag, CNT, fullerenes) for different regions. Environ. Sci. Technol. 43, 9216–9222. Available from: https://doi.org/10.1021/es9015553.

Gottschalk, F., Nowack, B., 2011. The release of engineered nanomaterials to the environment. J. Environ. Monit. 13, 1145–1155.

Groeneveld, E., de Mello Donegá, C., 2014. The challenge of colloidal nanoparticle synthesis. Nanoparticles. Springer Berlin Heidelberg, Berlin, Heidelberg, pp. 145–189. Available from: http://link.springer.com/10.1007/978-3-662-44823-6_6 (accessed 03.05.18.).

Hadioui, M., Merdzan, V., Wilkinson, K.J., 2015. Detection and characterization of ZnO nanoparticles in surface and waste waters using single particle ICPMS. Environ. Sci. Technol. 49 (10), 6141–6148. Available from: http://pubs.acs.org/doi/10.1021/acs.est.5b00681 (accessed 19.12.17.).

Hanna, S.K., et al., 2013. Accumulation and toxicity of metal oxide nanoparticles in a soft-sediment estuarine amphipod. Aquat. Toxicol. 142–143, 441–446. Available from: http://www.ncbi.nlm.nih.gov/pubmed/24121101 (accessed 01.06.18.).

Hanna, S.K., Miller, R.J., Lenihan, H.S., 2014. Deposition of carbon nanotubes by a marine suspension feeder revealed by chemical and isotopic tracers. J. Hazard. Mater. 279, 32–37. Available from: https://www.sciencedirect.com/science/article/pii/S0304389414005317 (accessed 01.06.18.).

Heinlaan, M., et al., 2008. Toxicity of nanosized and bulk ZnO, CuO and TiO_2 to bacteria *Vibrio fischeri* and crustaceans *Daphnia magna* and *Thamnocephalus platyurus*. Chemosphere 71 (7), 1308–1316. Available from: http://www.ncbi.nlm.nih.gov/pubmed/18194809 (accessed 01.06.18.).

Hendren, C.O., et al., 2013. Modeling nanomaterial fate in wastewater treatment: Monte Carlo simulation of silver nanoparticles (nano-Ag). Sci. Total Environ. 449, 418–425.

Hjorth, R., et al., 2017. The role of alternative testing strategies in environmental risk assessment of engineered nanomaterials. Environ. Sci.: Nano 4 (2), 292–301. Available from: http://xlink.rsc.org/?DOI = C6EN00443A (accessed 01.06.18.).

Hou, L., et al., 2012. Removal of silver nanoparticles in simulated wastewater treatment processes and its impact on COD and NH_4 reduction. Chemosphere 87, 248–252.

Kaegi, R., et al., 2011. Behavior of metallic silver nanoparticles in a pilot wastewater treatment plant. Environ. Sci. Technol. 45, 3902–3908. Available from: https://doi.org/10.1021/es1041892.

Kalatehjari, P., Yousefian, M., Khalilzadeh, M.A., 2015. Assessment of antifungal effects of copper nanoparticles on the growth of the fungus *Saprolegnia* sp. on white fish (*Rutilus frisii* kutum) eggs. Egypt. J. Aquat. Res. 41 (4), 303–306.

Kar, S., et al., 2014. Nano-quantitative structure–activity relationship modeling using easily computable and interpretable descriptors for uptake of magnetofluorescent engineered nanoparticles in pancreatic cancer cells. Toxicol. In Vitro 28 (4), 600–606. Available from: https://www.sciencedirect.com/science/article/pii/S0887233313003445?via%3Dihub (accessed 03.05.18.).

Kasemets, K., et al., 2009. Toxicity of nanoparticles of ZnO, CuO and TiO_2 to yeast *Saccharomyces cerevisiae*. Toxicol. in Vitro 23 (6), 1116–1122. Available from: http://www.ncbi.nlm.nih.gov/pubmed/19486936 (accessed 01.06.18.).

Keller, A.A., et al., 2010. Stability and aggregation of metal oxide nanoparticles in natural aqueous matrices. Environ. Sci. Technol. 44 (6), 1962–1967. Available from: https://doi.org/10.1021/es902987d.

Keller, A.A., et al., 2013. Global life-cycle emissions of engineered nanomaterials. J. Nanopart. Res. 15, 1692.

Keller, A.A., et al., 2014. Release of engineered nanomaterials from personal care products throughout their life cycle. J. Nanopart. Res. 16 (7), 2489. Available from: http://link.springer.com/10.1007/s11051-014-2489-9 (accessed 23.09.14.).

Keller, A.A., Lazareva, A., 2014. Predicted releases of engineered nanomaterials: from global to regional to local. Environ. Sci. Technol. Lett. 1 (1), 65–70. Available from: https://doi.org/10.1021/ez400106t.

Keller, A.A., Huang, Y., Nelson, J., 2018. Detection of nanoparticles in edible plant tissues exposed to nano-copper using single-particle ICP-MS. J. Nanopart. Res. 20 (4), 101. Available from: http://link.springer.com/10.1007/s11051-018-4192-8 (accessed 10.05.18.).

Kiser, M.A., et al., 2009. Titanium nanomaterial removal and release from wastewater treatment plants. Environ. Sci. Technol. 43, 6757–6763. Available from: https://doi.org/10.1021/es901102n.

Kuempel, E.D., Geraci, C.L., Schulte, P.A., 2012. Risk assessment and risk management of nanomaterials in the workplace: translating research to practice. Ann. Occup. Hyg. 56, 491–505. Available from: http://annhyg.oxfordjournals.org/content/56/5/491.abstract.

Kulkarni, N., Muddapur, U., 2014. Biosynthesis of metal nanoparticles: a review. J. Nanotechnol. 2014, 1–8. Available from: http://www.hindawi.com/journals/jnt/2014/510246/ (accessed 03.05.18.).

Kunhikrishnan, A., et al., 2015. Sources, distribution, environmental fate, and ecological effects of nanomaterials in wastewater streams. Crit. Rev. Environ. Sci. Technol. 45 (4), 277–318. Available from: http://www.tandfonline.com/doi/abs/10.1080/10643389.2013.852407 (accessed 03.05.18.).

Laborda, F., Bolea, E., Jiménez-Lamana, J., 2016. Single particle inductively coupled plasma mass spectrometry for the analysis of inorganic engineered nanoparticles in environmental samples. Trends Environ. Anal. Chem. 9, 15–23. Available from: https://www.sciencedirect.com/science/article/pii/S2214158815300155?via%3Dihub (accessed 19.12.17.).

Landvik, N.E., et al., 2018. Criteria for grouping of manufactured nanomaterials to facilitate hazard and risk assessment, a systematic review of expert opinions. Regul. Toxicol. Pharm. 95, 270–279. Available from: https://www.sciencedirect.com/science/article/pii/S0273230018301004 (accessed 01.06.18.).

Lazareva, A., Keller, A.A., 2014. Estimating potential life cycle releases of engineered nanomaterials from wastewater treatment plants. ACS Sustain. Chem. Eng. 2 (7), 1656–1665. Available from: https://doi.org/10.1021/sc500121w (accessed 23.09.14.).

Leso, V., Fontana, L., Mauriello, M.C., 2017. Occupational risk assessment of engineered nanomaterials: limits, challenges and opportunities. Curr. Nanosci. 13 (1), Available from: http://www.ingentaconnect.com/content/ben/cnano/2017/00000013/00000001/art00010 (accessed 01.06.18.).

Li, M., Zhu, L., Lin, D., 2011. Toxicity of ZnO nanoparticles to *Escherichia coli*: mechanism and the influence of medium components. Environ. Sci. Technol. 45 (5),

1977–1983. Available from: http://pubs.acs.org/doi/abs/10.1021/es102624t (accessed 01.06.18.).

Limbach, L.K., et al., 2008. Removal of oxide nanoparticles in a model wastewater treatment plant: influence of agglomeration and surfactants on clearing efficiency. Environ. Sci. Technol. 42, 5828–5833. Available from: https://doi.org/10.1021/es800091f.

Lin, D., et al., 2010. Fate and transport of engineered nanomaterials in the environment. J. Environ. Qual. 39 (6), 1896. Available from: https://www.agronomy.org/publications/jeq/abstracts/39/6/1896 (accessed 03.05.18.).

Liu, H.H., Cohen, Y., 2012. Multimedia environmental distribution of nanomaterials. In: Nanotech 2012 Vol. 3, Nanotechnology 2012: Bio Sensors, Instruments, Medical, Environment and Energy. pp. 304–306.

Liu, R., Zhang, H.Y., et al., 2013a. Development of structure–activity relationship for metal oxide nanoparticles. Nanoscale 5 (12), 5644. Available from: http://xlink.rsc.org/?DOI = c3nr01533e (accessed 03.05.18.).

Liu, R., Rallo, R., et al., 2013b. Nano-SAR development for bioactivity of nanoparticles with considerations of decision boundaries. Small 9 (9–10), 1842–1852. Available from: http://doi.wiley.com/10.1002/smll.201201903 (accessed 03.05.18.).

Lowry, G.V., Casman, E.A., 2009. Nanomaterial transport, transformation, and fate in the environment: a risk-based perspective on research needs. Nanomaterials: Risks and Benefits. Springer Netherlands, The Netherlands, pp. 125–137.

Luo, M., et al., 2017. Heteroaggregation of CeO_2 and TiO_2 engineered nanoparticles in the aqueous phase: application of Turbiscan stability index and fluorescence excitation-emission matrix (EEM) spectra. Colloids Surf. A 533, 9–19.

Ma, X., et al., 2010. Interactions between engineered nanoparticles (ENPs) and plants: phytotoxicity, uptake and accumulation. Sci. Total Environ. 408 (16), 3053–3061. Available from: https://www.sciencedirect.com/science/article/pii/S0048969710003025 (accessed 02.07.18.).

Mäkelä, J.M., et al., 2017. Liquid flame spray—a hydrogen-oxygen flame based method for nanoparticle synthesis and functional nanocoatings. KONA Powder Part. J. 34 (0), 141–154. Available from: https://www.jstage.jst.go.jp/article/kona/34/0/34_2017020/_article (accessed 03.05.18.).

Markus, A.A., et al., 2016. Modelling the transport of engineered metallic nanoparticles in the river Rhine. Water Res. 91, 214–224. Available from: https://www.sciencedirect.com/science/article/pii/S0043135416300033 (accessed 03.05.18.).

Meesters, J.A.J., et al., 2014. Multimedia modeling of engineered nanoparticles with SimpleBox4nano: model definition and evaluation. Environ. Sci. Technol. 48 (10), 5726–5736. Available from: http://pubs.acs.org/doi/10.1021/es500548h (accessed 11.01.18.).

Melegari, S.P., et al., 2013. Evaluation of toxicity and oxidative stress induced by copper oxide nanoparticles in the green alga *Chlamydomonas reinhardtii*. Aquat. Toxicol. 142–143, 431–440. Available from: https://www.sciencedirect.com/science/article/pii/S0166445X13002427 (accessed 02.07.18.).

Miao, A.-J., et al., 2010. Intracellular uptake: a possible mechanism for silver engineered nanoparticle toxicity to a freshwater alga *Ochromonas danica*. PLoS One 5 (12), e15196. Available from: http://dx.plos.org/10.1371/journal.pone.0015196 (accessed 02.07.18.).

Mitrano, D.M., et al., 2012a. Detecting nanoparticulate silver using single-particle inductively coupled plasma-mass spectrometry. Environ. Toxicol. Chem. 31 (1), 115–121.

Mitrano, D.M., et al., 2012b. Silver nanoparticle characterization using single particle ICP-MS (SP-ICP-MS) and asymmetrical flow field flow fractionation ICP-MS (AF4-ICP-MS). J. Anal. At. Spectrom. 27 (7), 1131. Available from: http://xlink.rsc.org/?DOI = c2ja30021d (accessed 19.12.17.).

Mueller, N.C., Nowack, B.B., Muller, N., 2008. Exposure modeling of engineered nanoparticles in the environment. Environ. Sci. Technol. 42, 4447–4453. Available from: http://pubs.acs.org/doi/abs/10.1021/es7029637.

NRC, 1983. In: National Academy of Sciences (Ed.), Risk Assessment in the Federal Government: Managing the Process. National Research Council, Division on Earth and Life Studies, Commission on Life Sciences, Committee on the Institutional Means for Assessment of Risks to Public Health, Washington, DC. Available from: <https://books.google.com/books?hl = en&lr = &id = eHTOLjpxSHkC&oi = fnd&pg = PT15&dq = %22human + health%22 + risk + assessment + process&ots = -1ZprOpc5c&sig = Tzz__uzPAvqfJ47-Gcoa9ThNrnY#v = onepage&q = %22humanhealth%22 risk assessment process&f = false> (accessed 03.05.18.).

O'Shaughnessy, P.T., 2013. Occupational health risk to nanoparticulate exposure. Environ. Sci. Processes Impacts 15 (1), 49–62. Available from: http://xlink.rsc.org/?DOI = C2EM30631J (accessed 01.06.18.).

OECD, 2018. Municipal Waste, Generation and Treatment. Available from: <https://stats.oecd.org/Index.aspx?DataSetCode = MUNW> (accessed 06.05.18.).

Oksel, C., et al., 2015. (Q)SAR modelling of nanomaterial toxicity: a critical review. Particuology 21, 1–19. Available from: https://www.sciencedirect.com/science/article/pii/S1674200115000358 (accessed 03.05.18.).

Peijnenburg, W.J.G.M., et al., 2015. A review of the properties and processes determining the fate of engineered nanomaterials in the aquatic environment. Crit. Rev. Environ. Sci. Technol. 45 (19), 2084–2134. Available from: http://www.tandfonline.com/doi/full/10.1080/10643389.2015.1010430 (accessed 03.05.18.).

Petersen, E.J., et al., 2016. Quantification of carbon nanotubes in environmental matrices: current capabilities, case studies, and future prospects. Environ. Sci. Technol. 50 (9), 4587–4605. Available from: http://pubs.acs.org/doi/10.1021/acs.est.5b05647 (accessed 01.06.18.).

Phenrat, T., et al., 2008. Partial oxidation ("aging") and surface modification decrease the toxicity of nanosized zerovalent iron. Environ. Sci. Technol. 43, 195–200. Available from: https://doi.org/10.1021/es801955n.

Praetorius, A., et al., 2014. Heteroaggregation of titanium dioxide nanoparticles with model natural colloids under environmentally relevant conditions. Environ. Sci. Technol. 48 (18), 10690–10698.

Praetorius, A., et al., 2017. Single-particle multi-element fingerprinting (spMEF) using inductively-coupled plasma time-of-flight mass spectrometry (ICP-TOFMS) to identify engineered nanoparticles against the elevated natural background in soils. Environ. Sci.: Nano 4 (2), 307–314. Available from: http://xlink.rsc.org/?DOI = C6EN00455E (accessed 19.12.17.).

Praetorius, A., Scheringer, M., Hungerbühler, K., 2012. Development of environmental fate models for engineered nanoparticles—a case study of TiO_2 nanoparticles in the Rhine River. Environ. Sci. Technol. 46 (12), 6705–6713. Available from: http://www.ncbi.nlm.nih.gov/pubmed/22502632 (accessed 01.06.14.).

Quik, J.T.K., et al., 2014. Heteroaggregation and sedimentation rates for nanomaterials in natural waters. Water Res. 48, 269–279.

Quik, J.T.K., de Klein, J.J.M., Koelmans, A.A., 2015. Spatially explicit fate modelling of nanomaterials in natural waters. Water Res. 80, 200–208. Available from: http://www.ncbi.nlm.nih.gov/pubmed/26001284 (accessed 03.05.18.).

Reed, R.B., et al., 2014. Characterization of nanomaterials in metal colloid-containing dietary supplement drinks and assessment of their potential interactions after ingestion. ACS Sustain. Chem. Eng. 2 (7), 1616–1624. Available from: http://pubs.acs.org/doi/10.1021/sc500108m (accessed 09.07.18.).

Savolainen, K., et al., 2010a. Nanotechnologies, engineered nanomaterials and occupational health and safety—a review. Saf. Sci. 48 (8), 957–963. Available from: https://www.sciencedirect.com/science/article/pii/S0925753510000767 (accessed 01.06.18.).

Savolainen, K., et al., 2010b. Risk assessment of engineered nanomaterials and nanotechnologies—a review. Toxicology 269, 92–104.

Schulte, P., et al., 2008. Occupational risk management of engineered nanoparticles. J. Occup. Environ. Hyg. 5 (4), 239–249. Available from: http://www.tandfonline.com/doi/abs/10.1080/15459620801907840 (accessed 01.06.18.).

Schulte, P.A., et al., 2010. Occupational exposure limits for nanomaterials: state of the art. J. Nanopart. Res. 12 (6), 1971–1987. Available from: http://link.springer.com/10.1007/s11051-010-0008-1 (accessed 01.06.18.).

Schwertfeger, D.M., et al., 2016. Single particle-inductively coupled plasma mass spectroscopy analysis of metallic nanoparticles in environmental samples with large dissolved analyte fractions. Anal. Chem. 88 (20), 9908–9914. Available from: http://pubs.acs.org/doi/10.1021/acs.analchem.6b02716 (accessed 19.12.17.).

SEDAC, 2018. Gridded Population of the World (GPW), v4. Available from: <http://sedac.ciesin.columbia.edu/data/collection/gpw-v4/documentation> (accessed 06.05.18.).

Shen, C., et al., 2014. Heteroaggregation of microparticles with nanoparticles changes the chemical reversibility of the microparticles' attachment to planar surfaces. J. Colloid Interface Sci. 421, 103–113.

Shepard, M.N., Brenner, S., 2013. An occupational exposure assessment for engineered nanoparticles used in semiconductor fabrication. Ann. Occup. Hyg. 58 (2), 251–265. Available from: https://academic.oup.com/annweh/article/58/2/251/199276/An-Occupational-Exposure-Assessment-for-Engineered (accessed 01.06.18.).

Song, R., et al., 2017. Dynamic model for the stocks and release flows of engineered nanomaterials. Environ. Sci. Technol. 51 (21), 12424–12433. Available from: http://pubs.acs.org/doi/10.1021/acs.est.7b01907 (accessed 18.12.17.).

Sun, T.Y., et al., 2014. Comprehensive probabilistic modelling of environmental emissions of engineered nanomaterials. Environ. Pollut. 185, 69–76. Available from: https://www.sciencedirect.com/science/article/pii/S0269749113005241 (accessed 01.06.18.).

Sun, T.Y., et al., 2017. Envisioning nano release dynamics in a changing world: using dynamic probabilistic modeling to assess future environmental emissions of engineered nanomaterials. Environ. Sci. Technol. 51 (5), 2854–2863. Available from: http://pubs.acs.org/doi/10.1021/acs.est.6b05702 (accessed 18.12.17.).

Suter, G.W.I., 2007. Ecological Risk Assessment, second ed. CRC Press, Boca Raton, FL. Available from: https://books.google.com/books?id = oEHMBQAAQBAJ&printsec = frontcover&dq = ecological + risk + assessment&hl = en&sa = X&ved = 0ahUKEwiw0eiNhuraAhXJxVQKHb6aB6kQ6AEIJzAA#v = onepage&q = ecologicalrisk assessment&f = false (accessed 03.05.18.).

Thio, B.J., et al., 2011. Mobility of capped silver nanoparticles under environmentally relevant conditions. Environ. Sci. Technol. 46 (13), 10. Available from: https://doi.org/10.1021/es203596w.

Tsang, M.P., et al., 2017. Modeling human health characterization factors for indoor nanomaterial emissions in life cycle assessment: a case-study of titanium dioxide. Environ. Sci.: Nano 4 (8), 1705–1721.

USEPA, 2018. Frequent Questions About Biosolids. Available from: <https://www.epa.gov/biosolids/frequent-questions-about-biosolids> (accessed 06.05.18.).

Vejerano, E.P., et al., 2014. Characterization of particle emissions and fate of nanomaterials during incineration. Environ. Sci.: Nano 1 (2), 133–143. Available from: http://xlink.rsc.org/?DOI = C3EN00080J (accessed 10.07.18.).

Verleysen, E., et al., 2015. TEM and SP-ICP-MS analysis of the release of silver nanoparticles from decoration of pastry. J. Agric. Food. Chem. 63 (13), 3570–3578. Available from: http://pubs.acs.org/doi/10.1021/acs.jafc.5b00578 (accessed 19.12.17.).

Walser, T., et al., 2012. Persistence of engineered nanoparticles in a municipal solid-waste incineration plant. Nat. Nanotechnol. 7 (8), 520–524. Available from: http://www.nature.com/articles/nnano.2012.64 (accessed 10.07.18.).

Wang, H., Dong, Y., et al., 2015a. Heteroaggregation of engineered nanoparticles and kaolin clays in aqueous environments. Water Res. 80, 130–138.

Wang, H., Adeleye, A.S., et al., 2015b. Heteroaggregation of nanoparticles with biocolloids and geocolloids. Adv. Colloid Interface Sci. 226, 24–36.

Wang, Z., et al., 2011. Toxicity and internalization of CuO nanoparticles to prokaryotic alga *Microcystis aeruginosa* as affected by dissolved organic matter. Environ. Sci. Technol. 45 (14), 6032–6040. Available from: http://pubs.acs.org/doi/abs/10.1021/es2010573 (accessed 02.07.18.).

Wang, Z., et al., 2013. Biological and environmental transformations of copper-based nanomaterials. ACS Nano 7 (10), 8715–8727. Available from: http://pubs.acs.org/doi/abs/10.1021/nn403080y (accessed 22.09.16.).

Westerhoff, P., et al., 2011. Occurrence and removal of titanium at full scale wastewater treatment plants: implications for TiO_2 nanomaterials. J. Environ. Monit. 13, 1195–1203.

Wong, S.W., et al., 2010. Toxicities of nano zinc oxide to five marine organisms: influences of aggregate size and ion solubility. Anal. Bioanal. Chem. 396 (2), 609–618. Available from: http://www.ncbi.nlm.nih.gov/pubmed/19902187 (accessed 01.06.18.).

Wu, B., et al., 2015. Copper oxide and zinc oxide nanomaterials act as inhibitors of multidrug resistance transport in sea urchin embryos: their role as chemosensitizers. Environ. Sci. Technol. 49 (9), 5760–5770. Available from: http://pubs.acs.org/doi/abs/10.1021/acs.est.5b00345 (accessed 29.09.16.).

Xia, T., et al., 2008. Comparison of the mechanism of toxicity of zinc oxide and cerium oxide nanoparticles based on dissolution and oxidative stress properties. ACS Nano 2, 2121–2134.

Xiong, D., et al., 2011. Effects of nano-scale TiO_2, ZnO and their bulk counterparts on zebrafish: acute toxicity, oxidative stress and oxidative damage. Sci. Total Environ. 409 (8), 1444–1452. Available from: http://www.ncbi.nlm.nih.gov/pubmed/21296382 (accessed 01.06.18.).

Yu, L., et al., 2011. Comparative toxicity of nano-ZnO and bulk ZnO suspensions to zebrafish and the effects of sedimentation, OH production and particle dissolution in distilled water. J. Environ. Monit. 13 (7), 1975. Available from: http://www.ncbi.nlm.nih.gov/pubmed/21611643 (accessed 01.06.18.).

Zhang, H., et al., 2012. Use of metal oxide nanoparticle band gap to develop a predictive paradigm for oxidative stress and acute pulmonary inflammation. ACS Nano 6 (5), 4349–4368. Available from: http://pubs.acs.org/doi/10.1021/nn3010087 (accessed 30.09.16.).

Zhao, C.-X., et al., 2011. Nanoparticle synthesis in microreactors. Chem. Eng. Sci. 66 (7), 1463–1479. Available from: https://www.sciencedirect.com/science/article/pii/S0009250910005142 (accessed 03.05.18.).

Zhao, L., Huang, Y., Keller, A.A., 2017. Comparative metabolic response between cucumber (*Cucumis sativus*) and corn (*Zea mays*) to a $Cu(OH)_2$ nanopesticide. J. Agric. Food. Chem. Available from: http://pubs.acs.org/doi/abs/10.1021/acs.jafc.7b01306 (accessed 19.06.17.).

Zhou, D., Abdel-Fattah, A.I., Keller, A.A.A., 2012. Clay particles destabilize engineered nanoparticles in aqueous environments. Environ. Sci. Technol. 46 (14), 7520–7526.

Zhu, X., et al., 2009. Acute toxicities of six manufactured nanomaterial suspensions to *Daphnia magna*. J. Nanopart. Res. 11 (1), 67–75. Available from: http://link.springer.com/10.1007/s11051-008-9426-8 (accessed 01.06.18.).

CHAPTER

Toxicology assessment of engineered nanomaterials: innovation and tradition

8

Marta Marmiroli[1], Elena Maestri[1], Luca Pagano[1], Brett H. Robinson[2], Roberta Ruotolo[1] and Nelson Marmiroli[1]

[1]*Department of Chemistry, Life Sciences and Environmental Sustainability, University of Parma, Parma, Italy*

[2]*Department of Chemistry, University of Canterbury, Christchurch, New Zealand*

8.1 Nanotoxicology and geno-nanotoxicology

Nanotoxicology is a research field investigating the interactions between nanomaterials (NMs) and biological entities at scales ranging from biological molecules, macromolecular complexes, organelles, cells, and whole organisms, including plants or animals. Understanding this bio−nano interaction is fundamental for the safe and efficient design of NMs not only for diagnostic, imaging, and drug delivery but also for other uses.

The unique physical and chemical properties of NMs may provide health benefits but may also be associated with deleterious effects on cells and tissues. ENMs can interfere with vital cell functions, resulting in toxicity. Some ENMs, such as metal-based nanoparticles (NPs) or fullerenes, can induce chromosomal fragmentation, DNA strand breaks or point mutations, and alterations in gene expression. The study of the genotoxic effects of ENM is defined as overall "geno-nanotoxicology." ENM toxicity is an emerging issue because the mechanisms and severity of nanotoxicity are not completely predictable or testable with current toxicology methods. Interactions between ENMs and cells, through nonspecific contacts or ligand−receptor interactions, as well as the intracellular mechanisms responsible for the trafficking of these materials in the cell, require better characterization. The physicochemical properties of ENMs and their interactions within an organism can lead to contrasting interactions. These interactions, therefore, determine the biocompatibility, stability, biological performance, and side effects of ENMs.

Exposure to Engineered Nanomaterials in the Environment. DOI: https://doi.org/10.1016/B978-0-12-814835-8.00008-X

8.2 Paradigm change in toxicity tests for engineered nanomaterials

The global market for nanotechnology in 2017 was estimated at around $39.2 billion and it is likely that it will rise to $90.5 billion by 2021 (McWilliams, 2016). In 2005 the Nanotechnology Consumer Product Inventory found 54 products that incorporated nanotechnology, this had increased by 2014 to more than 1800 products (Vance et al., 2015). Therefore it is necessary to minimize any risks posed by ENMs to maximize the benefits of this new technology (Maynard et al., 2006).

Despite the increased presence of ENMs in the marketplace, there are still uncertainties about their impacts on human health and the environment. At present, there is no evidence that ENMs are specifically damaging to human health; however, there is a range of in vivo (e.g., rodent), in vitro, and in silico tests that demonstrated their potential toxicity (Nel et al., 2013a). There is evidence for toxic effects of some ENMs, for example, metal oxide NMs, from studies on key environmental species in a range of taxa (insects, algae, plants, bacteria) as reviewed in Selck et al. (2016).

In 2009 it was estimated that the use of traditional approaches to evaluate the toxicity of the marketed ENMs would cost about $1.2 billion, take more than 53 years to be accomplished, and require an exceedingly large number of animals (Choi et al., 2009). It appears that there is a need of a paradigm shift in nanotoxicology, and in general in the whole field of toxicology, to enhance the use of alternative testing strategies (ATS), as advocated in 2007 by the National Academies of Sciences (Toxicity Testing for the 21st century, a vision and a strategy). In the same period the European Union (EU) legislation was promoting Intelligent or Integrated Testing Strategies (ITS) for chemicals and specifically for ENMs (REACH directive 2007). In general, toxicology for the 21st century promotes more efficient and more ethical tests, it encourages identification of toxicity mechanism to build evidence-based testing strategies and promotes the use of in vitro, high-throughput (HTP) systems which use cell line models mostly of human origin (Hartung, 2009, 2013).

8.2.1 The 3Rs principle: replace, reduce, refine

Among the key improvements in this new toxicology, vision is the widely accepted 3Rs principle, which aims to replace, reduce, and refine animal testing and is already implemented internationally. In particular, refine animal testing implies to use lower doses for shorter times (Russell and Burch, 1959). Recently, the need to better align nanotoxicology testing with the 3Rs principle has been raised (Burden et al., 2017). The EU regulation on the protection of animals used for scientific purposes (European Parliament, 2010) encourages the implementation of alternative approaches in scientific research. Thus the application of the

3Rs principles is embedded in regulations that are relevant to the safety assessment of ENMs in chemicals, cosmetics, food, pharmaceuticals, medical devices, and biocides (Rauscher et al., 2017).

Further criticisms directed toward standard guideline tests, using mice and rats, propound that they offer sparse information on the mechanism of toxicity of a substance thus providing little help in explaining why a substance might lead to an apical response (adverse effect) of regulatory concern (Gerloff et al., 2017). Therefore ITS or ATS is developed which enhances the use of biological information generated using in vitro and in silico methods to predict, through specific causes and molecular perturbations, whether any chemical, hence also any ENMs, would likely result in adverse outcomes (Burden et al., 2015).

8.2.2 Adverse outcome pathways

Since 2012 the OECD (Organisation for Economic Co-operation and Development) has developed new ITS to support the implementation of the 3Rs principles and to meet legislative requirements. This has resulted in the Integrated Approaches to Testing and Assessment (IATA) guidelines that combine and exploit existing information, in vitro assay data, and computational predictions to satisfy specific information requirements (OECD, IATA, 2013). Following this, the OECD launched a programme for the development of adverse outcome pathways (AOPs), which has taken up aspects of the World Health Organization and International Programme on Chemical Safety work on mode of action (MoA) of chemicals for a better evaluation and harmonization of the assessment of chemical risks (OECD, 2018). Initially, AOPs were described within the context of ecotoxicological risk assessment and defined as "conceptual construct that portrays existing knowledge concerning the linkage between a direct Molecular Initiating Event (MIE) and an Adverse Outcome (AO)," by capturing the sequential chain of causally linked key events (KEs) at different levels of biological organization (Ankley et al., 2010).

Subsequently, the AOP concept was extended to support the assessment of human health effects. AOPs aim to support regulatory decision-making by providing the knowledge base to support the development of novel test methods, test guidelines, quantitative structure–activity relationship tools, and IATA (Villeneuve et al., 2014). Of great importance are the links between KEs, described by key event relationships (KERs), because they complete the mechanistic view of the toxicity activity of a substance (Kleinstreuer et al., 2016). Fig. 8.1A represents a simple AOP as described in OECD, (2018), Fig. 8.1B represents the "Adverse Outcome Pathway on Protein Alkylation Leading to Liver Fibrosis" as described by Landesmann (2016).

At present, AOPs have been developed for chemical toxicology; nevertheless, there is increasing awareness of this concept also in the field of nanotoxicology. Recently, Vietti et al. (2016) published a review of the knowledge regarding the KEs and KERs involved in lung fibrosis development by carbon nanotubes

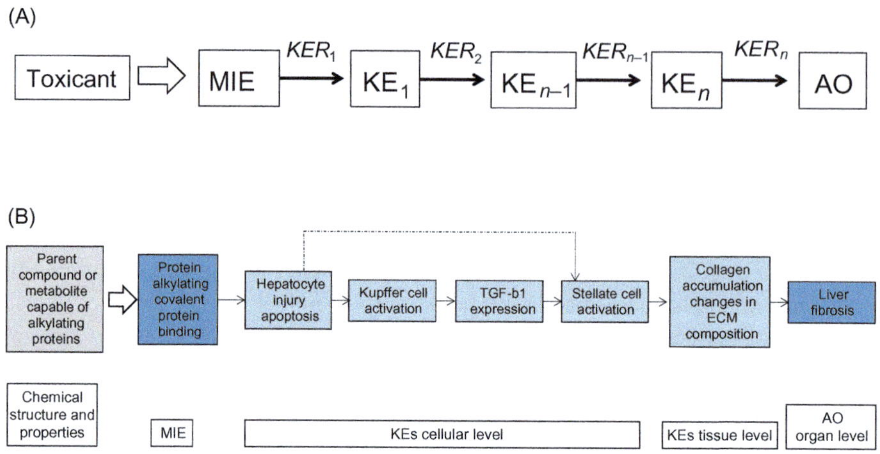

FIGURE 8.1

(A) A simple AOP as described in OECD (2018). (B) The "Adverse Outcome Pathway on Protein Alkylation Leading to Liver Fibrosis" as described by Landesmann (2016). *AOP*, adverse outcome pathway; *OECD*, Organisation for Economic Co-operation and Development.

(CNTs) with the intention to produce a draft of a likely AOP. Another work, by Labib et al., developed an AOP for multiwalled CNTs based on transcriptomic data from in vivo experiments. This work led to a linear AOP where biochemical toxicity end points are linked by well-identified cellular and organ response at a toxicogenomic level (Labib et al., 2016).

Two challenges when developing a draft for an AOP of a specific ENM are the physicochemical characterization of the ENM and the identification of a specific MIE (Nel et al., 2013a). An AOP by definition should be limited to the description of toxicodynamics, but the kinetics of the ENM can influence the occurrence and type of the initiating event (Johnston et al., 2018). In addition, it is likely that some ENMs interact with cells or cellular components not through specific molecular interaction, as seen for many chemicals (e.g., pharmaceuticals or pesticides), but by inducing mechanical/physical damage, for example, to the cell membrane or to the lysosome, which could not be described as a canonical "Molecular" Initiating Event (Gerloff et al., 2017).

8.2.3 Human cell lines for alternative in vitro and in vivo tests

An AOP can be understood and drafted utilizing in vitro cell or cell line cultures, which provide living systems for the investigation of toxicity. In the field of nanotoxicology the cultured cells from humans or other mammals of greatest relevance are those from lungs, liver, and immune systems (Nel et al., 2013b). Several examples can be, for instance, the trypan blue exclusion test of cell viability, which determines the number of viable cells on the base of the inclusion of trypan blue dye, depending on the state of the cell membrane, the clonogenic assay, which measure in vitro the ability of a single cell to grow into a colony or the measurements related to the activity of several enzymes involved in the primary metabolic functions (e.g., lactate dehydrogenase or glyceraldehyde 3-phosphate dehydrogenase).

To date, it has been established that ENMs cause adverse end points in the liver and lungs and cause the precipitation of immune reactions; however, the biochemical and molecular mechanisms involved are still under investigation. The first studies of particle toxicology were restricted to inhalation toxicology, meaning the accidental inhalation of particles in workplaces and the environment. Therefore the first experiment to establish ENM toxicity was conducted on the well-established models provided by lung tissues and cells (Donaldson and Seaton, 2012). Most of the toxicity to the lungs can involve cells located in the alveolar septa, which is the alveolar space surrounded by type I and type II epithelial cells. The latter is the most important cell type from a toxicology viewpoint. Type II alveolar cells (type II pneumocytes) have many crucial roles in the functioning of the lungs: (1) synthesis and secretion of lipoprotein surfactants to decrease surface tension at the air–liquid interface of the lungs resulting in increased alveolar stability and prevention of edema; (2) increase the ability of alveolar macrophages in phagocytosis of invading bacteria; (3) xenobiotic metabolism, including drugs and environmental pollutants, through the cytochrome P-450 (CYP450)-dependent monooxygenases system; and (4) extrude large amounts of Na and water across the plasma membrane and play a vital role in keeping the alveoli dry. This process involves the ascorbate cotransport system that aids in the accumulation of antioxidant vitamin C in these cells. Thus any toxic substance that could alter the potential of type II cells to synthesize and secrete surfactants impairs its capacity for xenobiotic metabolism, transepithelial Na transport or vitamin C accumulation may adversely affect lung function (Castranova et al., 1988). The main in vitro model for lung cell lines is the A549, which is a continuous human lung adenocarcinoma epithelial cell line initiated in 1972 by D.J. Giard through explant culture of lung carcinomatous tissue from a 58-year-old Caucasian male. Morphological, ultrastructural, and functional characteristics of A549 cells were like the in situ type II pneumocyte and were unchanged by the growth medium. A549 constitutes a well-established model for metabolic and molecular mechanisms specific to alveolar type II cells (Kermanizadeh et al., 2016). Although the gastrointestinal tract and the lungs are primary exposure systems for ENMs, it is now known that a proportion of ENMs entering the body through ingestion and/or inhalation reach secondary organs including the liver (Geiser and Kreyling, 2010). The liver, which is the primary organ of metabolism, is characterized by different types of cells, each with its own peculiar function and morphology. The hepatocytes, due to their abundance and their importance in the normal liver function, are of particular interest for (nano)-toxicology. Hepatocytes are also known to synthesize many hormones and cytokines including interleukin-8 and to activate in their endoplasmic reticulum xenobiotics and drugs detoxification through the CYP450 enzymes superfamily. The preferred human in vitro models are primary hepatocytes and hepatoma cell line HepG2, which was derived from an immortalized cell line of human liver carcinoma cells. HepG2 cells can be grown on a large scale, they can secrete plasma proteins, such as transferrin and albumin, they can also be transformed to express specific CYP450 enzymes (Kermanizadeh et al., 2016).

8.2.4 Model organisms for alternative in vitro and in vivo tests

To reduce the number of in vivo tests, not only human and murine cell lines can be utilized but also alternative whole model organisms.

Due to their versatility as molecular tools for toxicological screening, eukaryotic model systems widely used for nanotoxicity assessment are the yeast *Saccharomyces cerevisiae* and nematode *Caenorhabditis elegans*. Other models to assess toxicants activity are invertebrates, such as freshwater cnidarian *Hydra vulgaris*, the earthworm *Eisenia fetida*, the unicellular green alga *Chlamydomonas reinhardtii*, and the green alga *Chlorella vulgaris*. However, compared to humans, they differ in many aspects of their physiology and anatomy. Zebrafish (*Danio rerio*) are more similar to humans in their anatomy, physiology, and genome and thus offer many advantages over invertebrates as alternatives to rodent models (Lieschke and Currie, 2007; Davis et al., 2014). The main advantages of using zebrafish are (1) the ability to assess responses in a whole organism, (2) the small size and relative ease of maintenance, (3) the genetic similarity to mammals, (4) the high fecundity, (5) the low cost compared to rodent testing, (6) the potential to perform rapid toxicological screening, (7) the availability of genetically manipulated strains, and (8) the potential to generate HTP formats (Burns et al., 2005; Howe et al., 2013). Therefore zebrafish is a valuable addition for testing ENM toxicity with HTP platforms (Naatz et al., 2017). Model plants such as *Arabidopsis thaliana*, *Cucurbita pepo*, and *Oryza sativa* are used for ecotoxicological assessment (Marmiroli et al., 2015; Pagano et al., 2017).

In the future, in vitro studies or using alternative nonmammalian organisms may yield detailed information and novel insights regarding the toxicity mechanisms of ENMs. However, although animal testing should be kept to a minimum, in vivo studies are required to understand a set of phenomena related to the whole organism, such as (1) how ENMs distribute between different anatomical compartments, (2) whether the NPs accumulate or not in the organs and body, and (3) how NMs cross the biological barriers within the organism. Moreover, mammalian models are instrumental to establish any dosimetry limit.

8.2.5 High-content screening and high-throughput screening

To explore the mechanism of ENM toxicity, the approaches utilizing in vitro and in silico analyses can be improved by using high-content screening (HCS) and high-throughput screening (HTS) methods. Such integrated systems are defined as ATS (Nel and Malloy, 2017). A HCS refers to a cell-based HTS that uses microscopic images as assay readouts, for example, of phenotypic changes in a cell population after a treatment. Genomics was the first HTS technique applied in toxicology, this combined interaction, defined as toxicogenomics, elucidates how the entire genome is involved in biological responses of organisms exposed to environmental–chemical stressors. Toxicogenomics now incorporates the analyses of the proteome, transcriptome, and metabolome perturbations induced by environmental stress in disease initialization (Waters and Fostel, 2004). This approach, applied to in vivo organs and tissues extracts, is often the basis for the initial understanding of the mechanistic action of pollutants on specific genes and of the regulatory pathway through which toxicants influence cell homeostasis. The study of the mechanisms of ENM toxicity through this approach is known as nanotoxicogenomics (Labib et al., 2016).

8.3 Omics methods and system toxicology

The suffix "ome" is used in molecular biology to indicate a totality of some sort "omics" thus used to assess all the genes, proteins, transcript, metabolites, etc., globally when they are affected by a substance or condition. Omics tools, such as genomics, proteomics, transcriptomics, and metabolomics, can be coupled with computational approaches (bioinformatics) to identify pathways that can be quantitatively modeled. Systems biology aims at understanding in a holistic way, the output of several omics studies. Systems biology has recently been integrated with toxicology to produce "systems toxicology," which aims at integrating toxicology with new results from omics experiments when performed to evaluate the response of a living organism to a toxicological insult, among which ENMs are included (Sturla et al., 2014). Besides the ability to screen for multiple end points in a single experimental run, omics share the focus on changes at the molecular level. System toxicology is a multilevel screening, which integrates omics with more conventional, apical end points, such as histopathology and cytopathology, in the context of scientific integration (Fig. 8.2).

For example, transcriptomic profiling is useful in toxicogenomics as a discovery tool for identifying the biochemical mechanisms underlying compensatory responses to xenobiotics, the potential mechanisms of toxicity, and the biomarkers for drug safety evaluation. Microarrays are still an elective technology for transcriptome profiling, though emerging alternatives is the utilization of next-generation massively parallel sequencing (NGS) to sequence and count transcripts directly from samples (RNA-seq or digital transcriptomics) (Wang et al., 2009). In addition, the information content of RNA-seq exceeds mRNA abundance, including detection of alternative splicing and small interfering RNAs (Marguerat and Bahler, 2010). Proteomics provides important information about the cellular

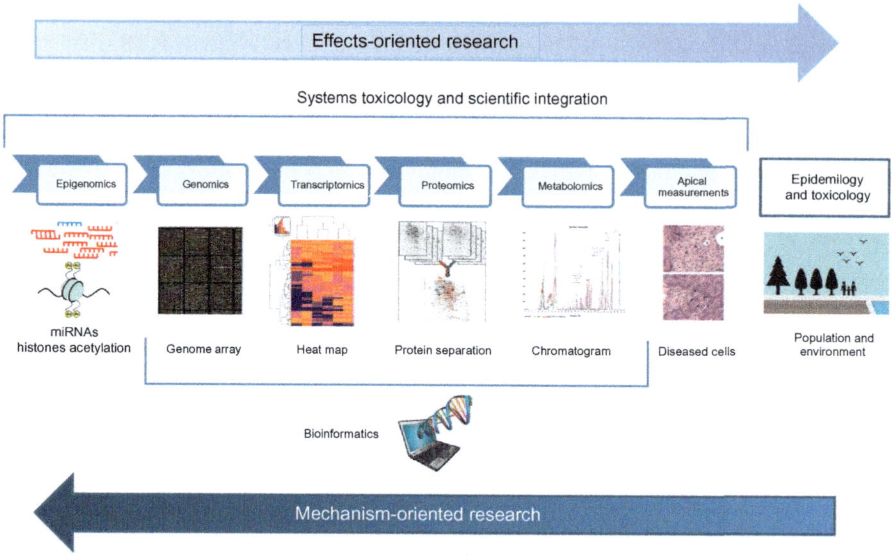

FIGURE 8.2

System toxicology flowchart: integrating omics with conventional end points.

response to toxicants, protein–protein interactions, and posttranslational modifications that are not detected by transcriptomics. Proteomics requires protein separation or fractionation through gel-based (two-dimensional polyacrylamide gel electrophoresis, two-dimensional differential in-gel electrophoresis), or gel-less (2D-liquid chromatography or Isobaric Tags for Relative and Absolute Quantitation) techniques. This is followed by the extraction peptides, which may be digested, ionized, and identified through mass fingerprinting or sequencing [time-of-flight mass spectrometry (MS), and MS/MS (tandem mass spectrometry)]. Proteomics analyses provide a snapshot of the proteins' abundance changed by the cells (in vitro) or organisms (in vivo) following exposure to a toxicant. In general the correlation between transcriptomics and proteomics results for the same toxicant can vary from 20% to 80% (Hajduch et al., 2010; Marmiroli et al., 2015). Metabolomics, the comprehensive study of the metabolites of a cell or an organism, aims at determining the dynamics in the production of metabolites following a physicochemical or environmental challenge. In toxicological terms, metabolomics studies the shifts in metabolites production in response to the toxicant exposure. Metabolites are chemically extracted from tissues or cells, isolated by high-performance liquid chromatography, gas chromatography, or electrospray ionization, then identified utilizing nuclear magnetic resonance or MS. In nanotoxicology, metabolomics can be useful for the identification of primary and secondary ROS (reactive oxygen species) produced within the cells because of the ENM reactivity. There is growing evidence that epigenetic modifications, such as DNA methylation and histone acetylation and tail modifications, may be caused by environmental factors (Jirtle and Skinner, 2007). In addition, further epigenetic posttranscriptional modification can be carried out by noncoding microRNAs (miRNAs), "which comprises species of short noncoding RNA that regulate gene expression posttranscriptionally" (Sato et al., 2011). There is increasing interest in the field of nanotoxicology in miRNAs expression and MoA: it has been shown that Ag ENMs can elicit an epigenetic response in Jurkat T cells through miRNAs action (Eom et al., 2014).

8.4 Engineered nanomaterial genotoxicity tests

The study of genotoxicity induced by ENMs can benefit from new, HTP testing strategies. NMs induce direct and indirect damages to DNA; in the former case, ENMs come in direct contact with the DNA material, in the latter, they induce the production of molecules and intermediates, such as ROS, that interacts negatively with the DNA structure and function. Due to the variety of mechanisms leading to ENM-induced DNA damage and the range of mutagenic events that may occur as a result, there should be several testing systems addressed to establish the genotoxic potential of ENMs; however, there is still a gray area. For pharmaceutical and chemical compounds, a battery of well-defined tests to assess various genotoxicity end points (point mutations, aneuploidy, and chromosomal fragmentation) is necessary for regulatory approval. The current guidelines include the in vitro bacterial reverse gene mutation test (Ames test; OECD 471), an in vitro mammalian cell gene mutation test (e.g., mouse lymphoma TK assay; OECD 476), and an in vitro mammalian cell chromosome aberration or micronucleus assay (OECD 473 or 487, respectively) (Doak et al., 2012). However, there is accumulating evidence demonstrating that ENMs interact with key genotoxicity assay components, including those that are fundamental to the tests mechanism, for example, dyes, which highlight the importance of generating well-validated assay protocols specifically for ENMs (Doak et al., 2009).

In nanogenotoxicology, assay refinement is vital for the development of testing strategies tailored to detect ENM genotoxicity, for example, new tests to identify hazards of mutation for chemicals may also prove valuable for ENM testing. The growth arrest and DNA damage 45 alpha (GADD45a)-GFP (GreenScreen) assay is a relatively new HTP test system effective in testing chemicals for genotoxic activity. The assay is based on a fluorescent reporter system that operates in a wide range of cell backgrounds (from in vitro and in vivo screenings) and is constructed on the observation that the upregulation of the GADD45a gene constitutes a sensitive "surrogate" marker of most types of DNA damages (Hastwell et al., 2006). The major benefit of this assay is its HTP nature, which allows multiple ENM variants to be tested in parallel for their genotoxic potential. Another genotoxicity test that is becoming widespread is the PIGa assay. The PIGa gene (which is X-chromosome linked) codes for a protein (phosphatidylinositol glycan complementation group A) heavily involved in the glycosylphosphatidylinositol anchors that attach cell markers to the cell membrane of blood cells. The PIGa mutation assay has been developed to identify cells that have lost these specific cell surface markers (e.g., CD59) through mutation of the PIGa gene, which leads to their loss of expression in the cell membrane. A combination of flow cytometry and fluorescent antibodies to detect the cell markers of interest is utilized to screen hundreds of thousands of cells searching for the relatively rare cells without expression of the cell marker. These are identified as the cells with PIGa mutations whose frequency informs on the mutation frequency induced by exposure to toxicants (Bryce et al., 2008). This test is currently under development for application on in vitro cell cultures. Nevertheless, it is a typical HCS test that can be used also to screen workers exposed in the sector of ENM production.

8.5 (Quantitative) Structure–activity relationships

Collectively, the use of AOP or MoA as the basis for toxicological analyses (possibly obtained by HTS or HCS assays) is a mechanistic toxicological approach, which has also the goal of establishing (Q)SARs [(quantitative) structure–activity relationships]. SARs are key aspects of nanotoxicology because they are useful for predicting the likelihood that a certain ENM with a specific structure, leading to unique physicochemical characteristics, would induce adverse effects in animals and humans. When the in vitro test and the connected SARs can establish effectively a pathogenic outcome in vivo, then the whole approach is called predictive toxicological platform (Nel et al., 2013a). The importance of a robust predictive toxicological platform is the bulk of the discovery and hazard ranking that can be performed in vitro, and are therefore useful for planning and prioritizing more complex and costly in vivo testing, according to the 3Rs principles (Gajewicz et al., 2018). In addition, any reliable (Q)SAR established for a specific ENM, based on strongly validated AOP, can be extremely useful to improve the ENM structure in accordance with the "safe-by-design" rule (Naatz et al., 2017).

8.6 Factors affecting engineered nanomaterial toxicity

Many different factors that affect ENM toxicity must be considered when designing ENMs for a specific application: size, shape, surface, charge, aggregation status, and crystallinity of the ENMs. The *size* of an ENM can affect both uptake and distribution within the organism. Small NPs, such as quantum dots (QDs), can influence subcellular trafficking in macrophages, enter the nucleus, and bind histones, thus interfering with critical biological functions. As evidenced by many authors, the smaller the size of an ENM, the more it is able to generate oxidative responses by formation of ROS, which can give risks to biological systems mainly through DNA damage, enzyme inactivation, lipid peroxidation and cause inflammatory responses (Gatoo et al., 2014). Smaller ENM accumulate more in rat liver and when they are retained in the respiratory tract for a longer duration, this leads to a greater translocation into the pulmonary interstitium with impaired alveolar macrophage function (Zhang et al., 2012; Gatoo et al., 2014).

Composition is a determining factor for ENM function, but some materials may contain toxic elements (such as heavy metals) and this toxicity may be reduced by adding a coating or nanoshell. In other cases, the surface coating also plays a role in ENM toxicity. Hydrophilic polymers, such as polyethylene glycol (PEG) conjugated to the surface of the particles, may extend the circulation time of NPs in vivo.

ENMs can be produced in various shapes including fibers, rings, tubes, spheres, and planes. Spherical NPs are relatively less toxic and their uptake mediated by endocytosis is facilitated (Gatoo et al., 2014). As for CNTs, concerns about its toxicity have emerged due to their shape, which is reminiscent of that of asbestos fibers. Toxicity of longer fibers is related to their plasma half-life.

In addition, the ENM charge significantly contributes to the toxicity of the material. Positively charged ENMs show significant cellular uptake compared to noncharged or negatively charged ENMs due to increased plasma protein opsonization. Furthermore, it has also been shown to induce hemolysis, platelet aggregation, and increased toxicity (Gatoo et al., 2014). The surface charge of the NPs alters the integrity of the blood–brain barrier and the transmembrane permeability.

ENMs, such as titanium dioxide (TiO_2), can cause toxicity based on their crystalline structure. Many studies have shown that the TiO_2 anatase form was more cytotoxic than the rutile form and that the toxicity was related to ROS generation and oxidative stress (Shah et al., 2017). The biodegradability of an ENM is another important parameter to consider for toxicity because ENMs can accumulate in organs and cells and exert toxic effects after long-term exposure. Biodegradable materials can also cause toxic effects if the degraded components of the material are toxic.

The results of toxicological studies in which ENMs of the same chemical nature are used can give different results depending on the exposure method of the ENM used. For example, TiO_2 NPs, whose use is widespread, even as a food additive, does not appear to cross the skin barrier in humans but may be slowly absorbed from both the lungs and the gastrointestinal tract and induce pulmonary tumors in rats if inhaled over a prolonged period. Some ENMs, such as fullerene, can easily cross the placental barrier and at high doses cause fetal toxicity and death.

8.6.1 Nano versus bulk versus ion

The particle size and form may profoundly affect the risk posed by ENMs, raising concerns about their environmental release and fate (Keller et al., 2013).

The physical and chemical stability of ENMs in the environment determines, to a large extent, their toxicological potential and fate (Gardea-Torresdey et al., 2014). Depending on the types of EMNs and their physicochemical properties, the effects observed can be similar to the respective bulk material (intended as raw or granular form), metal ions, or, in some cases, specific for the NM.

Examples for toxicity tests are reported in the literature, both for model organisms and species of economic interest, comprising bacteria (Wang et al., 2017), plants (Ma et al., 2015a), and animals (Brohi et al., 2017). A panel of 12 metal-based and metal oxide–based ENMs Pd, Al_2O_3, Co_3O_4, CuO, Fe_3O_4, MgO, Mn_3O_4, Sb_2O_3, SiO_2, TiO_2, ZnO, and WO_3 were tested on three bacterial species, *Escherichia coli*, *Staphylococcus aureus*, and *Vibrio fischeri*, and on the algae *Pseudokirchneriella subcapitata* and the protozoa *Tetrahymena thermophila*. Toxicity tests were also performed on the respective metal salts. ENM suspensions were characterized for their physicochemical properties (particle aggregation, solubility in different media) and for their effect on ROS formation. A cell viability test and growth inhibition test were performed on five species chosen for the study, estimating the EC_{50} values for each ENM tested. The results showed that *P. subcapitata* had the highest sensitivity to the ENM treatments. ENMs composed of CuO and ZnO were the most toxic, possibly due to their lower stability and higher dissolution rate, resulting in a higher concentration of ions in the medium (Aruoja et al., 2015).

In the case of simple eukaryotes, several formulations of functionalized gold NPs (Au NPs) were tested on the model organism *S. cerevisiae* (brewer's yeast). Functionalization was performed to test on the yeast model the differently charged particles, which showed that positively charged Au NPs reduced yeast survival. A screening of the yeast deletion mutant collection of nonessential genes revealed that of the 17 mutants resistant to Au NPs, several had changes related to mitochondrial function and organization (Smith et al., 2013). Similarly, the screening performed on the same mutant collection using cadmium sulfide-based QDs (CdS QDs) revealed 112 sensitive and 114 tolerant mutants. A comparison with Cd salts showed that only a few of these mutants to CdS QDs were also sensitive/resistant to Cd^{2+}, indicating that nano- and ionic Cd were elicited different response pathways. This study also showed that mitochondria were involved in the response to these ENMs (Marmiroli et al., 2016). Analyzing the ENM response in more complex eukaryotes, the effects of Ag NPs on the nematode *C. elegans* were reported, focusing on the roles of endocytosis and lysosomal function in uptake and in vivo toxicity (Maurer et al., 2016). Because of the intrinsic low stability in cellular environment, the results obtained from Ag NPs were compared with $AgNO_3$ and sparingly soluble citrate-coated Ag NPs (CIT-Ag NPs). The use of lysosomal- and endocytosis-deficient mutants carried out the importance of these mechanisms in the *C. elegans* uptake and response to Ag NPs. Furthermore, the pharmacological inhibition of endocytosis reduced the toxic effects of the stable CIT-Ag NP but not those related to $AgNO_3$. Likewise, toxicogenomic effects of low toxicity TiO_2 NPs (as anatase or rutile crystalline forms) and corresponding bulk forms on *C. elegans* were investigated. Microarray analysis revealed that the effects of the highly stable ENM and bulk were similar on the molecular level, involving mainly the oxidative stress response (Rocheleau et al., 2015).

Considering plants, studies were performed both on model organisms and crops: the response derived from CdS QD exposure of a panel of 400 *A. thaliana* (L.) Heynh. transposon and T-DNA-induced mutants was studied, isolating two specific QD-tolerant mutants named *atnp01* and *atnp02*. These mutants were characterized at genetic, transcriptomic, and physiological level and a comparison with Cd salt response evidenced the nano-specific nature of the molecular mechanisms and the pathways involved (Marmiroli et al., 2014). The specificity of the ENM response raises the possibility of monitoring the ENM exposure, using physiological and molecular biomarkers.

The effects of CuO and ZnO NPs on *Triticum aestivum* L. (wheat) grown in sand showed that the solubilization of Cu and Zn in the medium occurred at similar rates in both CuO and ZnO NPs, compared to their corresponding bulk forms (Dimkpa et al., 2012). Dissolved Cu derived from CuO NPs contributed to the phytotoxicity, whereas Zn release did not show any significant change in the plant growth. Total Cu and Zn levels in shoot were similar for NP or bulk forms, but oxidative stress was evident in the ENM-treated plants. The biotransformation of ENMs was observed in the wheat shoots, with speciation in the leaves as Cu(I)—sulfur complexes and Zn-phosphate. The effects of lanthanide-based ENMs (CeO_2 and La_2O_3) on *Cucumis sativus* L. (cucumber) plants indicated that CeO_2 NPs had no phytotoxic effect in *C. sativus*, while La_2O_3 NPs inhibited the elongation and biomass of roots and shoots, as well as inducing ROS (Ma et al., 2015b).

Through synchrotron analyses, La was found to combine with phosphate and carboxylic groups, whereas some Ce was transformed into Ce(III)—carboxyl complexes. This implied a higher dissolution rate of La_2O_3 as compared with CeO_2 NPs, which may explain their distinct behaviors in term of phytotoxicity and transport.

8.6.2 Bulk and ion in the engineered nanomaterial cocontamination

Another fundamental factor involved in the ENM environment are the interactions between ENMs and natural organic matter (NOM), which in the recent years were described in several different systems (Wang et al., 2016; Deng et al., 2017). Specifically, in the presence of NOM, soluble metal-based ENMs are more stable due to the increasing ionic strength on ENM suspensions. Humic acids present in the soil and pH of the suspension are fundamental factors in determining the speciation of metal-based ENMs and their degree of aggregation (Wang et al., 2016).

To date, few studies have investigated interactions between metal and metal oxide–based ENMs and their physiological and molecular effects on different organisms (e.g., protozoa, plants, human cells). The effects of the cocontamination have been studied also on a limited number of environmental systems such as their interactions with microbiota in the soil–plant system. Chemical interactions between NPs in aqueous medium indicate how stable ENMs or adsorbing surfaces, such as TiO_2 NPs, may contribute to determine the fate of ZnO NPs in water (Tong et al., 2015). The release of Zn^{2+} from ZnNPs can interact with the surfaces of TiO_2 NPs thus reducing the concentration of Zn^{2+} in water. Investigating the effect of the ENM cocontamination in *C. pepo* (zucchini) reported several interactions with metal oxide–based and rare-earth metal oxide–based ENMs when compared with the respective individual NM or bulk material treatments (Pagano et al., 2017).

The effects of ENMs as cocontaminants highlight the need for comprehensive methodologies like dosimetry, chronic, or long-term effects to elucidate every aspect of the phenomenon, for which a combination of biological and physicochemical data may be endorsed.

8.6.3 Off-target versus target

ENMs can enter cells through endocytosis via various pathways (Fig. 8.3), these include phagocytosis, clathrin, or caveolae-mediated endocytosis, clathrin/caveolae-independent endocytosis, and other types of pinocytosis, as macropinocytosis for large aggregates of ENMs. Phagocytosis mainly occurs in specialized "professional phagocyte" cells responsible for host defense and the uptake of dead cells, such as macrophages, neutrophils, and monocytes. However, cell internalization occurs in many cell types. Clathrin assembly units, known as the "triskelion," with other accessory proteins, stabilize membrane curvatures, inducing invagination of the membrane and the release of the vesicles with a diameter of 100–150 nm into the cytoplasm. Caveolae are flask-shaped membrane invagination of 50–80 nm induced by proteins called "caveolins" with the help of other proteins, such as dynamin. In these mechanisms, endocytosis may be mediated by membrane receptor–specific uptake, a process that is highly selective and specific, or by nonspecific adsorptive uptake, via nonspecific hydrophobic or electrostatic interactions. ENMs internalized via endocytosis remain trapped in the transport vesicles that travel through the endosomal scaffold, thus affecting subcellular organelles (Kafshgari et al., 2015). The intracellular traffic and the fate of ENMs are related to their physicochemical properties and to the endocytic pathways. For example, ENMs absorbed by clathrin-dependent receptor-mediated endocytosis (RME) are typically destined for lysosomal degradation, while internalization of RME-independent clathrin leads to accumulation and endosomal shunting in a nondegrading pathway. ENPs entering the cell by caveolae-

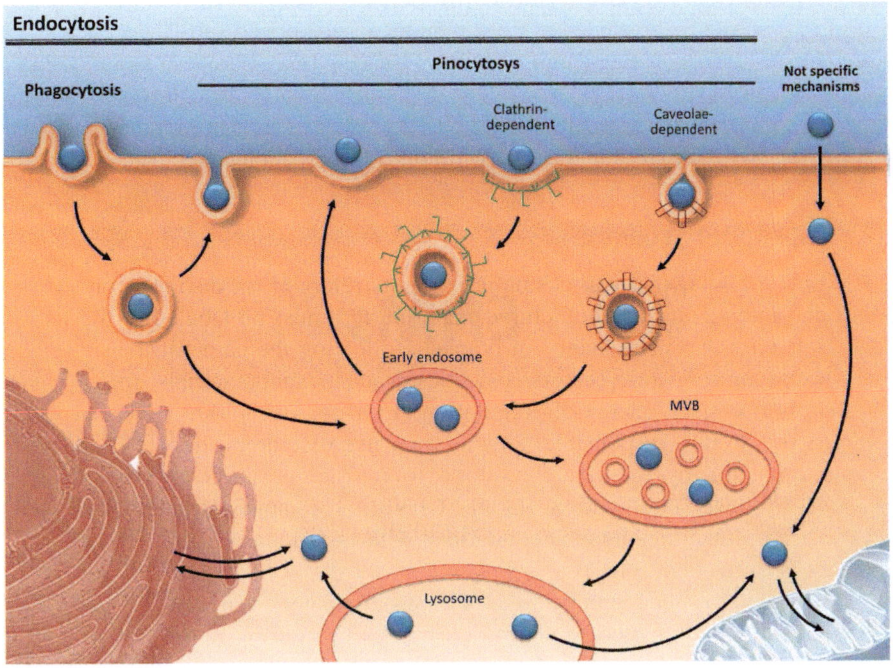

FIGURE 8.3

The main routes for intracellular uptake of ENMs. ENMs (indicated by dots) can enter the cell via different pathways of endocytosis: phagocytosis and clathrin/caveolae-mediated endocytosis. Upon cellular entry, ENMs are engulfed in membrane invaginations that are pinched off to form endosomes (or phagosomes in case of phagocytosis). These vesicles can become *MVB*/late endosomes and fuse with lysosomes. The ENMs may also be transported back to the cell surface or delivered to other organelles (e.g., mitochondria). *ENM*, Engineered nanomaterial; *MVB*, multivesicular bodies.

dependent mechanisms can sometimes escape lysosomal degradation. In addition to active mechanisms of ENM transport, ENMs may also enter cells by passive diffusion through the cell membrane. This occurs in cells lacking endocytosis machinery, such as red blood cells.

Inside the cell, ENMs may interact with various biological molecules (Fig. 8.4), including DNA and proteins. Despite the overall negative charge of DNA, it can form strong bonds with carbon-based NMs, such as fullerenes and CNTs. To design new NMs with reduced genotoxicity, it is essential to understand the mechanisms of interaction between DNA and ENMs. The physical interaction of ENMs with DNA can lead to potentially negative impacts on its structure, stability, and biological functions. The diameter of fullerenes and CNTs is close to the size of the major and minor grooves of DNA and RNA. Therefore it is likely that ENMs can bind and occupy these sites. Considering that the DNA

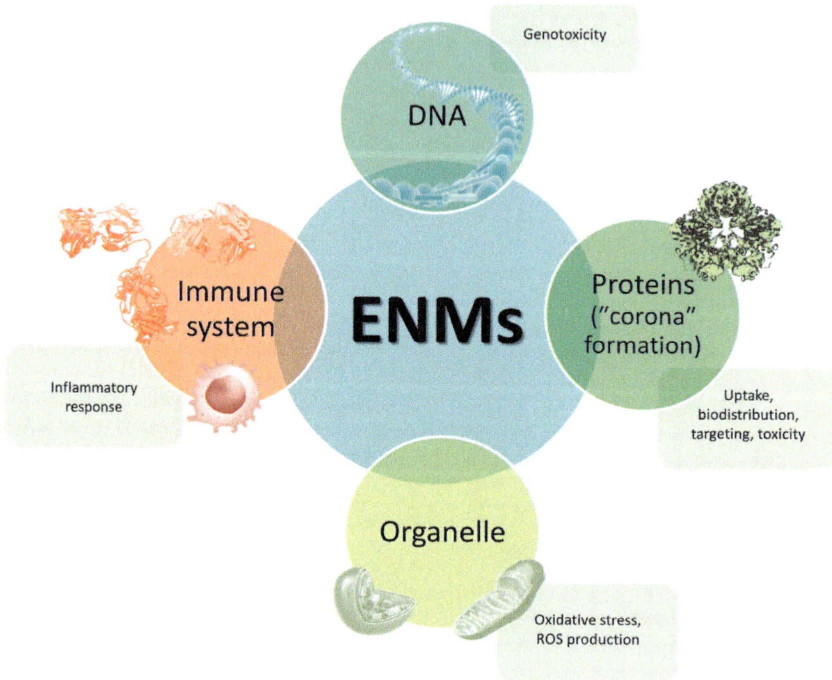

FIGURE 8.4

Schematic representation of possible biological targets for ENMs. Once in contact with a biological fluid, ENMs undergo the formation of a biomolecular corona made up primarily of proteins. These proteins may undergo conformational changes and expose cryptic epitopes, which can be recognized by different types of cells, such as immune cells, promoting inflammatory or autoimmune disorders. Protein–ENM interaction can also influence systemic circulation, biodistribution, bioavailability, and toxicity. Inside the cell, ENMs can be internalized in organelles, as mitochondria, in which the interaction with specific protein complexes can mediate ENM response (e.g., ROS production). ENMs can also interact with other biological molecules, as DNA, causing genotoxicity. *ENM*, Engineered nanomaterial; *ROS*, reactive oxygen species.

is wrapped around a histone complex to form chromatin with a diameter and a height of 6 nm containing positive charges, the charge on the ENM surface can also determine its ability to interfere with the chromatin structure.

Since 2000, most research has focused on *protein adsorption* of ENMs (Fig. 8.4). In contact with biological fluids (e.g., blood, interstitial fluid, or mucous secretions), ENPs are coated with macromolecules, *in primis* proteins that can modify surface charge and properties of these materials. This biological coating called the "protein corona" may subsequently lead to loss of ENM performance due to an increase in hydrodynamic or aggregation size or may in some cases modify its toxicological properties. The nature of the corona is a major determinant of ENM toxicity. The formation of the corona involves an element of competition between several proteins in a dynamic process that involves their adsorption/desorption on the ENP surface (Miclăuş et al., 2014; Docter et al., 2015). Corona proteins, which show a high affinity for the ENP surface, are exchanged slowly and these so-called hard proteins form the innermost layer, while the "soft" proteins are more loosely bound, possibly through protein–protein interactions, and are rapidly exchanged (Walczyk et al., 2010; Milani et al., 2012; Docter et al., 2015).

The proteins that adsorb onto the surface of the ENMs can influence their circulation and biodistribution. Several studies have shown that some plasma proteins are strongly bound to polymeric and metal-based NPs, liposomes, and CNTs, including albumin, immunoglobulins, fibrinogen, apolipoproteins, and proteins from the complement cascade (Zhang et al., 2012). The binding of complement proteins and immunoglobulin promotes, in many cases, the opsonization of ENPs, leading to their recognition via phagocytosis by the mononuclear phagocyte system (MPS) and by the rapid elimination from blood flow. It is known that ENPs are rapidly eliminated from the cells of the reticuloendothelial system/MPS in animals in organs, such as liver and spleen, where they can accumulate and cause toxicity. ENPs can, in fact, influence the immune system and induce an inflammatory response (Fig. 8.4). For example, it has been demonstrated that ZnO NPs can promote antigen absorption by antigen-presenting cells or promote targeting on specific cells, such as dendritic cells or macrophages. It has been discovered that the ZnO NPs are found specifically internalized in the immune cells, in which they induce toxicity. ENPs can also act as haptens and modify protein structures, increasing their potential for autoimmune effects.

The proteins that bind the ENMs often contain pockets that bind the ligands and of the catalytic sites that, despite their diversified functions, have dimensions similar to those of the ENMs. Some ENMs, such as fullerene (C60), CNTs, or QDs, can bind and suppress the activity of bound enzymes as in the case of the HIV-1 protease and glutathione-*s*-transferase (Yanamala et al., 2013; Ruotolo et al., 2018). In other cases, ENMs can bind proteins or pore channels formed by membrane proteins involved in the transport of molecules and ions and block their functionality. ENMs can also alter the characteristics of structural proteins that play crucial roles in maintaining the shape, size, morphology, and motility of cells, as observed in the case of actinic filaments treated with single walled carbon nano tubes (SWCNT). Interactions between ENMs and microtubule or centrosome proteins have been related to an interruption of the mitotic spindle, chromosomal rupture/fragmentation, and mutagenicity mediated by ENM treatment. These interactions of the ENMs can also indirectly influence the functioning of the organelles since their correct organization and distribution are mainly influenced by the cytoskeleton (Yanamala et al., 2013).

The ribosome is a molecular complex composed of proteins and different RNA molecules and ENMs, such as fullerenes and CNT, can bind to active sites on ribosomes, inhibiting or stopping protein synthesis. Other ENPs, such as silica or QDs, may interfere with protein synthesis by binding them to initiation or elongation factors (Klein et al., 2016; Ruotolo et al., 2018).

The use of the ENMs for therapeutic purposes can, therefore, be strongly influenced by the formation of the protein corona, because of the half-life of the ENM in the blood. The use of zwitterionic polymers, of PEG and its derivatives, has shown to drastically reduce the absorption of nonspecific proteins, in particular, apolipoprotein J and complement protein C3, thus prolonging the residence time in the blood.

ENM–protein interactions can be used to guide the distribution of ENMs to specific tissues. For example, the decoration of ENMs with specific proteins (e.g., formulations based on albumin, using the receptors of transferrin or epidermal growth factor receptor) prior to injection was exploited for particular targeting purposes. Transferrin is expressed in all types of tissue, is rapidly recycled to the cell surface after internalization, and is believed to facilitate the transport of macromolecules and nanoconstructs (coated with the specific receptor) through the blood–brain barrier (Ghadiri et al., 2017). Targeted ENMs can also identify molecular targets with good affinity and selectivity. These ENMs can bind to pathogens and biomarkers, amplifying their signal for molecular imaging and detection. Intracellular organelles can also be targeted. Direct administration of DNA to the mitochondrial matrix has been suggested for the treatment of genetic diseases associated with mitochondrial genome defects and for the development of direct mitochondrial gene therapy.

Since in animals, including humans, there is a certain interindividual variability in the circulating levels of various proteins (e.g., lipoproteins, immunoglobulins, cytokines), the differences in protein profiles may be the cause of variations in the formation of protein corona and be responsible for increased susceptibility to side effects. Physiological stress can trigger the overexpression of proteins in the acute phase and some of these proteins (e.g., C-reactive protein) can improve the activation of the complement proteins and the absorption of macrophages when fixed on the surface of pathogens and senescent cells (Fujita et al., 2010). The impact of these conditions on the fate of the ENMs must be carefully evaluated before using an ENM for therapeutic purposes.

Physical interactions between ENMs and organelles disrupt their normal functions in cells (Fig. 8.4). For example, mitochondrial damage is thought to be one of the probable mechanisms by which ENMs cause cytotoxicity by inducing oxidative stress through the destruction and distribution of normal electron transport by respiratory complexes. Furthermore, the voltage-dependent anionic channel (VDAC), located in the outer mitochondrial membrane, is the main component of its membrane and acts as a gatekeeper for the entry and exit of mitochondrial metabolites (Shoshan-Barmatz et al., 2017). Because mitochondria play an important role in maintaining energy balance in cells, binding and blocking of this channel by ENMs that bind VDAC can cause significant damage to cells (Wen et al., 2016).

8.7 Biomarkers for engineered nanomaterials

Considering the wide range of ENMs utilized, their physicochemical properties and their effects, there is still no consensus on consistent end points that are commonly shared in response to the different ENM classes (Fig. 8.5). Specific biomarkers for ENM exposure, effect, and susceptibility remain under investigation in plants, which exhibit contrasting responses from humans (Iavicoli et al., 2014). Biomarkers of exposure often involve the measurement of secondary metabolites or other physiological parameters able to reflect the biological dosage and effects, showing, directly or indirectly, the physiological implications of the exposure. Biomarkers of effects, generally, show changes at the cellular and molecular

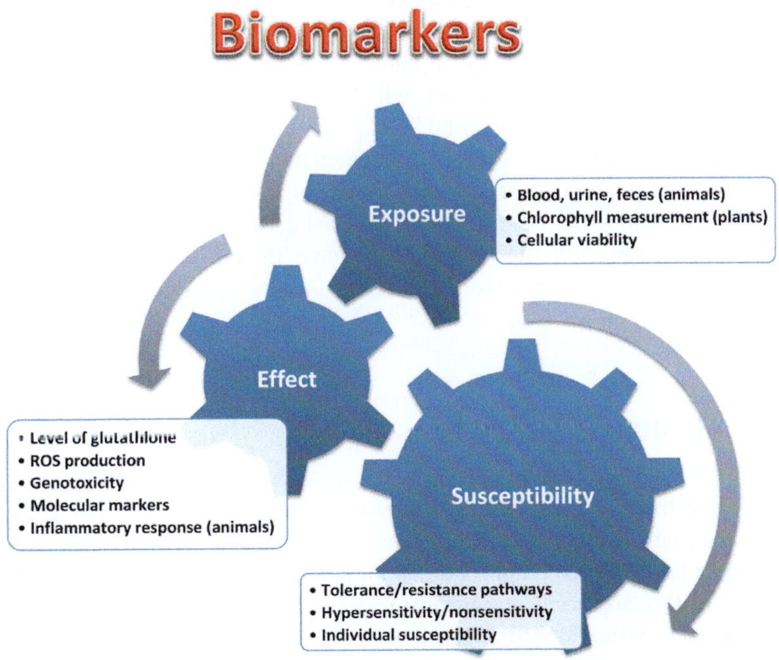

FIGURE 8.5

Schematic representation of the principal biomarkers of exposure, effect, and susceptibility that has been applied to ENMs. *ENM*, Engineered nanomaterial.

levels, reflecting the gene expression or the protein abundance under experimentally (controlled) conditions. These classes of biomarkers can be usefully applied as a tool for the in vivo toxicity or genotoxicity assessment (Brain and Cedergreen, 2009). Biomarkers of susceptibility indicate the constitutive responsiveness to the exposure of a contaminant, as such through tolerance and resistance pathways as described for hypersensitive and nonsensitive phenotypes (Iavicoli et al., 2016). Some of these categories of biomarkers reflect the complete organism response to ENM exposure, providing information related to the risk assessment.

Examples of exposure biomarkers in animals are the presence of ENMs in rats (Balasubramanian et al., 2013): blood, urine, and feces contained significantly higher amounts of Au after 15-day inhalation Au NPs exposure. Models were proposed in which biomarkers of effects might be ROS production (Tiwari et al., 2011), increased level of glutathione (GSH) (Genter et al., 2012), or the activation of the inflammatory response (Srinivas et al., 2012). Considering effects on the nucleus at the level of damage to DNA, a model for measuring genotoxicity can be found in a higher percentage of DNA strand breaks (Tiwari et al., 2011), detected by laboratory techniques as comet assay, microarray, polymerase chain reaction.

Ma et al. (2015a) described several examples of ENM-related physiological biomarkers in plants. These are related to photosynthetic activity, cellular viability, or oxidative stress. A model, which identifies molecular markers across different plant species, was proposed. Coupling metal uptake and physiological measurements, it was possible to give a mechanistic insight of the ENM response, through interatomic networks that represent molecular pathways, cellular components, and biological processes involved in ENM exposure and varying according to particle type, size, and plant species (Pagano et al., 2016). This systematic approach can be useful in order to characterize the risk associated to ENMs.

8.8 Conclusion

The use of ENMs is increasing rapidly and this raises concerns about their potential diffusion in environment and food chains. To avoid repeating past mistakes when new materials were introduced, which were subsequently shown to be harmful, a precautionary approach should be adopted where ENMs are "safe by design." In this approach, a unifying role is played by genotoxicology and by all the "omics" applications to toxicology. These new technologies allow for a "trace back" into the appearance of negative events, which are triggered by exposure to ENMs. Molecular interactions are so far the very first "omic" science able to elucidate the response to ENM exposure. Biomarkers of exposure and/or effect can be profitably found by tagging molecular events, offering also the possibility of finding similar effects in different cases like ENM type, way of exposure, type of organism.

Key points

- The chapter provides insights on traditional toxicology end points in new model organisms, as well as innovative approaches to nanotoxicology.
- The effects related to engineered nanomaterial (ENM) exposure, compared to the relative ions and bulk materials, are also considered.
- Critical end points related to molecular trafficking are highlighted, as well as potential molecular tools for early detection and monitoring of ENM exposure and effects.

Acknowledgments

Authors acknowledge the support of the project INTENSE, grant no. 652515 and the support of FIL (Fondi Locali per la Ricerca) provided by the University of Parma.

References

Ankley, G.T., Bennett, R.S., Erickson, R.J., Hoff, D.J., Hornung, M.W., Johnson, R.D., et al., 2010. Adverse outcome pathways: a conceptual framework to support ecotoxicology research and risk assessment. Environ. Toxicol. Chem. 29, 730–741.

Aruoja, V., Pokhrel, S., Sihtmäe, M., Mortimer, M., Mädler, L., Kahru, A., 2015. Toxicity of 12 metal-based nanoparticles to algae, bacteria and protozoa. Environ. Sci.: Nano 2, 630.

Balasubramanian, S.K., Poh, K.W., Ong, C.N., Kreyling, W.G., Ong, W.Y., Yu, L.E., 2013. The effect of primary particle size on biodistribution of inhaled gold nano-agglomerates. Biomaterials 34, 5439–5452.

Brain, R.A., Cedergreen, N., 2009. Biomarkers in aquatic plants: selection and utility. Rev. Environ. Contam. Toxicol. 198, 49–109.

Brohi, R.D., Wang, L., Talpur, H.S., Wu, D., Khan, F.A., Bhattarai, D., et al., 2017. Toxicity of nanoparticles on the reproductive system in animal models: a review. Front. Pharmacol. 8, 606.

Bryce, S.M., Bemis, J.C., Dertinger, S.D., 2008. In vivo mutation assay based on the endogenous Pig-a locus. Environ. Mol. Mutagen. 49, 256–264.

Burden, N., Sewell, F., Andersen, M.E., Boobis, A., Chipman, J.K., Cronin, M.T.D., et al., 2015. Adverse outcome pathways can drive non-animal approaches for safety assessment. J. Appl. Toxicol. 35, 971–975.

Burden, N., Aschberger, K., Chaudhry, Q., Clift, M.J.D., Doak, S.H., Fowler, P., et al., 2017. The 3Rs as a framework to support a 21st century approach for nanosafety assessment. Nano Today 12, 10–13.

Burns, C.G., Milan, D.J., Grande, E.J., Rottbauer, W., MacRae, C.A., Fishman, M.C., 2005. High-throughput assay for small molecules that modulate zebrafish embryonic heart rate. Nat. Chem. Biol. 1, 263–264.

Castranova, V., Rabovsky, J., Tucker, J.H., Miles, P.R., 1988. The alveolar type II epithelial cell: a multifunctional pneumocyte. Toxicol. Appl. Pharmacol. 93, 472–483.

Choi, J.Y., Ramachandran, G., Kandlikar, M., 2009. The impact of toxicity testing costs on nanomaterial regulation. Environ. Sci. Technol. 43, 3030–3034.

Davis, E.E., Frangakis, S., Katsanis, N., 2014. Interpreting human genetic variation with in vivo zebrafish assays. Biochim. Biophys. Acta 1842, 1960–1970.

Deng, R., Lin, D., Zhu, L., Majumdar, S., White, J.C., Gardea-Torresdey, J.L., et al., 2017. Nanoparticle interactions with co-existing contaminants: joint toxicity, bioaccumulation and risk. Nanotoxicology 11, 591–612.

Dimkpa, C.O., McLean, J.E., Latta, D.E., Manangon, E., Britt, D.W., Johnson, W.P., et al., 2012. CuO and ZnO nanoparticles: phytotoxicity, metal speciation and induction of oxidative stress in sand-grown wheat. J. Nanopart. Res. 14, 1125–1140.

Doak, S.H., Griffiths, S.M., Manshian, B., Singh, N., Williams, P.M., Brown, A.P., et al., 2009. Confounding experimental considerations in nanogenotoxicology. Mutagenesis 24, 285–293.

Doak, S.H., Manshian, B., Jenkins, G.J., Singh, N., 2012. In vitro genotoxicity testing strategy for nanomaterials and the adaptation of current OECD guidelines. Mutat. Res. 745, 104–111.

Docter, D., Westmeier, D., Markiewicz, M., Stolte, S., Knauer, S.K., Stauber, R.H., 2015. The nanoparticle biomolecule corona: lessons learned—challenge accepted? Chem. Soc. Rev. 44, 6094–6121.

Donaldson, K., Seaton, A., 2012. A short history of the toxicology of inhaled particles. Part. Fibre Toxicol. 9, 13.

Eom, H.J., Chatterjee, N., Lee, J., Choi, J., 2014. Integrated mRNA and micro RNA profiling reveals epigenetic mechanism of differential sensitivity of Jurkat T cells to AgNPs and Ag ions. Toxicol. Lett. 229, 311–318.

European Parliament, 2010. Directive 2010/63/EU of the European Parliament and of the Council of 22 September 2010 on the Protection of Animals Used for Scientific Purposes.

Fujita, Y., Kakino, A., Harada-Shiba, M., Sato, Y., Otsui, K., Yoshimoto, R., et al., 2010. C-reactive protein uptake by macrophage cell line via class-A scavenger receptor. Clin. Chem. 56, 478–481.

Gajewicz, A., Puzyn, T., Odziomek, K., Urbaszek, P., Haase, A., Riebeling, C., et al., 2018. Decision tree models to classify nanomaterials according to the DF4nanoGrouping scheme. Nanotoxicology 12, 1–17.

Gardea-Torresdey, J.L., Rico, C.M., White, J.C., 2014. Trophic transfer, transformation, and impact of engineered nanomaterials in terrestrial environments. Environ. Sci. Technol. 48, 2526–2540.

Gatoo, M.A., Naseem, S., Arfat, M.Y., Dar, A.M., Qasim, K., Zubair, S., 2014. Physicochemical properties of nanomaterials: implication in associated toxic manifestations. Biomed. Res. Int. 2014, 498420.

Geiser, M., Kreyling, W.G., 2010. Deposition and biokinetics of inhaled nanoparticles. Part. Fibre Toxicol. 7, 2.

Genter, M.B., Newman, N.C., Shertzer, H.G., Ali, S.F., Bolon, B., 2012. Distribution and systemic effects of intranasally administered 25 nm silver nanoparticles in adult mice. Toxicol. Pathol. 40, 1004–1013.

Gerloff, K., Landesmann, B., Worth, A., Munn, S., Palosaari, T., Whelan, M., 2017. The adverse outcome pathway in nanotoxicology. Comput. Toxicol. 1, 3–11.

Ghadiri, M., Vasheghani-Farahani, E., Atyabi, F., Kobarfard, F., Mohamadyar-Toupkanlou, F., Hosseinkhani, H., 2017. Transferrin-conjugated magnetic dextran–spermine nanoparticles for targeted drug transport across blood–brain barrier. J. Biomed. Mater. Res. A 105, 2851–2864.

Hajduch, M., Hearne, L.B., Miernyk, J.A., Casteel, J.E., Joshi, T., Agrawal, G.K., et al., 2010. Systems analysis of seed filling in *Arabidopsis*: using general linear modeling to assess concordance of transcript and protein expression. Plant Physiol. Biochem. 152, 2078–2087.

Hartung, T., 2009. Toxicology for the twenty-first century. Nature 460, 208–212.

Hartung, T., 2013. Look back in anger—what clinical studies tell us about preclinical work. ALTEX 30, 275–291.

Hastwell, P.W., Chai, L.L., Roberts, K.J., Webster, T.W., Harvey, J.S., Rees, R.W., et al., 2006. High-specificity and high-sensitivity genotoxicity assessment in a human cell line: validation of the GreenScreen HC GADD45a-GFP genotoxicity assay. Mutat. Res. 607, 160–175.

Howe, K., Clark, M.D., Torroja, C.F., Torrance, J., Berthelot, C., Muffato, M., et al., 2013. The zebrafish reference genome sequence and its relationship to the human genome. Nature 496, 498–503.

Iavicoli, I., Leso, V., Manno, M., Shutle, P.A., 2014. Biomarkers of nanomaterial exposure and effect: current status. J. Nanopart. Res. 16, 2302.

Iavicoli, I., Leso, V., Schulte, P.A., 2016. Biomarkers of susceptibility: state of the art and implications for occupational exposure to engineered nanomaterials. Toxicol. Appl. Pharmacol. 299, 112–124.

Jirtle, R.L., Skinner, M.K., 2007. Environmental epigenomics and disease susceptibility. Nat. Rev. Genet. 8, 253–262.

Johnston, H.J., Verdon, R., Gillies, S., Brown, D.M., Fernandes, T.F., Henry, T.B., et al., 2018. Adoption of in vitro systems and zebrafish embryos as alternative models for reducing rodent use in assessments of immunological and oxidative stress responses to nanomaterials. Crit. Rev. Toxicol. 48, 252–271.

Kafshgari, M.H., Harding, F.J., Voelcker, N.H., 2015. Insights into cellular uptake of nanoparticles. Curr. Drug Deliv. 12, 63–77.

Keller, A., McFerran, S., Lazareva, A., Suh, S., 2013. Global life cycle releases of engineered nanomaterials. Nanopart. Res. 15, 1692–1709.

Kermanizadeh, A., Gosens, I., MacCalman, L., Johnston, H., Danielsen, P.H., Jacobsen, N. R., et al., 2016. A multilaboratory toxicological assessment of a panel of 10 engineered nanomaterials to human health—ENPRA project—the highlights, limitations, and current and future challenges. J. Toxicol. Environ. Health B Crit. Rev. 19, 1–28.

Klein, G., Mathe, C., Biola-Clier, M., Devineau, S., Drouineau, E., Hatem, E., et al., 2016. RNA-binding proteins are a major target of silica nanoparticles in cell extracts. Nanotoxicology 10, 1555–1564.

Kleinstreuer, N.C., Sullivan, K., Allen, D., Edwards, S., Mendrick, D.L., Embry, M., et al., 2016. Adverse outcome pathways: from research to regulation scientific workshop report. Regul. Toxicol. Pharmacol. 76, 39–50.

Labib, S., Williams, A., Yauk, C.L., Nikota, J.K., Wallin, H., Vogel, U., et al., 2016. Nano-risk Science: application of toxicogenomics in an adverse outcome pathway framework for risk assessment of multi-walled carbon nanotubes. Part. Fibre Toxicol. 13, 15.

Landesmann, B., 2016. Adverse Outcome Pathway on Protein Alkylation Leading to Liver Fibrosis. OECD Series on Adverse Outcome Pathways, No. 2. OECD Publishing.

Lieschke, G.J., Currie, P.D., 2007. Animal models of human disease: zebrafish swim into view. Nat. Rev. Genet. 8, 353–367.

Ma, C., White, J.C., Dhankher, O.P., Xing, B., 2015a. Metal-based nanotoxicity and detoxification pathways in higher plants. Environ. Sci. Technol. 49, 7109–7122.

Ma, Y., Zhang, P., Zhang, Z., He, X., Li, Y., Zhang, J., et al., 2015b. Origin of the different phytotoxicity and biotransformation of cerium and lanthanum oxide nanoparticles in cucumber. Nanotoxicology 9, 262–270.

Marguerat, S., Bahler, J., 2010. RNA-seq: from technology to biology. Cell. Mol. Life Sci. 67, 569–579.

Marmiroli, M., Pagano, L., Savo Sardaro, M.L., Villani, M., Marmiroli, N., 2014. Genome-wide approach in *Arabidopsis thaliana* to assess the toxicity of cadmium sulfide quantum dots. Environ. Sci. Technol. 48, 5902–5909.

Marmiroli, M., Imperiale, D., Pagano, L., Villani, M., Zappettini, A., Marmiroli, N., 2015. The proteomic response of *Arabidopsis thaliana* to cadmium sulfide quantum dots, and its correlation with the transcriptomic response. Front. Plant Sci. 6, 1104.

Marmiroli, M., Pagano, L., Pasquali, F., Zappettini, A., Tosato, V., Bruschi, C.V., et al., 2016. A genome-wide nanotoxicology screen of *Saccharomyces cerevisiae* mutants reveals the basis for cadmium sulphide quantum dot tolerance and sensitivity. Nanotoxicology 10, 84–93.

Maurer, L.L., Yang, X., Schindler, A.J., Taggart, R.K., Jiang, C., Hsu-Kim, H., et al., 2016. Intracellular trafficking pathways in silver nanoparticle uptake and toxicity in *Caenorhabditis elegans*. Nanotoxicology 10, 831–835.

Maynard, A.D., Aitken, R.J., Butz, T., Colvin, V., Donaldson, K., Oberdorster, G., et al., 2006. Safe handling of nanotechnology. Nature 444, 267–269.

McWilliams, A., 2016. The Maturing Nanotechnology Market: Products and Applications. BCC Research, Wellesley, MA.

Miclăuş, T., Bochenkov, V.E., Ogaki, R., Howard, K.A., Sutherland, D.S., 2014. Spatial mapping and quantification of soft and hard protein coronas at silver nanocubes. Nano Lett. 14, 2086–2093.

Milani, S., Bombelli, F.B., Pitek, A.S., Dawson, K.A., Radler, J., 2012. Reversible versus irreversible binding of transferrin to polystyrene nanoparticles: soft and hard corona. ACS Nano 6, 2532–2541.

Naatz, H., Lin, S., Li, R., Jiang, W., Ji, Z., Chang, C.H., et al., 2017. Safe-by-design CuO nanoparticles via Fe-doping, Cu–O bond length variation, and biological assessment in cells and zebrafish embryos. ACS Nano 11, 501–515.

Nel, A.E., Malloy, T.F., 2017. Policy reforms to update chemical safety testing. Science 355, 1016–1018.

Nel, A., Xia, T., Meng, H., Wang, X., Lin, S., Ji, Z., et al., 2013a. Nanomaterial toxicity testing in the 21st century: use of a predictive toxicological approach and high-throughput screening. Acc. Chem. Res. 46, 607–621.

Nel, A.E., Nasser, E., Godwin, H., Avery, D., Bahadori, T., Bergeson, L., et al., 2013b. A multi-stakeholder perspective on the use of alternative test strategies for nanomaterial safety assessment. ACS Nano 7, 6422–6433.

OECD, 2013. Revised guidance document on developing and assessing adverse outcome pathways. In: Environment, Health and Safety Publications Series on Testing and Assessment No. 184. ENV/JM/MONO(2013)6.

OECD, 2018. Users' handbook supplement to the guidance document for developing and accessing AOPs. In: Environment, Health and Safety Publications Series on Testing and Assessment No. 233 ENV/JM/MONO(2016)12.

Pagano, L., Serving, A.D., De La Torre-Roche, R., Mukherjee, A., Majumdar, S., Hawthorne, J., et al., 2016. Molecular response of crop plants to engineered nanomaterials. Environ. Sci. Technol. 50, 7198–7207.

Pagano, L., Pasquali, F., Majumdar, S., De La Torre-Roche, R., Zueva-Mena, N., Villani, M., et al., 2017. Exposure of *Cucurbita pepo* to binary combinations of engineered nanomaterials: physiological and molecular response. Environ. Sci.: Nano 4, 1579–1590.

Rauscher, H., Rasmussen, K., Sokull-Klüttgen, B., 2017. Regulatory aspects of nanomaterials in the EU. Chem. Ing. Tech. 89, 224–231.

Rocheleau, S., Arbour, M., Elias, M., Sunahara, G.I., Masson, L., 2015. Toxicogenomic effects of nano- and bulk-TiO_2 particles in the soil nematode *Caenorhabditis elegans*. Nanotoxicology 9, 502–512.

Ruotolo, R., Pira, G., Villani, M., Zappettini, A., Marmiroli, N., 2018. Ring-shaped corona proteins influence the toxicity of engineered nanoparticles to yeast. Environ. Sci.: Nano 5, 1428–1440.

Russell, W.M.S., Burch, R.L., 1959. The Principles of Humane Experimental Technique. Universities Federation for Animal Welfare, Wheathampstead, UK.

Sato, F., Tsuchiya, S., Meltzer, S.J., Shimizu, K., 2011. MicroRNAs and epigenetics. FEBS J. 278, 1598–1609.

Selck, H., Handy, R.D., Fernandes, T.F., Klaine, S.J., Petersen, E.J., 2016. Nanomaterials in the aquatic environment: A European Union–United States perspective on the status of ecotoxicity testing, research priorities, and challenges ahead. Environ Toxicol Chem 35, 1055–1067. Available from: https://doi.org/10.1002/etc.3385.

Shah, S.N., Shah, Z., Hussain, M., Khan, M., 2017. Hazardous effects of titanium dioxide nanoparticles in ecosystem. Bioinorg. Chem. Appl. 2017, 4101735.

Shoshan-Barmatz, V., Krelin, Y., Shteinfer-Kuzmine, A., Arif, T., 2017. Voltage-dependent anion channel 1 as an emerging drug target for novel anti-cancer therapeutics. Front. Oncol. 7, 154.

Smith, M.R., Boenzli, M.G., Hindagolla, V., Ding, J., Miller, J.M., Hutchison, J.E., et al., 2013. Identification of gold nanoparticle-resistant mutants of *Saccharomyces cerevisiae* suggests a role for respiratory metabolism in mediating toxicity. Appl. Environ. Microbiol. 79, 728–733.

Srinivas, A., Rao, P.J., Selvam, G., Goparaju, A., Murthy, P.B., Reddy, P.N., 2012. Oxidative stress and inflammatory responses of rat following acute inhalation exposure to iron oxide nanoparticles. Hum. Exp. Toxicol. 31, 1113–1131.

Sturla, S.J., Boobis, A.R., FitzGerald, R.E., Hoeng, J., Kavlock, R.J., Schirmer, K., et al., 2014. Systems toxicology: from basic research to risk assessment. Chem. Res. Toxicol. 27, 314–329.

Tiwari, D.K., Jin, T., Behari, J., 2011. Dose-dependent in-vivo toxicity assessment of silver nanoparticle in Wistar rats. Toxicol. Mech. Methods 21, 13–24.

Tong, T., Wilke, C.M., Wu, J., Binh, C.T., Kelly, J.J., Gaillard, J.F., et al., 2015. Combined toxicity of nano-ZnO and nano-TiO_2: from single-to-multinanomaterial systems. Environ. Sci. Technol. 49, 8113–8123.

Vance, M.E., Kuiken, T., Vejerano, E.P., McGinnis, S.P., Hochella Jr., M.F., et al., 2015. Nanotechnology in the real world: redeveloping the nanomaterial consumer products inventory. Beilstein J. Nanotechnol. 6, 1769–1780.

Vietti, G., Lison, D., van den Brule, S., 2016. Mechanisms of lung fibrosis induced by carbon nanotubes: towards an adverse outcome pathway (AOP). Part. Fibre Toxicol. 13, 11.

Villeneuve, D.L., Crump, D., Garcia-Reyero, N., Hecker, M., Hutchinson, T.H., LaLone, C.A., et al., 2014. Adverse outcome pathway development II: best practices. Toxicol. Sci. 142, 321–330.

Walczyk, D., Bombelli, F.B., Monopoli, M.P., Lynch, I., Dawson, K.A., 2010. What the cell "sees" in bionanoscience. J. Am. Chem. Soc. 132, 5761–5768.

Wang, Z., Gerstein, M., Snyder, M., 2009. RNA-Seq: a revolutionary tool for transcriptomics. Nat. Rev. Genet. 10, 57–63.

Wang, Z., Zhang, L., Zhao, J., Xing, B., 2016. Environmental processes and toxicity of metallic nanoparticles in aquatic systems as affected by natural organic matter. Environ. Sci.: Nano 3, 240.

Wang, L., Hu, C., Shao, L., 2017. The antimicrobial activity of nanoparticles: present situation and prospects for the future. Int. J. Nanomed. 12, 1227–1249.

Waters, M.D., Fostel, J.M., 2004. Toxicogenomics and systems toxicology: aims and prospects. Nat. Rev. Genet. 5, 936–948.

Wen, R., Banik, B., Pathak, R.K., Kumar, A., Kolishetti, N., Dhar, S., 2016. Nanotechnology inspired tools for mitochondrial dysfunction related diseases. Adv. Drug Deliv. Rev. 99, 52–69.

Yanamala, N., Kagan, V.E., Shvedova, A.A., 2013. Molecular modeling in structural nano-toxicology: interactions of nano-particles with nano-machinery of cells. Adv. Drug Deliv. Rev. 65, 2070–2077.

Zhang, X.Q., Xu, X., Bertrand, N., Pridgen, E., Swami, A., Farokhzad, O.C., 2012. Interactions of nanomaterials and biological systems: implications to personalized nanomedicine. Adv. Drug Deliv. Rev. 64, 1363–1384.

CHAPTER

Innovation in nanomedicine and engineered nanomaterials for therapeutic purposes

9

Maricla Galetti[1,2], Stefano Rossi[2], Cristina Caffarra[2], Amparo Guerrero Gerboles[2] and Michele Miragoli[2,3]

[1]*Center of Excellence for Toxicological Research, INAIL, University of Parma, Parma, Italy*
[2]*Department of Medicine and Surgery, University of Parma, Parma, Italy*
[3]*Institute of Genetic and Biomedical Research, IRGB-CNR, Milan, Italy*

9.1 Introduction

9.1.1 Evolution of engineered nanomaterials for therapeutic purposes

In the last decades, great efforts have been made for proving novel carriers able to transport specific molecules in targeted organs. Nanomedicine is considered a revolution process for reinventing pharmacology and medicine with an incredible potential in the development, treatment, and following noncommunicable and communicable diseases [e.g., cancer, cardiovascular diseases (CDs), and infection] (Coty and Vauthier, 2018). It is logic that the production and the market of engineered nanomaterials (ENMs) take particular attention in the medical field with a perspective in reaching 350.8 billion dollars in 2025 (https://www.grandviewresearch.com/press-release/global-nanomedicine market).

The ENM workforce is estimated to grow ~15% annually (Schubauer-Berigan et al., 2011). This is because a "nanocarrier" denotes several innovations compared to conventional pharmacological treatment: (1) can be either invisible to the immune system (Corbo et al., 2016); (2) can cross several physiological barriers [pulmonary (Choi et al., 2010), blood–brain (Garcia-Garcia et al., 2005), gastrointestinal tract (Pietroiusti et al., 2017)]; and (3) can protect the molecular compound without special excipients that can be contraindicated in the presence of chronic pathologies (e.g., sugar for diabetes). In several cases the ENMs are themselves drugs in producing antibacterial or anticancer therapeutic effects [e.g., Ag^+ nanoparticles (NPs) (Siddiqi et al., 2018)]. Still, we have to keep in mind that ENMs are foreign (nano)bodies in our organisms; therefore a

Exposure to Engineered Nanomaterials in the Environment. DOI: https://doi.org/10.1016/B978-0-12-814835-8.00009-1

nanotoxicological assessment is always imperative before reaching a preclinical or clinical investigation using ENMs.

In this chapter, we highlighted what we need to take care of when ENMs are using in medicine. We move then toward the innovations in nanomedicine, in the field of cancer, CDs, regenerative medicine, and medical imaging. Finally, we will touch the upcoming and promising directions of personalized nanomedicine and how this field can be enlarged in the context of engineered medicine.

9.1.2 Engineered nanoparticles for nanomedicine

NPs are the armamentarium of therapeutics aiming to fight a plethora of diseases; however, still NPs are foreign bodies, some of them are not biodegradable and as such need to be deeply investigated. Sa et al. (2012) labeled micelles, NPs, and mesoporous silica loaded with different drugs (tamoxifen, hydroxyapatite); they found a heterogeneous biodistribution after injection of ENMs that strictly depend on their nature and cargo. Thus understanding the occupational, safety, and risk of ENMs is imperative. ENM toxicity has been also attributed to residual metals, but more importantly, the physicochemical properties of ENMs and their transformation in the body may affect their uptake. When ENMs are not sufficiently characterized or tested in vitro and in vivo, the prediction of toxicity is undervalued (Liu et al., 2008).

9.1.2.1 What we need to take care of?

The architecture of the ENMs plays a role in cellular uptake, flow, and redistribution. Chitiriani et al. (2006) showed that spherical NPs are better internalized from HeLa cells. On the other hand, rod-shaped mesoporous silica NPs have better uptake rate for human melanoma cells (Huang et al., 2010). The shape of the ENMs is important when injected intravenously; the flow dynamics of different-sized spherical, elliptical, circular, or rod-shaped particles has been studied by Doshi et al. (2010). The rod-shaped particles, for instance, appear to have higher adhesion properties. The effect of the particle's shape is prominent when you compare with large particles or ENMs with different electrical surface charge.

Recently, we showed an interaction between the electrical surface charged of the NPs and excitable cells, such as cardiomyocyte, possibly reflecting the excitable nature of the bioelectricity passing through the heart (Savi et al., 2014). NPs that denote a positive Z-potential and thus positively charged are highly cytotoxic, disrupt the cell membrane, and induce apoptosis in cardiomyocytes (Miragoli et al., 2013). This has been observed by amino-modified NPs (Fig. 9.1) in both adult and neonatal rat ventricular myocytes. Before the death of the cells, arrhythmia was also induced during early exposure (2 hours, Fig. 9.1). This scenario is accompanied by a significant dose–response increment on percentage of intracellular calcium transient alternance.

On the other hand the negative charged NPs can be gentler or ruder to the cells depending on the dose and exposure. The carboxyl-modified polystyrene

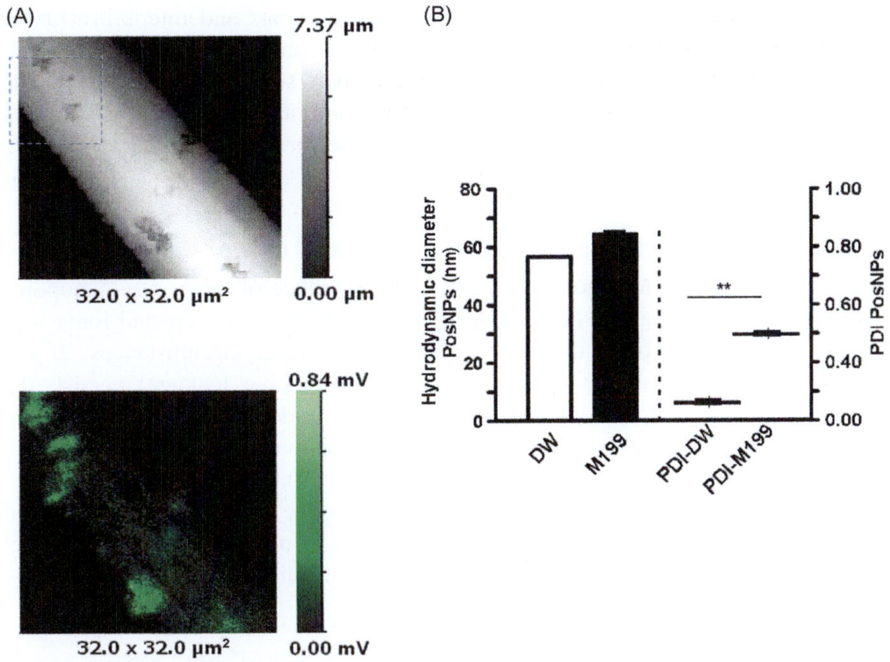

FIGURE 9.1 Physicochemical characteristics of positively charged nanoparticles and their interactions with cardiac cell membrane.

(A) (Top) SICM topography of an adult rat ventricular myocyte subjected to 50 nm amino-modified positive charged nanoparticles. (Bottom) Same as (A) topography obtained by surface confocal SICM. Nanoparticles are labeled in *green* (gray in print version). To note the cell membrane is disrupted. (B) Hydrodynamic diameters and polydispersion index of the same NPs. *DW,* Distilled water; *M199,* medium; *SICM,* scanning ion conductance microscopy.

Modified from Miragoli et al., 2013 (14) with permission.

latex NPs induced temporary nanopores (same diameter as amino-modified NPs) which are compatible with life (Miragoli et al., 2013).

We also showed that size plays a pivotal role in this context. While the internalization is always dependent by clathrin and dynamin (Novak et al., 2014) for particle with aerodynamic diameter <200 nm, the internalization is clathrin-mediated pinocytosis. Thanks to the scanning ion conductance microscopy (Miragoli et al., 2011; Shevchuk et al., 2011) and other techniques (Miragoli et al., 2013; Novak et al., 2014), we showed that in vitro negatively charged NPs are internalized in a couple of minutes thanks to the life-compatible nanopores.

Moreover, in biological systems, NPs are covered by biological molecules from the bloodstream (DNA, metabolites, and proteins) leading to the formation of "protein corona." These biologically modified NPs play an important role in nano–bio interactions, and they affect cell binding events and internalization of NPs. Therefore NPs are formed not only by polymers and drugs but are also associated with biological molecules adsorbed onto their surfaces. Thus beyond the intrinsic architecture of ENMs, it is important to study and take into consideration these interactions for their safety functionality performances (Pederzoli et al., 2017).

9.1.2.2 Negatively charged nanoparticles: friend or foe?

Recently, we showed (Miragoli et al., 2013) the presence of temporary nanopores made by negatively charged NPs; such membrane perturbation created ionic leak that can be restored from the membrane under certain circumstances. If the nanopores-caused membrane opening (and therefore ionic leakage) persist, the ionic imbalance will determine the cell death. We understood this phenomenon when we tracheally instilled rats with TiO_2 NPs (Savi et al., 2014). Titanium dioxide NPs exhibit a negative *Z*-potential (c. − 19 mV), while TiO_2 NPs are normally considered inert; we showed that TiO_2 can be found in the heart and increase the susceptibility to arrhythmias (Fig. 9.2). Moreover, if we can control the physicochemical characteristics, the biocompatibility and the nanotoxicity,

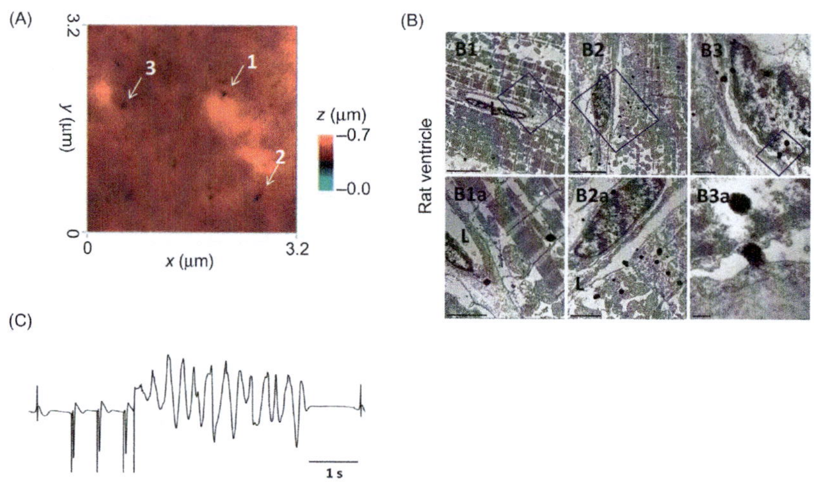

FIGURE 9.2 Physicochemical characteristics of negatively charged nanoparticles and their interaction with cell membrane.

(A) SICM images of a ventricular myocyte subjected to 50 nm carboxyl-modified negative charged. (B) Transmission electron microscopy images of rare cardiac tissue subjected by negatively charged TiO_2 NPs after tracheal instillation. B1a, B2a, B3a referred to the areas zoomed in B1, B2, and B3, respectively. Scale bar 100 nm. (C) Induced ventricular fibrillation in vivo in TiO_2 NPs exposed rats using S1–S2 stimulation protocol. Scale bar: 1 s. *NPs*, Nanoparticles; *SICM*, scanning ion conductance microscopy.

Modified from (13) and (14) with permission.

negatively charged NPs can be used as carriers for treating CDs (Di Mauro et al., 2016; Miragoli et al., 2018) (Section 9.1.6).

When Paracelsus said "is the dose that make[s] the poison" he was certainly right; in nanomedicine also the morphology, and the behavior of the nanocarrier (loaded/unloaded) within the body as well as the nanocarrier–tissue interaction play a key role.

9.1.3 Organ specificity

Injectable NPs in the bloodstream can theoretically reach all organs; however, it was observed that the biodistribution is not uniform among anatomical districts (Di Mauro et al., 2016; Li et al., 2016; Spigoni et al., 2015). Taking into account the physicochemical characteristics for ENMs, there are other two main reasons, by which intravenous (i.v.) administration of ENMs targets specific organs: (1) the presence of physiological barrier and (2) the organ specificity.

In light to protect itself from diverse pathogenic substrate, the body has developed several physiological barriers that limit the entry of potential hazardous foreign bodies (Braakhuis et al., 2015). This needs to be taken into consideration when novel ENMs for nanomedicine are produced with specific aim. We propose that for cardiomyocytes the Coulomb force, that is, the attraction between two charged bodies, may occur especially from excitable cells and charged NPs. This can be explained as charged NPs, like i.v. administered calcium phosphate (CaP), can attract preferentially to the heart and to the kidney but cannot be detected in the brain due to the blood–brain barrier (Miragoli et al., 2018). Similarly, discoidal NPs seem to accumulate less in the liver but can reach the lungs and other organs (Morachis et al., 2012). The organ specificity should also be important in the context of medical imaging. Paramagnetic NPs are widely used as an agent for magnetic resonance imaging (MRI), but if they can be labeled properly, they can be used for the characterization of specific region of the organ such as pediatric brain tumors (Dumont et al., 2014) or Vx2 tumors (Schmieder et al., 2013). The cargo itself can contribute to the organ specificity; this can be obvious for drugs or molecule that can exert an effect only into a specific organ. It is less obvious if a given peptide concentration can influence the organ specificity. Harris et al. (2010) demonstrate that particle coated with 2.5:1 peptide:DNA weight/ratio in the presence of serum facilitates the release of the gene to the liver; on the other hand the same particle but coated with 20:1 peptide:DNA weight/ratio facilitates the delivery to the spleen and to the bone marrow.

9.1.4 Nanomedicine in cancer therapy

The majority of nanomedicines currently marketed or in development are for cancer indications (Hare et al., 2017; Shi et al., 2017; Wicki et al., 2015), although nanomedicines to treat different diseases are being placed successfully, for example, CD, diabetes and other immune disorders, including rheumatoid arthritis, and asthma and allergic conditions (Fonseca-Santos et al., 2015; Godin et al., 2010; Schiener et al., 2014; Takedatsu et al., 2015; Veiseh et al., 2015).

In conventional oral or i.v. drug delivery of chemotherapy the drug is distributed indiscriminately throughout the body, with different concentrations reaching both the disease site and the healthy tissue. Consequently, chemotherapy causes adverse effects on healthy tissue and requires a trade-off between optimal disease treatment and patient's quality of life. In contrast, NP-based drug-delivery systems offer the possibility to develop highly effective target therapy with improved circulation half-life, bioavailability, biodistribution, pharmacokinetic, and safety profiles. The pharmacokinetics and biodistribution of nanomedicines are independent from the molecular structure of the payload, overcoming some of the limitations of traditional medicines. Moreover, NPs are indispensable in maintaining synergistic drug ratios in combination therapy and offer the first possibility of delivering therapeutic agents such as nucleic acids and unstable proteins.

9.1.4.1 Anticancer nanomedicines in preclinical and clinical development

Nanotechnology has made important contributions to oncology over the past several decades. Although only a relatively small number of nanosized drug-delivery carriers have been approved for human use, nanotechnologies will likely constitute a growing share of the oncologist's therapeutic arsenal over the next years to come. In parallel, nanotechnologies will also be appealing because they could facilitate the combination regiments that are routinely used in cancer therapy. The timely codelivery to cancer cells of drugs combined in a single NP carrier could provide improved synergistic action among the multiple active pharmaceutical ingredients (Valencia et al., 2013; Xu et al., 2013).

Currently, there are considerable arguments in favor of developing nanosized therapeutics. First, NPs may help to overcome problems of solubility and chemical stability of anticancer drugs. For example, many promising drugs such as Wortmannin failed clinical development because they did not meet solubility and chemical requirements, but with the new advancement in nanomedicine, this problem was resolved using a lipid-based nanocarrier (Karve et al., 2012).

Second, a nanocarrier can protect anticancer drugs from biodegradation or excretion, changing the bioavailability and pharmacokinetic of the therapeutic agents. Indeed, nanocarriers provide an additional protection of anticancer agent, because the drug encapsulation protects them from the environment and the immune system.

Third, nanotechnology could improve distribution and targeting of anticancer drugs, increasing drug penetration into tumor tissue. We will discuss in detail this argument in Section 9.1.4.2.

Fourth, nanocarriers can release their payload only upon a trigger in the neoplastic tissue, minimizing systemic exposure to the compound. Triggers can be divided into internal (pathophysiological/pathochemical condition) and external (physical stimuli such as temperature, light, magnetic force, and electric field) stimuli. For example, drugs whose delivery is not pH dependent, such as doxorubicin, can be conjugated with a pH sensitive NP to increase cellular drug uptake (Du et al., 2011).

Finally, another important benefit is the ability to achieve the "right safety" profile. Actually, the maximum tolerated dose of the active agents could be increase by avoiding the tolerability problems cause by the solubilizing excipients. For example, high dose of paclitaxel can be administrated to patients using Abraxane (Celgene) or the polymeric micelle formulation Genexol-PM (Samyang Biopharmaceuticals), because these formulations avoid the use of Cremophor (polyoxyethylated castor oil), the lipid-based solvent used as a vehicle to formulate Taxol (Green et al., 2006).

Anticancer nanomedicines in clinical development comprise mainly polymer therapeutics (polymer–protein and polymer–drug conjugates), where the drug is covalently bound or conjugated to a polymeric structure, and particulate drug nanocarriers, where the drug is physically entrapped. In particular, in the last class, drugs are entrapped within molecular assemblies with different structures made from different materials, such as polymers (polymeric micelles, dendrimers, and polymeric NP), lipids (liposome), or organometallic compounds [carbon nanotubes (CNTs)].

Polymer therapeutics are defined as nanoscale linear water-soluble polymeric molecules conjugated to antitumor proteins or small molecule anticancer drugs, via cleavable linkers that are stable during the transport of the drugs and release these compounds in the tumors (Duncan, 2006). Polymer proteins have the capability to protect drugs (small molecules, proteins, nucleic acids, or peptides) from hepatic inactivation, enzymatic degradation, and rapid clearance in vivo, increasing their stability and time circulation in blood. Otherwise, in the case of polymer–drug conjugation, polymers give cytotoxic drugs an increased circulation time in blood, an improved aqueous solubility, a passively delivery to tumor areas, and a reduced toxicity, improving the therapeutic effect (Bildstein et al., 2011; Duncan, 2006). Both of them are considered as new chemicals rather than drug carriers, because they have a reduced payload and a restricted capacity of active targeting, for the reason that they presented a limited number of conjugation sites.

The most well established polymer is poly(ethylene glycol) (PEG). One polymer–protein conjugates, which has been FDA-approved since 1994, is the extremely popular PEG-L-asparaginase, used for the treatment of acute lymphoblastic leukemia (Dinndorf et al., 2007; Masetti and Pession, 2009). From this approval, numerous clinical trials of polymer–drug conjugates have been or are being performed. Those include enzymes [arginase (Cheng et al., 2007; Yau et al., 2013), glutaminase] and biological response modifiers [interleukin-2 (Bukowski et al., 1993)], interferon-α, but there are only few FDA approvals for cancer treatment. They are an interesting group of nanosized (5–20 nm) drug-delivery systems, because the change of the pharmacokinetic profile of a drug (Duncan, 2006). Today there are at least 20 conjugates currently in clinical trials, based on traditional cytotoxic drugs [doxorubicin, paclitaxel/docetaxel (Langer et al., 2008)]. The *N*-(2-hydroxypropyl)methacrylamide (HPMA)−doxorubicin conjugate was the first synthetic polymer−drug conjugate to enter in clinical trials in 1994, and today, a phase II study show promising signs of activity of HPMA copolymer−doxorubicin conjugate PK1 in breast cancer and non−small-cell lung cancer (NSCLC) (Seymour et al., 2009).

Otherwise, particulate drug nanocarriers are nanosized molecular structures that provide an additional protection of the anticancer agents over the polymeric conjugates since the anticancer drugs are isolated from the environment and immune system. In addition, respect to polymer therapeutics, they provide the possibility to control drug release. Particulate drug nanocarriers include liposomes, CNTs, and polymeric nanocarriers (polymeric micelles, dendrimers, and polymeric NP).

Liposomes are self-assembled colloidal vesicles with a closed phospholipid bilayers, which not only allow the encapsulation on different hydrophilic anticancer drugs and siRNA in their aqueous core but also can host hydrophobic cytotoxic agents in their hydrophobic membrane. Lipid nanocarriers exhibit longtime circulation in blood, which can be enhanced by conjugation of PEG to the liposome surface (Infante et al., 2012), and can be used as active targeting carriers by bounding monoclonal antibody to the surface of liposome [immunoliposomes (ILs)] (Carter, 2001; Kontermann, 2006). PEGylated liposomal doxorubicin (Doxil) was the first nanocarrier approved by FDA in 1995 (Harrison et al., 1995) and it is a well-studied example of altered pharmacokinetic and toxicity: the liposomal formulation of doxorubicin markedly improves the critical cardiotoxicity of free doxorubicin.

Despite the abovementioned advantages of liposomal nanocarriers, only five lipid nanocarriers are approved for the clinical use: non-PEG liposomal doxorubicin (Myocet) (Chan et al., 2004), non-PEG liposomal daunorubicin (DaunoXome) (Gill et al., 1996), non-PEG liposomal cytarabine (DepoCyt) (Gokbuget et al., 2011), vincristine sulfate liposomes (Marqibo) (O'Brien et al., 2013), and liposomal mifamurtide (Mepact) (Frampton, 2010). Also in these cases, the nanoformulations of doxorubicin, daunorubicin, and vincristine prolonged the half-life of the cytotoxic compounds and profoundly improved their toxicity profiles.

Furthermore, liposomes are also being investigated for systemic delivery of nucleic acids or synthetic short interfering siRNA that silence certain genes involved in cancer progression or plasmid coding cDNAs for the recovery of tumor suppressor gene (Li and Rana, 2014; Pecot et al., 2011). Although none of these nanoformulation drugs have been approved for clinical use yet, efficacy of several siRNA-loaded liposomal formulations has been demonstrated in delivery in phase I clinical trials (Coelho et al., 2013; Tabernero et al., 2013).

CNTs are tubular hydrophobic networks of carbon atoms with a diameter of approximately 1−4 nm; 1−100 μm in length; and unique structural, electronic, optimal, and mechanical properties (Sinha and Yeow, 2005). Different anticancer drugs can be included in their inner cavity or in their surface, with a high payload due to their ultrahigh surface area (Ajima et al., 2008). No FDA approvals are up to date using CNTs, although preclinical results in vitro and in vivo encourage cancer treatment with these nanoformulation drugs. For example, using a first line cytotoxic agent (cisplatin) and a targeting moiety epidermal growth factor (EGF), conjugated to the sidewall of single-walled nanotubes (SWNTs), it has been reported an enhanced efficacy to target squamous cancer cells overexpressing EGF receptor (EGFR). Moreover, an increasing uptake of nanotubes by the tumor areas and an augmented efficacy to reduce the size of head and neck tumors, compared to nanotubes without EGF molecules, were reported (Bhirde et al., 2010). This work is an example of multiple functionalization of the sidewall of SWNTs that permit to use several molecules with different functions in the same carriers, being very promising for cancer treatment.

Despite the numerous advantages and clinical benefits of liposomes, they showed problems of stability and limited control of drug release. Otherwise, polymeric nanocarriers overcome these problems, since they are stable in vivo, show increased circulation time in blood, higher loading capacities, and produce more controlled drug release profile. Polymeric nanocarriers are polymer-based nanodrugs with different possible structures: amphiphilic core/shell (polymeric micelles), hyperbranched macromolecules (dendrimers), or capsule/particles (polymeric NPs). Polymeric micelles are composed of amphiphilic block copolymers that form nanosized micellar structures with a hydrophobic core (with inside anticancer drugs) and a hydrophilic shell, which provides stability to the micelle. In spite of the promising characteristics of the polymeric micelles, there are only eight polymeric micelle-based formulations in clinical trials (Gong et al., 2012), because of insufficient stability in systemic circulation and the premature drug leakage.

Polymeric NPs are biodegradable colloidal systems where cytotoxic drug can be encapsulated within a polymeric matrix (nanospheres) or entrapped into a cavity surrounded by a polymeric membrane (nanocapsules). Because of their chemical versatility, they are a promising tool for nanometric therapeutics. Moreover, respect to other nanocarriers, they show more homogenous size distribution, better controllable physicochemical properties, higher drug payload, and more controlled drug release (Hu et al., 2010). However, there are few polymeric NPs available on the market and used for the therapy of breast cancer, NSCLC and pancreatic cancer, such as albumin NP–bound paclitaxel (nab-paclitaxel, Abraxane) (Desai et al., 2006). By incorporation of paclitaxel (a cytotoxic compound poorly water-soluble) into an albumin NP, drug solubility is improved; moreover, albumin may mediate transcytosis of the compound on the endothelium and thus enhance its delivery to tumors.

However, there remain much to be learned in the emerging field of nanomedicine and it is necessary to develop a nanocarrier that can effectively deliver a payload into the tumors with clinically relevant results.

9.1.4.2 The enhanced permeability and retention effect on nanotherapeutic efficacy

The preferential accumulation of NPs in solid tumors is generally ascribed to defective tumor vessels, to an extensive amounts of various vascular permeability factors (such as VPF/VEGF) (Baban and Seymour, 1998) and to a poor lymphatic drainage system (Fang et al., 2011; Maeda et al., 2013; Perez-Herrero and Fernandez-Medarde, 2015) (cf. Fig. 9.3). Overall, the general effect is that macromolecular particles or nanomedicines are entrapped in tumors and their clearance is delaying. This important phenomenon, called the enhanced permeability and

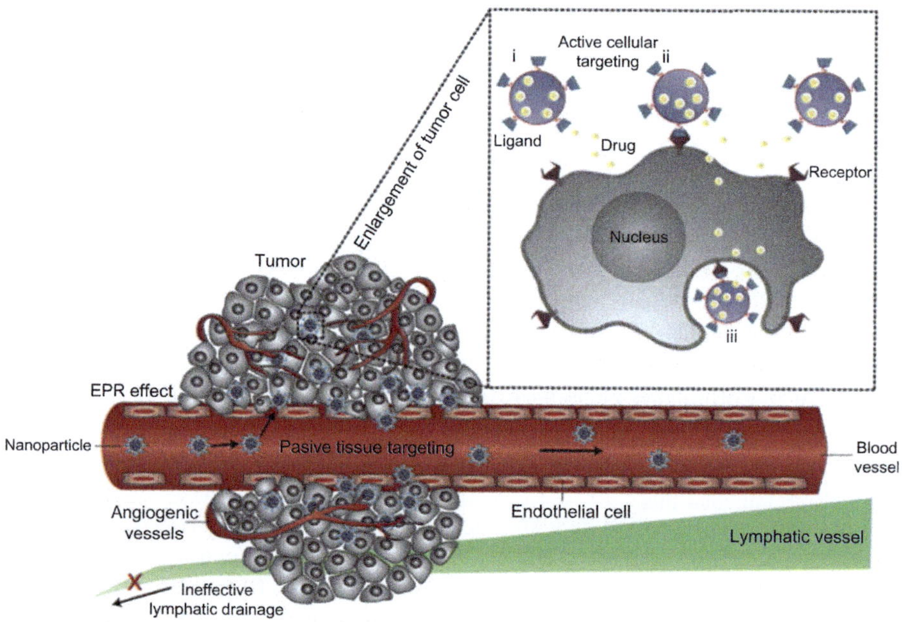

FIGURE 9.3

Nanocarriers as promising transporters of anticancer drugs to tumors by passive tissue targeting and active cellular targeting. Passive tissue targeting uses the increased permeability of tumor vasculature and the poor lymphatic drainage of tumors (EPR effect), allowing the release of chemotherapeutic agents near the tumor. Active cellular targeting is achieved by functionalization of the surface of nanocarriers, containing chemotherapy drugs, with targeting moieties that provide selective recognition of different receptors or antigens overexpressed in cancerous cells, increase their therapeutic efficacy, and overcomes the multiple drug resistance. Nanocarriers, once in the vicinity of the tumor, can: (1) release their cytotoxic content next to the cancer cells; (2) bind to the membrane of the cancer cells and release their content in a sustained way; (3) be internalized into the cells. *EPR*, Enhanced permeability and retention.

Reprinted from (75).

retention (ERP) effect, is becoming an increasingly promising paradigm for anticancer development. It was first proposed in 1986 as a new concept of delivery of drugs in tumors (Matsumura and Maeda, 1986), relative to normal tissue, and it elegantly explains the fact that solid tumors retain more polymeric NPs, proteins, liposomes, and micelles than other tissues, after i.v. administration.

While in normal vasculature tight junctions prevent particles larger than 2 nm to extravasate near the interstitium, in tumor tissue the permeable vasculature and disordered basement membrane lead to preferential accumulation of entities of 10–500 nm in size near the tumors. This passive tissue targeting is used to increase the release of nanotherapeutics in tumors, achieving a cytotoxic concentration several folds higher in tumors with a reduced toxicity for the rest of the body (Bertrand et al., 2014; Torchilin, 2011). This size effect has also been associated with improved efficacy (Cabral et al., 2011). Therefore an ideal nanocarrier must not exceed 500 nm to escape from mononuclear phagocyte system and the reticuloendothelial system (mainly liver and spleen) and to achieve the extravasation into tumors by the EPR effect, which is more effective with diameters below 200 nm (Maeda, 2012). In addition, because blood vessels and cells contains negatively charged surface constituents that might reject nanocarriers with negatively charged surface, hydrophilic and neutral surface modifications must be considered during nanotherapeutic drug development.

Nevertheless, it is too premature to generalize this size dependency, which is likely tumor and nanomedicine dependent. Furthermore, it is increasingly clear that EPR effect varies substantially between patients and also between tumor types. Many experimental and clinical data have demonstrated that tumor tissue changes in pathological characteristics, sizes (from size less than 1 mm to more than 10 cm), etiologies, and vascular densities (Bertrand et al., 2014). For example, hepatocellular carcinoma and renal cell carcinoma have a high vascular density, so they tend to have a reasonably augmented EPR effect. Otherwise, pancreatic and prostate cancer have a low vascular density and this condition affects negatively the EPR effect.

Therefore if until now tumor enhanced permeability and retention (ERP) is a driving principle in the design characteristics of a nanomedicine, then it will be essential to treat tumors that present an EPR effect modulating the biological conditions driving EPR variability. This is particularly important since only a few first-generation anticancer medicines really improve overall survival in patient cohorts, mainly because subpopulation with higher susceptibility to NPs may be masked by nonresponsive patients (Prabhakar et al., 2013). If such subpopulations exist, defining predictive biomarkers for EPR would significantly affect the clinical outcome and must be considered when optimizing nanomedicine systems.

In addition, most macromolecular drugs (including nucleic acids and some proteins) cannot really permeate through the cell membrane and reach their pharmacological target. For these reasons, it is possible to achieve an improved therapeutic efficacy respect to passive tissue targeting (EPR effect) by the conjugation of different moieties, such as monoclonal antibodies, antibody fragments, peptides, small molecules, and aptamers, to the surface of nanocarriers containing chemotherapy drugs (Fig. 9.3). This active cellular targeting is being explored as a method to concentrate a therapeutic nanomedicine at an active diseased site, producing a preferred distribution profile. Targeted molecules can be proteins, sugar, or lipid proteins present in diseased organs or on the surface of cells (Yu et al., 2010). The specific ligand–receptor interaction can be utilized increase interactions between NPs and cells, enhancing internalization of drugs without altering the overall biodistribution.

In several preclinical experiments the surface of liposome was coated with anti-HER2, anti-EGFR, anti-VEGFR2, or other antibodies. The tumor masses treated with anti-VEGFR2 and doxorubicin-loaded ILs, shrank to one-sixth of those treated with the same dose of nontarget liposomal doxorubicin (Mamot et al., 2012). This illustrates the potency of target nanodrugs compared to their nontarget counterparts.

One example in clinical trials are doxorubicin-loaded target ILs against the EGFR (Mamot et al., 2005, 2012). Anti-EGFR-ILs were given intravenously every 4 weeks for a total of six cycles or until toxicity or progression appeared. In this study, they observed some preliminary signs of clinical activity, including a patient with a complete remission among 24 evaluated patients. Coating of liposomes with antibody did not exhibit any cardiac effect, a critical side effect observed in doxorubicin treated patients; concurrently, with regard to toxicity, anti-EGFR-ILs mimic the favorable cardiotoxicity profile of liposomal doxorubicin.

At the same time, decoration of nanocarriers with target moieties overcomes the multidrug resistance in cancer cells, a situation where transporter proteins, overexpressed on the surface of cancer cells, expel drugs from cells. Expelling drugs inevitably lowers the therapeutic effect and cancer cells soon develop resistance to a variety of drugs. On the contrary, it has been reported that membrane transporter such as *P*-glycoprotein or multidrug resistant protein 1 (MRP-1), which expel anticancer agents outside from the cells, are bypassed by target ILs (Huwyler et al., 2002).

Currently, actively targeted NPs are envisioned as a promising complementary strategy to EPR to further achieve the efficacy of cancer nanotherapy. Nevertheless, the design of actively target NP drug carriers is complex because NP architecture, physicochemical properties (e.g., ligand density and size of the NPs), and the choice of the targeting ligand may affect the efficacy of the active targeting strategy both in vitro and in vivo. In addition, measuring the impact of active targeting in humans is not simple, requires time and resources, and the parallel evaluation of matching targeted and nontargeted NP remains therefore unlike.

All together, these elements suggest a better understanding of how these nanomedicines work to maximize the benefits for patients and to positively impact the same way as nanotechnology has influenced other scientific fields, such as electronic, energy, and material sciences.

9.1.5 Image-guided therapeutic delivery

Great strides have been done in medicine in the past few decades. We are in the era of precision medicine, where research disciplines and clinical practices are integrated to guide individualized patient care. Medical imaging and targeting therapy are very important components of medicine care. On one side, medical imaging can be used in the diagnosis and monitoring of several diseases (i.e., cancer), and on the other, targeting therapy can offer precise disease treatment, maximize intervention efficacy, and reduce the side effects of the drug. Thanks to their unique size-dependent properties and to their unique quantized photophysical characteristics, several NP-based products have been approved for diagnostic and therapeutic approaches, and a lot of them are under clinical trials (Ehlerding et al., 2018). Recently, the concept of "NPs multifunctionality" grew up, that is, the idea to have one single NP platform for different functions. In this context the idea of image-guided therapeutic delivery emerged, which had functions, both imaging and therapeutic, of the NPs to guide disease interventions according to individual patients' pathological conditions and responses to the treatment (Bao et al., 2013). This kind of multifunctional NPs should have both the capabilities to be visualized with medical tools and to reach the diseased tissue and release the drug. Among the broad spectrum of NPs, inorganic NPs (i.e., gold, silica, iron oxide, and quantum dots) are the suitable candidates for image-guided therapeutic delivery because they often possess unique electric, magnetic, optical, and plasmonic properties due to the quantum mechanical effects at nanometer scales (Hosoya et al., 2016). These intrinsic quantum mechanical properties make them remarkable contrast agent for optical imaging, MRI, nuclear imaging, computed tomography, surface-enhanced Raman spectroscopy, and ultrasound. However, inorganic particles tend to agglomerate in the bloodstream, limit nonspecific cell interaction, and compare to their organic counterpart, they can load less amount of drug. NPs can be coated with stealth layer of polymer or carbohydrates to protect them against agglomeration, to enhance their stability, biocompatibility, and safety; limit nonspecific cell interaction; and provide chemical handles for the conjunctions of drug molecules (Ehlerding et al., 2018). Superparamagnetic iron oxide (SPIO) NPs are a good example of NPs used in image-guided therapeutic delivery. SPIOs have excellent MRI contrast and biocompatibility. Micelles, liposomes, and mesoporous silica containing single or multiple SPIOs have been loaded with chemotherapeutic drugs such as doxorubicin and paclitaxel (Kim et al., 2008; Sinha et al., 2014). The resulting SPIO/drug complex demonstrated to achieve high drug payload while maintaining good MRI-T2 contrast. MRI/NIR (near-infrared) dual-modality SPIOs have been also used to image in vivo siRNA delivery. In this study, siRNA was conjugated to dextran-coated SPIOs, to which an NIR dye, Cy5.5, was also attached. The accumulation of siRNA in the tumor was visualized with both MRI and in vivo NIR optical imaging (Medarova et al., 2007). Multifunctional NPs, in particular combining imaging and therapeutic functions into a single NP platform, offer several advantages: (1) disease diagnosis at their very early stage, (2) noninvasive monitoring of the disease, (3) the possibility to intervene and to check the status of the disease treatment in real time, and (4) delivery of the drugs specifically to diseased tissue decreasing the side effects. This new tool seems to be promising and could contribute to help clinicians to suite a personal treatment for each patient and checking real time the effectiveness of the treatment.

9.1.6 Cardiac nanomedicine and safety

9.1.6.1 Inhaled nanoparticles for cardiac nanomedicine

The incidence of CD has increased over the last decades and still remains a major cause of death worldwide, claiming 17.1 million lives a year and accounting for estimated 31% of all deaths globally. Treatment of CD costs 139 billion euros for the EU annually, with hospitalizations comprising 60%–70% of direct treatment costs. Up to 40% of all deaths occur among the elderly (65 years of age or older), who are expected to reach approximately 20% of the whole population by the year 2030. In addition, the cost to treat CD will triple by that time. The economic impact of CD on the overall European healthcare currently stands at 192 billion euros annually, and this figure continues to rise every year. While largely investigated in the cancer field, the development and use of efficient NPs for the therapeutic treatment of CD are still in its infancy. The idea that Coulomb forces may better guide the negatively charged NPs to the heart via inhalation route has inspired a recent Horizon2020 project named "Cupido" (www.cupidoproject.eu). The negatively charged NPs candidate and the way of administration were the two main obstacles to overcome. In fact, to the best of our knowledge, only dendrimers, liposomes, and NPs based on synthetic polymers have so far been investigated for the in vivo delivery of various therapeutic molecules to myocardial cells with limited biodegradability and less efficiency.

9.1.6.2 Bioinspired and bioresorbable engineered nanomaterial: a new frontier for cardiac nanomedicine

During respiration the oxygenated blood moves from the pulmonary circulation first to the heart via the pulmonary vein. In line with this, inhaled CaP particles carrying a cardiospecific peptide able to cross the alveolar–capillary barrier in the lung would rapidly reach the myocardium (Miragoli et al., 2018) (Fig. 9.4). Indeed, we provide evidences that diabetic mice subjected to inhaled CaP recovers completely from the diabetic cardiac failure conditions (Fig. 9.4A and B). For achieving the results, CaP NPs have passed a severe screening for toxicological investigation in several organs including the heart for mice, rats, and pigs (Table 9.1). In addition, mice subjected to CaP inhalation (single or multiple one-a-day administration protocol) show no alterations of cardiac excitability parameters compared to the toxic outcome we previously obtained with the inhalation of titanium dioxide NPs (Miragoli et al., 2016a; Savi et al., 2014), these data

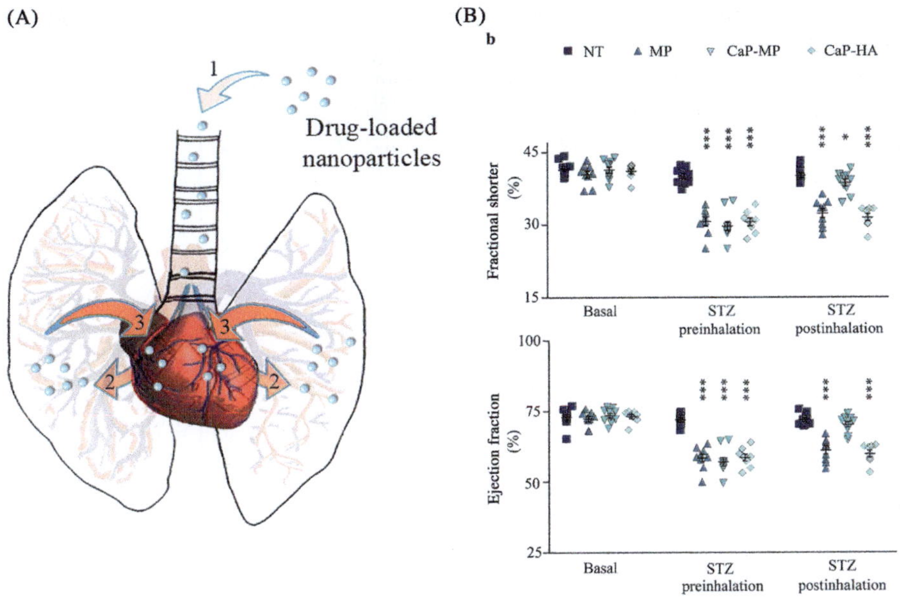

FIGURE 9.4 Inhaled peptide-loaded NPs for improving heart failure.

(A) Schematic representation of the pathway for inhaled NPs reaching the heart. (B) Cardiac function assessed by echocardiography in diabetic mice showing heart failure. *CaP*, Calcium phosphate nanoparticles; *MP*, mimetic peptide; *NPs*, nanoparticles; *Nt*, not treated; *STZ*, streptozotocin.

Modified from (18) with permission.

provide a preliminary safe profile for CaP administration. Definitely, we did not observe any signs of functional, metabolical, or toxicological alteration for the CaP-treated hearts (Miragoli et al., 2018). We think that reduced first-pass metabolism, that is liver barrier, may help the delivery of inhaled CaPs to the heart. Recently also metal ENMs such as gold NPs pave the way for potential cardiac applications (Ahmed et al., 2017). Ahmed et al. demonstrated the efficacy of this intervention by administering for 14 days 400 μg/kg in rat model expressing myocardial infarction. The main finding relies to the reduction in cardiac myocyte disarray, while cardiac biomarkers (creatine kinase and cardiac troponin T) did not change compared to sham rats. Adenosine triphosphate (ATP) is a molecule that tends to reduce during ischemia and it is possible to be encapsulated in a nanovector (Verma et al., 2006). Given that ATP levels are highly reduced during ischemia (Fassina et al., 2017), exogenous ATP supplement could provide a better recovery during reperfusion. To this end, ATP was encapsulated in liposomes with a diameter of 190 nm coated with the 2G4 antibody. They infuse the nanocarrier for 1 minute before the ischemia and fascinatingly; this resulted in a fourfold higher left ventricular developed pressure and threefold lower left ventricular

Table 9.1 Thiobarbituric acid reactive substance (TBARS) assays in several organs after calcium phosphate (CaP) exposure.

	LV			RV			Septum			Kidney		
	TBARS	**Proteins**	**TBARS/prot**	**TBARS**	**Proteins**	**TBARS/prot**	**TBARS**	**Proteins**	**TBARS/prot**	**TBARS**	**Proteins**	**TBARS/ prot**
	(nM)	(g/L)	(nmol/g)	(nM)	(g/L)	(nmol/g)	(nM)	(g/L)	(nmol/g)	(nM)	(g/L)	(nmol/g)
NT	429 ± 125	6.75 ± 3.99	80.66 ± 43.61	663 ± 109	3.64 ± 1.69	212.07 ± 96.90	507 ± 112	4.69 ± 2.66	131.90 ± 63.52	154 ± 114	29.60 ± 1156	5.57 ± 3.66
MP	499 ± 107	12.25 ± 3.80	41.66 ± 4.73	661 ± 45	3.71 ± 1.11	192.98 ± 76.09	611 ± 86	3.60 ± 1.69	195.38 ± 90.89	118 ± 62	30.43 ± 10.28	4.11 ± 2.27
CaP–HA	485 ± 197	7.82 ± 1.60	65.69 ± 37.04	563 ± 184	3.08 ± 0.50	188.97 ± 76.22	586 ± 141	4.54 ± 1.08	134.74 ± 47.89	91 ± 54	26.88 ± 5.47	3.75 ± 2.91
CaP–MP	507 ± 103	6.81 ± 1.24	78.41 ± 32.55	667 ± 24	3.53 ± 0.42	191.20 ± 27.80	532 ± 58	3.61 ± 1.32	167.17 ± 84.18	95 ± 84	19.25 ± 4.00	5.34 ± 2.93

	Lung			Liver			Spleen			Serum
	TBARS	**Proteins**	**TBARS/prot**	**TBARS**	**Proteins**	**TBARS/prot**	**TBARS**	**Proteins**	**TBARS/prot**	**TBARS**
	(nM)	(g/L)	(nmol/g)	(nM)	(g/L)	(nmol/g)	(nM)	(g/L)	(nmol/g)	(nM)
NT	77 ± 50	7.83 ± 2.60	9.90 ± 4.71	67 ± 27	26.63 ± 17.37	4.27 ± 4.01	214 ± 131	12.12 ± 6.27	19.04 ± 14.40	0.26 ± 0.05
MP	37 ± 10	4.80 ± 0.87	7.87 ± 2.09	72 ± 25	18.75 ± 11.75	5.27 ± 3.46	122 ± 62	7.33 ± 2.01	19.41 ± 15.96	0.24 ± 0.02
CaP–HA	40 ± 15	4.64 ± 2.24	9.88 ± 3.80	45 ± 19	15.38 ± 6.13*	3.65 ± 3.02	128 ± 79	4.38 ± 2.63	29.46 ± 12.44	0.23 ± 0.02
CaP–MP	93 ± 37	8.66 ± 1.23	10.66 ± 3.36	68 ± 21	14.50 ± 8.49*	6.19 ± 4.41	168 ± 69	8.17 ± 2.33	23.55 ± 16.56	0.25 ± 0.04

Data were obtained after collection of organs in the four group of mice (n = 3 each). All data are expressed as mean ± SEM (two-way ANOVA test). HA, *Hemagglutinin;* LV, *left ventricle;* MP, *mimetic peptide;* NT, *not treated;* RV, *right ventricle;* *$^*p < 0.05$ vs NT.*

Source: *Modified from (18) with permission.*

end systolic pressure compared to control. ENMs gain attraction for tissue engineering. For example, nanosheet, graphene, or CNTs have been using for controlling tissue conductivity (You et al., 2011). Those results suggest that we can intend cardiac nanomedicine not only from the pharmacological side but also ENMs can bionically integrate with the cardiac tissue for assuring the normal electrical and mechanical function of the heart.

Recently, Lozano et al. (2018) have fascinatingly summarized where we are in cardiac nanomedicine, encompassing a various range of application including biomaterials, metal, and carbon-based NPs. Interestingly, the review describes how NPs can target either cardiac sarcoplasmic reticulum, or mitochondria, or nucleus, depending on their physicochemical characteristics. For example, during myocardial infarction, cyclosporine A (CsA), a direct mitochondrial permeability transition pore inhibitor (Miragoli et al., 2016b), has been administered at the onset of reperfusion in mice during ischemia/reperfusion (I/R) protocol via poly(lactic-*co*-glycolic acid) (PLGA) NPs (Ikeda et al., 2016). They found a reduction of c.40% of infarct area 10 times higher than the normal administration of CsA. Recently, also chronic cardiomyopathy has been treated, experimentally, with NPs (Maxwell et al., 2015). Polyketal NPs decorated with *N*-acetylglucosamine have been loaded with S100A1 (a calcium binding protein that regulates calcium release machinery via SERCa2a, RyR2, and phospholamban (PLB)). The beneficial response was measured in restoration of sarcomeric shortening, Ca^{2+} transient amplitude and reducing the frequency of Ca^{2+} sparks.

NPs, mainly iron oxide NPs, have been also selected as a tool for the cardiac imaging field. Verma et al. (2015) created fluorescent iron oxide magnetic NPs for (1) cell selection and imaging with (2) direct translational therapeutic purpose. Other colleagues take advantage of the same NPs for imaging cardiac inflammation (Podrouzkova et al., 2015). To further confirm that this is an expanded field with "plenty of space at the bottom," we took advantage with the superparamagnetic characteristics of iron oxide by decorating during synthesis CaP NPs. After tracheal instillation, we observed those particles in the heart, and via magnetic field stimulation, we were able to control their cargo delivery (Marrella et al., 2018).

9.1.7 Engineered nanomaterials in regenerative medicine

The theoretical and experimental potential of nanotechnology to the fundamental developments in regenerative medicines for treatments of injuries and degenerative diseases are being increasingly recognized (Khang et al., 2010; Smith and Ma, 2010). Since cell function occurs at the nanometer scale, nanotechnology can influence and even alter cellular behavior, which ultimately promotes the functions of various cell types and, therefore, tissue or organ. In this light, different approaches have been proposed to apply nanotechnology in regenerative medicine; most of them can be related to (1) NPs; (2) scaffolds with nanofibers; (3) scaffolds with nanotopographic modifications; (4) drug/gene delivery;

(5) extracellular matrix patterning (Chaudhury et al., 2014). In all these cases the integration of nanotechnology to regenerative medicine allows better control over physical and biological properties of a biomaterial than conventional technologies. Moreover, these different approaches accelerate various regenerative therapies, such as those for bone (Oryan et al., 2014), vascular (Miller et al., 2004), heart (Cicha et al., 2016), cartilage (Guasti et al., 2014), bladder (Miller et al., 2002), and brain tissues (Malmo et al., 2013). For example, recent advancements in nanotechnology-based matrices demonstrated greater bone-forming cell (osteoblast) functions on various nanomaterials, such as nano-hydroxyapatite (Sato et al., 2005), electrospun silk (Jin et al., 2004), anodized titanium (Yao et al., 2008), and nanostructured titanium surfaces as compared to conventional orthopedic implant materials (Khang et al., 2008). In addition to these observed greater initial responses of osteoblasts, long-term functions of bone cells are promoted on nanostructured biomaterials, having larger consequences toward calcium crystallization and, therefore, bone regeneration (Ergun et al., 2008). Therefore it has been demonstrated that nanostructured implant surface promotes bone cell responses, improving integration with surrounding bone compared to conventional surface.

Furthermore, vascular graft (PLGA) and stent (titanium) surfaces with nanometers surface roughness values improve endothelial (inner vessel cells) cell functions as compared to nanosmooth polymer and titanium surface (Choudhary et al., 2007).

The multidisciplinary applications of nanotechnologies for discovering new molecules and tailoring those could improve human health, increasing the quality of life and the initial formation of tissue necessary to prolong implant lifetime.

9.1.8 Where do we go from here?

9.1.8.1 Personalized (nano)medicine

Personalized medicine is a healthcare strategy that aims at the development of specific treatments for each patient or cohort of patients, taking into account their biological "personality" such as phenotype, coding and noncoding genome, as well as epigenetic landscape. All these factors could influence the outcome (efficacy and safety) of the therapy with a personalized footprint (Fornaguera and Garcia-Celma, 2017). Nanomedicine fits well in the concept of personalization especially if we introduce the organ specificity. Because the "personalization" can be defined in the target organs, in the drugs and in the carrier, nanomedicine has been embraced with great enthusiasm in the medical and research community. For example, specific ENMs carrying multiple drugs specifically designed toward the patient's diseases or ENMs able to return diagnostically imaging and therapeutic results: herewith, we may introduce the polypharmaco therapy in personalized nanomedicine (Elsabahy et al., 2015). Another step where ENMs can be a disruptive medical technology is the combination with genome editing technology for DNA repairing (Yin et al., 2017). The therapeutic potential of clustered regularly interspaced short palindromic repeats (CRISPR)-based genome is higher and, depending by the efficiency, the optimal expectation would suggest that the "cut-and-replace" work needs to be done in a given amount of time. Thus long-term expression of Cas9–sgRNA in vivo has been shown using virus, including adeno-associated virus (Swiech et al., 2015; Tabebordbar et al., 2016). Ideally, limiting the exposure from the editing machinery and resorbable NPs or nanobots (see next) would be crucial in this passage (Schumann et al., 2015).

9.1.8.2 Engineering nanomedicine: nanobots and nanosponge

ENMs can also be dynamic and producing the effect on-site by changing its mechanical and structural conformation. The terms of nanobots do not rely for a little Transformers robot swimming in our body but to a chain of atoms that can be activate externally from light or other energetical sources. Robert Pal's lab pioneering described how mechanical atomic machine can act as a drill in cancer cells (Garcia-Lopez et al., 2017) where a nanomechanical action is provided onto the tumor cell membrane and the light-driven motors inactivated by UV. The same photodynamic therapy can be adopted for hybrid nano(bot)system based on compatible silicon carbide nanowires conjugated with porphyrin derivative that release singlet oxygen when irradiated by the X-ray (Rossi et al., 2015). Nanomachine can also be used to simulate some in vivo function, for example, muscle contraction. Nobel laureate Bernard Feringa recently proposed that artificial muscle–like function using nanofibers assembled into a linear bundle that makes strings centimeter long. When illuminate with light (so energy needs to be transferred somehow), the artificial muscle contracts toward the light source in a controlled and defined manner (Chen et al., 2018).

A last fascinating aspect of ENMs is related to the possibility to create a "nanosponge" able to clean the district where it will be delivered or stick onto a given target for delivery their cargo. Recently, a nanosponge therapy has been proposed in the context of intraocular infection where cytolysin produced by bacteria can be absorbed by the nanosponge and reducing the infection and the hemolityic activity by 70% (LaGrow et al., 2017). The material of miracles, that is, graphene, can also be adopted as nanosponge; Su et al. (2016) demonstrate that lipid bilayer-capped ultrasmall graphene nanosponge (a sponge-like carbon material on graphene nanosheet) is an excellent delivery platform for overcoming the multiple obstacles of drug delivery to tumors because of the biocompatibility, tumor penetration, large-carrier payload, and NIR response.

So, ENMs for therapeutic purpose are certainty on its infancy requiring a multidisciplinary background for growing; the medical community aspect this therapeutic and healthcare revolution with open arms (Ventola, 2012a) but with a technical, safety, and security aspects that needs to be controlled and assessed rigorously (Ventola, 2012b,c).

Acknowledgments

We thank the Department of Medicine and Surgery at University of Parma and the CERT (Center of Excellence for Toxicological Research) for the financial support; we also thank INAIL-DIMEILA for the collaboration and contributions to nanotoxicology data, CNR-IRGB Humanitas Research Center (Milan) for providing published experimental data and Cupido Consortium from the EU project CUPIDO (www.cupidoproject.eu) for allowing the possibility to discuss published data in this chapter. We also thank Prof. Alberico Borghetti and Prof. Aderville Cabassi for providing salary for S.R. by means of "Fondi per la Ricerca Guido Erluison."

References

Ahmed, S.M., Abdelrahman, S.A., Salama, A.E., 2017. Efficacy of gold nanoparticles against isoproterenol induced acute myocardial infarction in adult male albino rats. Ultrastruct. Pathol. 41 (2), 168–185.

Ajima, K., Murakami, T., Mizoguchi, Y., Tsuchida, K., Ichihashi, T., Iijima, S., et al., 2008. Enhancement of in vivo anticancer effects of cisplatin by incorporation inside single-wall carbon nanohorns. ACS Nano 2 (10), 2057–2064.

Baban, D.F., Seymour, L.W., 1998. Control of tumour vascular permeability. Adv. Drug Deliv. Rev. 34 (1), 109–119.

Bao, G., Mitragotri, S., Tong, S., 2013. Multifunctional nanoparticles for drug delivery and molecular imaging. Annu. Rev. Biomed. Eng. 15, 253–282.

Bertrand, N., Wu, J., Xu, X., Kamaly, N., Farokhzad, O.C., 2014. Cancer nanotechnology: the impact of passive and active targeting in the era of modern cancer biology. Adv. Drug Deliv. Rev. 66, 2–25.

Bhirde, A.A., Patel, S., Sousa, A.A., Patel, V., Molinolo, A.A., Ji, Y., et al., 2010. Distribution and clearance of PEG-single-walled carbon nanotube cancer drug delivery vehicles in mice. Nanomedicine (Lond.) 5 (10), 1535–1546.

Bildstein, L., Dubernet, C., Couvreur, P., 2011. Prodrug-based intracellular delivery of anticancer agents. Adv. Drug Deliv. Rev. 63 (1–2), 3–23.

Braakhuis, H.M., Kloet, S.K., Kezic, S., Kuper, F., Park, M.V., Bellmann, S., et al., 2015. Progress and future of in vitro models to study translocation of nanoparticles. Arch. Toxicol. 89 (9), 1469–1495.

Bukowski, R.M., Young, J., Goodman, G., Meyers, F., Issell, B.F., Sergi, J.S., et al., 1993. Polyethylene glycol conjugated interleukin-2: clinical and immunologic effects in patients with advanced renal cell carcinoma. Invest. New Drugs 11 (2–3), 211–217.

Cabral, H., Matsumoto, Y., Mizuno, K., Chen, Q., Murakami, M., Kimura, M., et al., 2011. Accumulation of sub-100 nm polymeric micelles in poorly permeable tumours depends on size. Nat. Nanotechnol. 6 (12), 815–823.

Carter, P., 2001. Improving the efficacy of antibody-based cancer therapies. Nat. Rev. Cancer 1 (2), 118–129.

Chan, S., Davidson, N., Juozaityte, E., Erdkamp, F., Pluzanska, A., Azarnia, N., et al., 2004. Phase III trial of liposomal doxorubicin and cyclophosphamide compared with epirubicin and cyclophosphamide as first-line therapy for metastatic breast cancer. Ann. Oncol. 15 (10), 1527–1534.

Chaudhury, K., Kumar, V., Kandasamy, J., RoyChoudhury, S., 2014. Regenerative nanomedicine: current perspectives and future directions. Int. J. Nanomed. 9, 4153–4167.

Chen, J., Leung, F.K., Stuart, M.C.A., Kajitani, T., Fukushima, T., van der Giessen, E., et al., 2018. Artificial muscle-like function from hierarchical supramolecular assembly of photoresponsive molecular motors. Nat. Chem. 10 (2), 132–138.

Cheng, P.N., Lam, T.L., Lam, W.M., Tsui, S.M., Cheng, A.W., Lo, W.H., et al., 2007. PEGylated recombinant human arginase (rhArg-peg5,000mw) inhibits the in vitro and in vivo proliferation of human hepatocellular carcinoma through arginine depletion. Cancer Res. 67 (1), 309–317.

Chithrani, B.D., Ghazani, A.A., Chan, W.C., 2006. Determining the size and shape dependence of gold nanoparticle uptake into mammalian cells. Nano Lett. 6 (4), 662–668.

Choi, H.S., Ashitate, Y., Lee, J.H., Kim, S.H., Matsui, A., Insin, N., et al., 2010. Rapid translocation of nanoparticles from the lung airspaces to the body. Nat. Biotechnol. 28 (12), 1300–1303.

Choudhary, S., Haberstroh, K.M., Webster, T.J., 2007. Enhanced functions of vascular cells on nanostructured Ti for improved stent applications. Tissue Eng. 13 (7), 1421–1430.

Cicha, I., Singh, R., Garlichs, C.D., Alexiou, C., 2016. Nano-biomaterials for cardiovascular applications: clinical perspective. J. Control. Release 229, 23–36.

Coelho, T., Adams, D., Silva, A., Lozeron, P., Hawkins, P.N., Mant, T., et al., 2013. Safety and efficacy of RNAi therapy for transthyretin amyloidosis. N. Engl. J. Med. 369 (9), 819–829.

Corbo, C., Molinaro, R., Parodi, A., Toledano Furman, N.E., Salvatore, F., Tasciotti, E., 2016. The impact of nanoparticle protein corona on cytotoxicity, immunotoxicity and target drug delivery. Nanomedicine (Lond.) 11 (1), 81–100.

Coty, J.B., Vauthier, C., 2018. Characterization of nanomedicines: a reflection on a field under construction needed for clinical translation success. J. Control. Release 275, 254–268.

Desai, N., Trieu, V., Yao, Z., Louie, L., Ci, S., Yang, A., et al., 2006. Increased antitumor activity, intratumor paclitaxel concentrations, and endothelial cell transport of cremophor-free, albumin-bound paclitaxel, ABI-007, compared with cremophor-based paclitaxel. Clin. Cancer Res. 12 (4), 1317–1324.

Di Mauro, V., Iafisco, M., Salvarani, N., Vacchiano, M., Carullo, P., Ramirez-Rodriguez, G.B., et al., 2016. Bioinspired negatively charged calcium phosphate nanocarriers for cardiac delivery of MicroRNAs. Nanomedicine (Lond.) 11 (8), 891–906.

Dinndorf, P.A., Gootenberg, J., Cohen, M.H., Keegan, P., Pazdur, R., 2007. FDA drug approval summary: pegaspargase (oncaspar) for the first-line treatment of children with acute lymphoblastic leukemia (ALL). Oncologist 12 (8), 991–998.

Doshi, N., Prabhakarpandian, B., Rea-Ramsey, A., Pant, K., Sundaram, S., Mitragotri, S., 2010. Flow and adhesion of drug carriers in blood vessels depend on their shape: a study using model synthetic microvascular networks. J. Control. Release 146 (2), 196–200.

Du, J.Z., Du, X.J., Mao, C.Q., Wang, J., 2011. Tailor-made dual pH-sensitive polymer-doxorubicin nanoparticles for efficient anticancer drug delivery. J. Am. Chem. Soc. 133 (44), 17560–17563.

Dumont, M.F., Yadavilli, S., Sze, R.W., Nazarian, J., Fernandes, R., 2014. Manganese-containing Prussian blue nanoparticles for imaging of pediatric brain tumors. Int. J. Nanomed. 9, 2581–2595.

Duncan, R., 2006. Polymer conjugates as anticancer nanomedicines. Nat. Rev. Cancer 6 (9), 688–701.
Ehlerding, E.B., Grodzinski, P., Cai, W., Liu, C.H., 2018. Big potential from small agents: nanoparticles for imaging-based companion diagnostics. ACS Nano 12 (3), 2106–2121.
Elsabahy, M., Heo, G.S., Lim, S.M., Sun, G., Wooley, K.L., 2015. Polymeric nanostructures for imaging and therapy. Chem. Rev. 115 (19), 10967–11011.
Ergun, C., Liu, H., Webster, T.J., Olcay, E., Yilmaz, S., Sahin, F.C., 2008. Increased osteoblast adhesion on nanoparticulate calcium phosphates with higher Ca/P ratios. J. Biomed. Mater. Res. A 85 (1), 236–241.
Fang, J., Nakamura, H., Maeda, H., 2011. The EPR effect: unique features of tumor blood vessels for drug delivery, factors involved, and limitations and augmentation of the effect. Adv. Drug Deliv. Rev. 63 (3), 136–151.
Fassina, L., Rozzi, G., Rossi, S., Scacchi, S., Galetti, M., Lo Muzio, F.P., et al., 2017. Cardiac kinematic parameters computed from video of in situ beating heart. Sci. Rep. 7, 46143.
Fonseca-Santos, B., Gremiao, M.P., Chorilli, M., 2015. Nanotechnology-based drug delivery systems for the treatment of Alzheimer's disease. Int. J. Nanomed. 10, 4981–5003.
Fornaguera, C., Garcia-Celma, M.J., 2017. Personalized nanomedicine: a revolution at the nanoscale. J. Pers. Med. 7 (4). Available from: https://doi.org/10.3390/jpm7040012.
Frampton, J.E., 2010. Mifamurtide: a review of its use in the treatment of osteosarcoma. Paediatr. Drugs 12 (3), 141–153.
Garcia-Garcia, E., Gil, S., Andrieux, K., Desmaele, D., Nicolas, V., Taran, F., et al., 2005. A relevant in vitro rat model for the evaluation of blood-brain barrier translocation of nanoparticles. Cell. Mol. Life Sci. 62 (12), 1400–1408.
Garcia-Lopez, V., Chen, F., Nilewski, L.G., Duret, G., Aliyan, A., Kolomeisky, A.B., et al., 2017. Molecular machines open cell membranes. Nature 548 (7669), 567–572.
Gill, P.S., Wernz, J., Scadden, D.T., Cohen, P., Mukwaya, G.M., von Roenn, J.H., et al., 1996. Randomized phase III trial of liposomal daunorubicin versus doxorubicin, bleomycin, and vincristine in AIDS-related Kaposi's sarcoma. J. Clin. Oncol. 14 (8), 2353–2364.
Godin, B., Sakamoto, J.H., Serda, R.E., Grattoni, A., Bouamrani, A., Ferrari, M., 2010. Emerging applications of nanomedicine for the diagnosis and treatment of cardiovascular diseases. Trends Pharmacol. Sci. 31 (5), 199–205.
Gokbuget, N., Hartog, C.M., Bassan, R., Derigs, H.G., Dombret, H., Greil, R., et al., 2011. Liposomal cytarabine is effective and tolerable in the treatment of central nervous system relapse of acute lymphoblastic leukemia and very aggressive lymphoma. Haematologica 96 (2), 238–244.
Gong, J., Chen, M., Zheng, Y., Wang, S., Wang, Y., 2012. Polymeric micelles drug delivery system in oncology. J. Control. Release 159 (3), 312–323.
Green, M.R., Manikhas, G.M., Orlov, S., Afanasyev, B., Makhson, A.M., Bhar, P., et al., 2006. Abraxane, a novel Cremophor-free, albumin-bound particle form of paclitaxel for the treatment of advanced non-small-cell lung cancer. Ann. Oncol. 17 (8), 1263–1268.
Guasti, L., Vagaska, B., Bulstrode, N.W., Seifalian, A.M., Ferretti, P., 2014. Chondrogenic differentiation of adipose tissue-derived stem cells within nanocaged POSS-PCU scaffolds: a new tool for nanomedicine. Nanomedicine 10 (2), 279–289.

Hare, J.I., Lammers, T., Ashford, M.B., Puri, S., Storm, G., Barry, S.T., 2017. Challenges and strategies in anti-cancer nanomedicine development: an industry perspective. Adv. Drug Deliv. Rev. 108, 25–38.

Harris, T.J., Green, J.J., Fung, P.W., Langer, R., Anderson, D.G., Bhatia, S.N., 2010. Tissue-specific gene delivery via nanoparticle coating. Biomaterials 31 (5), 998–1006.

Harrison, M., Tomlinson, D., Stewart, S., 1995. Liposomal-entrapped doxorubicin: an active agent in AIDS-related Kaposi's sarcoma. J. Clin. Oncol. 13 (4), 914–920.

Hosoya, H., Dobroff, A.S., Driessen, W.H., Cristini, V., Brinker, L.M., Staquicini, F.I., et al., 2016. Integrated nanotechnology platform for tumor-targeted multimodal imaging and therapeutic cargo release. Proc. Natl. Acad. Sci. U.S.A. 113 (7), 1877–1882.

Hu, C.M., Aryal, S., Zhang, L., 2010. Nanoparticle-assisted combination therapies for effective cancer treatment. Ther. Deliv. 1 (2), 323–334.

Huang, X., Teng, X., Chen, D., Tang, F., He, J., 2010. The effect of the shape of mesoporous silica nanoparticles on cellular uptake and cell function. Biomaterials 31 (3), 438–448.

Huwyler, J., Cerletti, A., Fricker, G., Eberle, A.N., Drewe, J., 2002. By-passing of P-glycoprotein using immunoliposomes. J. Drug Target. 10 (1), 73–79.

Ikeda, G., Matoba, T., Nakano, Y., Nagaoka, K., Ishikita, A., Nakano, K., et al., 2016. Nanoparticle-mediated targeting of cyclosporine a enhances cardioprotection against ischemia-reperfusion injury through inhibition of mitochondrial permeability transition pore opening. Sci. Rep. 6, 20467.

Infante, J.R., Keedy, V.L., Jones, S.F., Zamboni, W.C., Chan, E., Bendell, J.C., et al., 2012. Phase I and pharmacokinetic study of IHL-305 (PEGylated liposomal irinotecan) in patients with advanced solid tumors. Cancer Chemother. Pharmacol. 70 (5), 699–705.

Jin, H.J., Chen, J., Karageorgiou, V., Altman, G.H., Kaplan, D.L., 2004. Human bone marrow stromal cell responses on electrospun silk fibroin mats. Biomaterials 25 (6), 1039–1047.

Karve, S., Werner, M.E., Sukumar, R., Cummings, N.D., Copp, J.A., Wang, E.C., et al., 2012. Revival of the abandoned therapeutic wortmannin by nanoparticle drug delivery. Proc. Natl. Acad. Sci. U.S.A. 109 (21), 8230–8235.

Khang, D., Park, G.E., Webster, T.J., 2008. Enhanced chondrocyte densities on carbon nanotube composites: the combined role of nanosurface roughness and electrical stimulation. J. Biomed. Mater. Res. A 86 (1), 253–260.

Khang, D., Carpenter, J., Chun, Y.W., Pareta, R., Webster, T.J., 2010. Nanotechnology for regenerative medicine. Biomed. Microdevices 12 (4), 575–587.

Kim, J., Kim, H.S., Lee, N., Kim, T., Kim, H., Yu, T., et al., 2008. Multifunctional uniform nanoparticles composed of a magnetite nanocrystal core and a mesoporous silica shell for magnetic resonance and fluorescence imaging and for drug delivery. Angew. Chem. Int. Ed. Engl. 47 (44), 8438–8441.

Kontermann, R.E., 2006. Immunoliposomes for cancer therapy. Curr. Opin. Mol. Ther. 8 (1), 39–45.

LaGrow, A.L., Coburn, P.S., Miller, F.C., Land, C., Parkunan, S.M., Luk, B.T., et al., 2017. A novel biomimetic nanosponge protects the retina from the *Enterococcus faecalis* cytolysin. mSphere 2 (6), e00335–17.

Langer, C.J., O'Byrne, K.J., Socinski, M.A., Mikhailov, S.M., Lesniewski-Kmak, K., Smakal, M., et al., 2008. Phase III trial comparing paclitaxel poliglumex (CT-2103,

PPX) in combination with carboplatin versus standard paclitaxel and carboplatin in the treatment of PS 2 patients with chemotherapy-naive advanced non-small cell lung cancer. J. Thorac. Oncol. 3 (6), 623–630.

Li, Z., Rana, T.M., 2014. Therapeutic targeting of microRNAs: current status and future challenges. Nat. Rev. Drug Discov. 13 (8), 622–638.

Li, D., Morishita, M., Wagner, J.G., Fatouraie, M., Wooldridge, M., Eagle, W.E., et al., 2016. In vivo biodistribution and physiologically based pharmacokinetic modeling of inhaled fresh and aged cerium oxide nanoparticles in rats. Part. Fibre Toxicol. 13 (1), 45.

Liu, X., Guo, L., Morris, D., Kane, A.B., Hurt, R.H., 2008. Targeted removal of bioavailable metal as a detoxification strategy for carbon nanotubes. Carbon N.Y. 46 (3), 489–500.

Lozano, O., Torres-Quintanilla, A., Garcia-Rivas, G., 2018. Nanomedicine for the cardiac myocyte: where are we? J. Control. Release 271, 149–165.

Maeda, H., 2012. Macromolecular therapeutics in cancer treatment: the EPR effect and beyond. J. Control. Release 164 (2), 138–144.

Maeda, H., Nakamura, H., Fang, J., 2013. The EPR effect for macromolecular drug delivery to solid tumors: improvement of tumor uptake, lowering of systemic toxicity, and distinct tumor imaging in vivo. Adv. Drug Deliv. Rev. 65 (1), 71–79.

Malmo, J., Sandvig, A., Varum, K.M., Strand, S.P., 2013. Nanoparticle mediated P-glycoprotein silencing for improved drug delivery across the blood-brain barrier: a siRNA-chitosan approach. PLoS One 8 (1), e54182.

Mamot, C., Drummond, D.C., Noble, C.O., Kallab, V., Guo, Z., Hong, K., et al., 2005. Epidermal growth factor receptor-targeted immunoliposomes significantly enhance the efficacy of multiple anticancer drugs in vivo. Cancer Res. 65 (24), 11631–11638.

Mamot, C., Ritschard, R., Wicki, A., Stehle, G., Dieterle, T., Bubendorf, L., et al., 2012. Tolerability, safety, pharmacokinetics, and efficacy of doxorubicin-loaded anti-EGFR immunoliposomes in advanced solid tumours: a phase 1 dose-escalation study. Lancet Oncol. 13 (12), 1234–1241.

Marrella, A., Iafisco, M., Adamiano, A., Rossi, S., Aiello, M., Barandalla-Sobrados, M., et al., 2018. A combined low-frequency electromagnetic and fluidic stimulation for a controlled drug release from superparamagnetic calcium phosphate nanoparticles: potential application for cardiovascular diseases. J. R. Soc. Interface 15 (144), 20180236.

Masetti, R., Pession, A., 2009. First-line treatment of acute lymphoblastic leukemia with pegasparaginase. Biologics 3, 359–368.

Matsumura, Y., Maeda, H., 1986. A new concept for macromolecular therapeutics in cancer chemotherapy: mechanism of tumoritropic accumulation of proteins and the antitumor agent smancs. Cancer Res. 46 (12 Pt 1), 6387–6392.

Maxwell, J.T., Somasuntharam, I., Gray, W.D., Shen, M., Singer, J.M., Wang, B., et al., 2015. Bioactive nanoparticles improve calcium handling in failing cardiac myocytes. Nanomedicine (Lond.) 10 (22), 3343–3357.

Medarova, Z., Pham, W., Farrar, C., Petkova, V., Moore, A., 2007. In vivo imaging of siRNA delivery and silencing in tumors. Nat. Med. 13 (3), 372–377.

Miller, D.C., Thapa, A., Haberstroh, K.M., Webster, T.J., 2002. Enhanced functions of vascular and bladder cells on poly-lactic-*co*-glycolic acid polymers with nanostructured surfaces. IEEE Trans. Nanobioscience 1 (2), 61–66.

Miller, D.C., Thapa, A., Haberstroh, K.M., Webster, T.J., 2004. Endothelial and vascular smooth muscle cell function on poly(lactic-*co*-glycolic acid) with nano-structured surface features. Biomaterials 25 (1), 53–61.

Miragoli, M., Moshkov, A., Novak, P., Shevchuk, A., Nikolaev, V.O., El-Hamamsy, I., et al., 2011. Scanning ion conductance microscopy: a convergent high-resolution technology for multi-parametric analysis of living cardiovascular cells. J. R. Soc. Interface 8 (60), 913–925.

Miragoli, M., Novak, P., Ruenraroengsak, P., Shevchuk, A.I., Korchev, Y.E., Lab, M.J., et al., 2013. Functional interaction between charged nanoparticles and cardiac tissue: a new paradigm for cardiac arrhythmia? Nanomedicine (Lond.) 8 (5), 725–737.

Miragoli, M., Rossi, S., Savi, M., Goldoni, M., Pinelli, S., Alinovi, R., et al., 2016a. Nanoparticles exposure: in-vitro and in-vivo investigation to evaluate cardiovascular risk factors in normal and failing cardiac tissue. G. Ital. Med. Lav. Ergon. 38 (3), 146–150.

Miragoli, M., Sanchez-Alonso, J.L., Bhargava, A., Wright, P.T., Sikkel, M., Schobesberger, S., et al., 2016b. Microtubule-dependent mitochondria alignment regulates calcium release in response to nanomechanical stimulus in heart myocytes. Cell Rep. 14 (1), 140–151.

Miragoli, M., Ceriotti, P., Iafisco, M., Vacchiano, M., Salvarani, N., Alogna, A., et al., 2018. Inhalation of peptide-loaded nanoparticles improves heart failure. Sci. Transl. Med. 10 (424).

Morachis, J.M., Mahmoud, E.A., Almutairi, A., 2012. Physical and chemical strategies for therapeutic delivery by using polymeric nanoparticles. Pharmacol. Rev. 64 (3), 505–519.

Novak, P., Shevchuk, A., Ruenraroengsak, P., Miragoli, M., Thorley, A.J., Klenerman, D., et al., 2014. Imaging single nanoparticle interactions with human lung cells using fast ion conductance microscopy. Nano Lett. 14 (3), 1202–1207.

O'Brien, S., Schiller, G., Lister, J., Damon, L., Goldberg, S., Aulitzky, W., et al., 2013. High-dose vincristine sulfate liposome injection for advanced, relapsed, and refractory adult Philadelphia chromosome-negative acute lymphoblastic leukemia. J. Clin. Oncol. 31 (6), 676–683.

Oryan, A., Alidadi, S., Moshiri, A., Maffulli, N., 2014. Bone regenerative medicine: classic options, novel strategies, and future directions. J. Orthop. Surg. Res. 9 (1), 18.

Pecot, C.V., Calin, G.A., Coleman, R.L., Lopez-Berestein, G., Sood, A.K., 2011. RNA interference in the clinic: challenges and future directions. Nat. Rev. Cancer 11 (1), 59–67.

Pederzoli, F., Tosi, G., Vandelli, M.A., Belletti, D., Forni, F., Ruozi, B., 2017. Protein corona and nanoparticles: how can we investigate on? Wiley Interdiscip. Rev. Nanomed. Nanobiotechnol. 9 (6), e1467.

Perez-Herrero, E., Fernandez-Medarde, A., 2015. Advanced targeted therapies in cancer: drug nanocarriers, the future of chemotherapy. Eur. J. Pharm. Biopharm. 93, 52–79.

Pietroiusti, A., Bergamaschi, E., Campagna, M., Campagnolo, L., De Palma, G., Iavicoli, S., et al., 2017. The unrecognized occupational relevance of the interaction between engineered nanomaterials and the gastro-intestinal tract: a consensus paper from a multidisciplinary working group. Part. Fibre Toxicol. 14 (1), 47.

Podrouzkova, H., Feitova, V., Panovsky, R., Meluzin, J., Orban, M., 2015. Superparamagnetic iron oxide-enhanced magnetic resonance for imaging cardiac inflammation. A minireview. Biomed. Pap. Med. Fac. Univ. Palacky Olomouc Czech Repub. 159 (3), 378–381.

Prabhakar, U., Maeda, H., Jain, R.K., Sevick-Muraca, E.M., Zamboni, W., Farokhzad, O. C., et al., 2013. Challenges and key considerations of the enhanced permeability and retention effect for nanomedicine drug delivery in oncology. Cancer Res. 73 (8), 2412–2417.

Rossi, F., Bedogni, E., Bigi, F., Rimoldi, T., Cristofolini, L., Pinelli, S., et al., 2015. Porphyrin conjugated SiC/SiO_x nanowires for X-ray-excited photodynamic therapy. Sci. Rep. 5, 7606.

Sa, L.T., Albernaz Mde, S., Patricio, B.F., Falcao Jr., M.V., Coelho, B.F., Bordim, A., et al., 2012. Biodistribution of nanoparticles: initial considerations. J. Pharm. Biomed. Anal. 70, 602–604.

Sato, M., Slamovich, E.B., Webster, T.J., 2005. Enhanced osteoblast adhesion on hydrothermally treated hydroxyapatite/titania/poly(lactide-*co*-glycolide) sol-gel titanium coatings. Biomaterials 26 (12), 1349–1357.

Savi, M., Rossi, S., Bocchi, L., Gennaccaro, L., Cacciani, F., Perotti, A., et al., 2014. Titanium dioxide nanoparticles promote arrhythmias via a direct interaction with rat cardiac tissue. Part. Fibre Toxicol. 11, 63.

Schiener, M., Hossann, M., Viola, J.R., Ortega-Gomez, A., Weber, C., Lauber, K., et al., 2014. Nanomedicine-based strategies for treatment of atherosclerosis. Trends Mol. Med. 20 (5), 271–281.

Schmieder, A.H., Winter, P.M., Williams, T.A., Allen, J.S., Hu, G., Zhang, H., et al., 2013. Molecular MR imaging of neovascular progression in the Vx2 tumor with alphavbeta3-targeted paramagnetic nanoparticles. Radiology 268 (2), 470–480.

Schubauer-Berigan, M.K., Dahm, M.M., Yencken, M.S., 2011. Engineered carbonaceous nanomaterials manufacturers in the United States: workforce size, characteristics, and feasibility of epidemiologic studies. J. Occup. Environ. Med. 53 (6 Suppl.), S62–S67.

Schumann, K., Lin, S., Boyer, E., Simeonov, D.R., Subramaniam, M., Gate, R.E., et al., 2015. Generation of knock-in primary human T cells using Cas9 ribonucleoproteins. Proc. Natl. Acad. Sci. U.S.A. 112 (33), 10437–10442.

Seymour, L.W., Ferry, D.R., Kerr, D.J., Rea, D., Whitlock, M., Poyner, R., et al., 2009. Phase II studies of polymer-doxorubicin (PK1, FCE28068) in the treatment of breast, lung and colorectal cancer. Int. J. Oncol. 34 (6), 1629–1636.

Shevchuk, A.I., Novak, P., Takahashi, Y., Clarke, R., Miragoli, M., Babakinejad, B., et al., 2011. Realizing the biological and biomedical potential of nanoscale imaging using a pipette probe. Nanomedicine (Lond.) 6 (3), 565–575.

Shi, J., Kantoff, P.W., Wooster, R., Farokhzad, O.C., 2017. Cancer nanomedicine: progress, challenges and opportunities. Nat. Rev. Cancer 17 (1), 20–37.

Siddiqi, K.S., Husen, A., Rao, R.A.K., 2018. A review on biosynthesis of silver nanoparticles and their biocidal properties. J. Nanobiotechnol. 16 (1), 14.

Sinha, A., Chakraborty, A., Jana, N.R., 2014. Dextran-gated, multifunctional mesoporous nanoparticle for glucose-responsive and targeted drug delivery. ACS Appl. Mater. Interfaces 6 (24), 22183–22191.

Sinha, N., Yeow, J.T., 2005. Carbon nanotubes for biomedical applications. IEEE Trans. Nanobioscience 4 (2), 180–195.

Smith, I.O., Ma, P.X., 2010. Cell and biomolecule delivery for regenerative medicine. Sci. Technol. Adv. Mater. 11 (1), 014102.

Spigoni, V., Cito, M., Alinovi, R., Pinelli, S., Passeri, G., Zavaroni, I., et al., 2015. Effects of TiO_2 and Co_3O_4 nanoparticles on circulating angiogenic cells. PLoS One 10 (3), e0119310.

Su, Y.L., Chen, K.T., Sheu, Y.C., Sung, S.Y., Hsu, R.S., Chiang, C.S., et al., 2016. The penetrated delivery of drug and energy to tumors by lipo-graphene nanosponges for photolytic therapy. ACS Nano 10 (10), 9420−9433.

Swiech, L., Heidenreich, M., Banerjee, A., Habib, N., Li, Y., Trombetta, J., et al., 2015. In vivo interrogation of gene function in the mammalian brain using CRISPR-Cas9. Nat. Biotechnol. 33 (1), 102−106.

Tabebordbar, M., Zhu, K., Cheng, J.K.W., Chew, W.L., Widrick, J.J., Yan, W.X., et al., 2016. In vivo gene editing in dystrophic mouse muscle and muscle stem cells. Science 351 (6271), 407−411.

Tabernero, J., Shapiro, G.I., LoRusso, P.M., Cervantes, A., Schwartz, G.K., Weiss, G.J., et al., 2013. First-in-humans trial of an RNA interference therapeutic targeting VEGF and KSP in cancer patients with liver involvement. Cancer Discov. 3 (4), 406−417.

Takedatsu, H., Mitsuyama, K., Torimura, T., 2015. Nanomedicine and drug delivery strategies for treatment of inflammatory bowel disease. World J. Gastroenterol. 21 (40), 11343−11352.

Torchilin, V., 2011. Tumor delivery of macromolecular drugs based on the EPR effect. Adv. Drug Deliv. Rev. 63 (3), 131−135.

Valencia, P.M., Pridgen, E.M., Perea, B., Gadde, S., Sweeney, C., Kantoff, P.W., et al., 2013. Synergistic cytotoxicity of irinotecan and cisplatin in dual-drug targeted polymeric nanoparticles. Nanomedicine (Lond.) 8 (5), 687−698.

Veiseh, O., Tang, B.C., Whitehead, K.A., Anderson, D.G., Langer, R., 2015. Managing diabetes with nanomedicine: challenges and opportunities. Nat. Rev. Drug Discov. 14 (1), 45−57.

Ventola, C.L., 2012a. The nanomedicine revolution: Part 1: Emerging concepts. P T 37 (9), 512−525.

Ventola, C.L., 2012b. The nanomedicine revolution: Part 2: Current and future clinical applications. P T 37 (10), 582−591.

Ventola, C.L., 2012c. The nanomedicine revolution: Part 3: Regulatory and safety challenges. P T 37 (11), 631−639.

Verma, D.D., Levchenko, T.S., Bernstein, E.A., Mongayt, D., Torchilin, V.P., 2006. ATP-loaded immunoliposomes specific for cardiac myosin provide improved protection of the mechanical functions of myocardium from global ischemia in an isolated rat heart model. J. Drug Target. 14 (5), 273−280.

Verma, V.K., Kamaraju, S.R., Kancherla, R., Kona, L.K., Beevi, S.S., Debnath, T., et al., 2015. Fluorescent magnetic iron oxide nanoparticles for cardiac precursor cell selection from stromal vascular fraction and optimization for magnetic resonance imaging. Int. J. Nanomed. 10, 711−726.

Wicki, A., Witzigmann, D., Balasubramanian, V., Huwyler, J., 2015. Nanomedicine in cancer therapy: challenges, opportunities, and clinical applications. J. Control. Release 200, 138−157.

Xu, X., Xie, K., Zhang, X.Q., Pridgen, E.M., Park, G.Y., Cui, D.S., et al., 2013. Enhancing tumor cell response to chemotherapy through nanoparticle-mediated codelivery of siRNA and cisplatin prodrug. Proc. Natl. Acad. Sci. U.S.A. 110 (46), 18638−18643.

Yao, C., Slamovich, E.B., Webster, T.J., 2008. Enhanced osteoblast functions on anodized titanium with nanotube-like structures. J. Biomed. Mater. Res. A 85 (1), 157–166.

Yau, T., Cheng, P.N., Chan, P., Chan, W., Chen, L., Yuen, J., et al., 2013. A phase 1 dose-escalating study of pegylated recombinant human arginase 1 (Peg-rhArg1) in patients with advanced hepatocellular carcinoma. Invest. New Drugs 31 (1), 99–107.

Yin, H., Song, C.Q., Suresh, S., Wu, Q., Walsh, S., Rhym, L.H., et al., 2017. Structure-guided chemical modification of guide RNA enables potent non-viral in vivo genome editing. Nat. Biotechnol. 35 (12), 1179–1187.

You, J.O., Rafat, M., Ye, G.J., Auguste, D.T., 2011. Nanoengineering the heart: conductive scaffolds enhance connexin 43 expression. Nano Lett. 11 (9), 3643–3648.

Yu, B., Tai, H.C., Xue, W., Lee, L.J., Lee, R.J., 2010. Receptor-targeted nanocarriers for therapeutic delivery to cancer. Mol. Membr. Biol. 27 (7), 286–298.

CHAPTER

10 Evaluation of potential engineered nanomaterials impacts on human health: from risk for workers to impact on consumers

Massimiliano G. Bianchi[1], Ovidio Bussolati[1], Martina Chiu[1], Giuseppe Taurino[1] and Enrico Bergamaschi[2]

[1]*Laboratory of General Pathology, Department of Medicine and Surgery, University of Parma, Parma, Italy*

[2]*Department of Public Health Science and Pediatrics, University of Turin, Turin, Italy*

10.1 Introduction—what does this contribution deal with?

In this chapter we make an introductory appraisal of the impact of engineered nanomaterials (ENM) on human health. We speak of "potential ENM impacts on human health" because no specific toxic effect of ENM on humans has been demonstrated yet, in contrast with other unintentionally generated ultrafine particles. Moreover, as concluded by Krug (2014), "many good studies demonstrate, through careful analysis of the dose–response relationship, that we are operating in a safe area, since neither the effects shown nor the predicted environmental concentrations lead one to expect any impact on human health or the environment." To date, the situation does not seem substantially changed (Warheit, 2018).

Experimental studies on the effects of ENM on animals or in vitro models are being published at the impressing rate of more than 2000 per year, and an attempt to perform a critical review would be unreasonable. Moreover, the findings of these studies, especially those published some years ago, are often highly questionable, since many contributions do not take into account the "fundamental rules as are applied to toxicology" (Krug, 2014) and actually "contribute to the Babylonian plethora of low-value results that exists today" (Krug, 2014). However, those studies have fed the perception of an increased risk potential in the public opinion. At the same time, we have excluded from our contribution any discussion about the intrinsic limitations of in vitro studies and how these limitations can be overcome by the most advanced and sometimes sophisticated

Exposure to Engineered Nanomaterials in the Environment. DOI: https://doi.org/10.1016/B978-0-12-814835-8.00010-8

developments in the models adopted, such as 3D cultures, cocultures of different cell lines or primary cells, organoids, or organ-on-a-chip microdevices.

Our contribution will focus, instead, on two limited, yet critical, aspects of the rapidly developing field of nanotoxicology. We try to answer to the following questions. First, is there any documented risk for workers who produce, modify, handle, or in any sense exploit ENM at their workplace? Second, does the exposure to ENM in our everyday life of consumers have any potential consequence? Moreover, three subjects of great current interest, such as safety-by-design (SbD) strategies, adverse outcome pathway (AOP) approaches, and the relationship between synthetic and biological identities of ENM, will be also considered.

Emphasis is given to ENM widely present in industry or on the market, such as multiwalled carbon nanotubes (MWCNT) or TiO_2 and SiO_2 nanoparticles (NPs). Instead, intentional exposure to nanobiomedical devices, which are increasingly exploited or proposed in a variety of diagnostic and therapeutic applications, will not be considered.

10.2 Are engineered nanomaterial workers at risk?

The occupational setting represents perhaps the most likely situation in which chronic exposure to a specific ENM may occur. This exposure usually concerns the respiratory system where, due to their size, ENM can reach the alveoli (Borm et al., 2006). In animal models exposure to ENM often causes inflammation (Chou et al., 2008), intuitively related to the production of oxidant compounds, directly or indirectly dependent on the interaction between ENM and the tissue (Xia et al., 2006). On a chronic time scale, fibrosis commonly ensues (Sanchez et al., 2009).

In particular fiber-like materials, such as carbon nanotubes, behave like asbestos fibers in animals, and the mechanisms involved are actively investigated (Wang et al., 2017). Even though asbestos reminds of carcinogenic risks, available literature on fibrous ENM as possible carcinogens is not conclusive. In its 2017 monograph on carcinogenicity of some ENM and fibers (IARC, 2017), the IARC concluded that no relevant data on human carcinogenicity of MWCNT were available to the ad hoc constituted working group. However, several studies documented oncogenic activity of MWCNT in experimental animals. Intrascrotal injection of nanotubes in male rats caused peritoneal mesothelioma, the typical asbestos-related cancer (Sakamoto et al., 2009). These results were consistent with a previous contribution, reporting mesothelioma development in male $p53^{+/-}$ mice intraperitoneally injected with MWCNT (Takagi et al., 2008) and were successively confirmed in either male or female rats exposed through a single intraperitoneal injection of two distinct, although similar, MWCNT preparations (Nagai et al., 2011). A possible limit of these studies consists in the route of administration adopted, since exposure to MWCNT is expected to occur through the airways. However, inhaled MWCNT had a promoting effect on 3-methylcholanthrene-initiated bronchioloalveolar adenoma and lung adenocarcinoma in male mice (Sargent et al., 2014). On the contrary, negative results were reported for subcutaneous administration of MWCNT in male mice (Takanashi et al., 2012). Analysis of mechanistic data available in literature led the IARC Working Group to conclude for a moderate mechanistic evidence for mesothelioma-related end-points for MWCNT and for a weak evidence for single-walled carbon nanotubes (SWCNT), due to the lack of data. The IARC Working Group underlined that "the mechanistic events relevant to genotoxicity, lung inflammation, and fibrosis as well as translocation to the pleura, are liable to occur in humans exposed to Carbon NanoTubes (CNT) by inhalation." It should be stressed that most of the positive studies exploited a specific MWCNT preparation (MWCNT-7 from Mitsui Ltd) and that IARC pointed to the existence of significant gaps in the comprehension of the mechanisms underlying carcinogenicity due to CNT heterogeneity and limited number of long-term studies. Thus, while concluding that evidence for the carcinogenicity of carbon nanotubes in humans is inadequate, IARC considered sufficient the available evidence to define a particular MWCNT preparation (MWCNT-7) carcinogenic in experimental animals. The more limited availability of studies led IARC to define inadequate the evidence for the carcinogenicity of SWCNT. These elements led to the definition of MWCNT-7 as "possibly carcinogenic to humans (Group 2B)," according to the IARC grouping of carcinogens, while other MWCNT and SWCNT have been defined as not classifiable as to their carcinogenicity to humans (Group 3).

Another type of ENM produced in thousands of tons per year in several manufacturing processes is represented by NP of TiO_2. In a 2010 IARC monography TiO_2 is also categorized as a Group 2B carcinogen for humans (IARC, 2010); however, the monography concern both bulk and nanosized TiO_2. Moreover, none of the studies concerning human carcinogenicity considered the possible impact of particle size. IARC concluded that these studies (three epidemiological cohort studies and a single population-based case–control study) do not suggest an association between occupational exposure and risk for cancer (either in the lung or in other sites). In contrast, evidence for carcinogenicity in animals is stronger, although somewhat inconsistent. Indeed, while two inhalation studies in rats and one in mice were negative, two other studies reported an increased incidence of tumors in rats exposed to nanosized TiO_2. Conversely, oral, subcutaneous, and intraperitoneal exposures were all negative. On the basis of these data, IARC concluded that "there is inadequate evidence in humans for the carcinogenicity of titanium dioxide" but that "there is sufficient evidence in experimental animals for the carcinogenicity of titanium dioxide" and that "titanium dioxide is possibly carcinogenic to humans (Group 2B)." Again, these conclusions are not specifically applied to nanosized or bulk TiO_2.

It should be recalled that possible consequences of ENM inhalation may not be limited to local lung effects. Indeed, the large alveolar surface constitutes a possible way of ENM absorption and translocation to distant organs. Although this mechanism is quantitatively poorly efficient (Kreyling et al., 2017a), it has been documented in experimental models and may be of great pathophysiological significance. Moreover, local inflammation may lead to the release of inflammatory mediators, in particular cytokines, in the blood. Recent evidence indicates that this indirect mechanism may be more important than direct effects of the ENM translocated from the lungs to the rest of the body (Ganguly et al., 2017).

The relevance of the respiratory route explains why the presence of ENM in aerosols or in dust derived from any phase of the productive process is considered an important parameter (Savolainen and Pietroiusti, 2017; Kuhlbusch et al., 2011). Experimental evidence has indicated that in a real-life scenario nanomaterials often agglomerate or adsorb to background particles, suggesting that they interact with the respiratory system under these forms (Brouwer, 2010; Ding et al., 2017). This peculiarity should be adequately considered when experimental conditions for in vitro testing of biological effects are planned. On the contrary, a common tendency in the field is to disperse the ENM as much as possible (ideally obtaining monodispersed suspensions) using natural or synthetic dispersing agents, which, on the other hand, may also have unwanted effects. On the other hand, it is possible that the surfactant produced by alveolar cells acts as a natural dispersant, thus increasing the fraction of monodispersed ENM. To complicate further the situation, the biological effects of monodispersed or agglomerated ENM may be different, and ENM agglomerates may be endowed with peculiar toxic effects not detected with the monodispersed counterparts (Rotoli et al., 2015).

In the workplace environment ENM may be present under different forms, monodispersed or included in homo- or hetero-agglomerates. This fact, along with the high possibility of confounding effects from the NPs present in the environment independently from the productive process, renders exposure assessment a very challenging issue. Thus, robust exposure limits based on experimental evidence have not yet been established, and the recommended preventive approaches are still based on precautionary principles. The approaches proposed include modifying production processes, enforcing administrative means, and adopting personal protective equipment [see Savolainen and Pietroiusti, (2017) for an extensive discussion]. In particular, in the more general context of Safety-by-Design (SbD) strategies, modifications of the production process should be conceived and implemented, on the basis of ENM physicochemical properties, to lower the risk acting on either the exposure potential or the ENM-associated hazards.

However, occupational exposure limits (OELs) have been proposed for some ENM (TiO_2 NPs, carbon nanotubes, and nanofibers, silver NPs, cellulose nanocrystals) (Schulte et al., 2018). Given the hundreds of different ENM that are being produced and marketed, substance-by-substance efforts to establish specific OEL are not feasible, and the OELs proposed should be considered as prototypes for risk assessment in the perspective of grouping and categorization of ENM (Schulte et al., 2018).

Lack of clear cut adverse effects of ENM upon occupational respiratory exposure has not prevented the search for exposure biomarkers. The first wave of studies, reviewed by Liou et al. (2015), did not offer consistent results. More recently, in a series of contributions (Pelclova et al., 2016a,b,c, 2017), Pelclova et al. have demonstrated that inflammatory markers, markers of DNA and protein oxidative damage, and lipid oxidative markers are all increased in the exhaled breath condensate of workers exposed to NP of TiO_2. The authors report that the "median particle number concentration in the production line ranged from 1.98×10^4 to 2.32×10^4 particles/cm^3 with approximately 80% of the particles <100 nm in diameter," thus pointing to an effective exposure to nanosized titanium dioxide, and a mass concentration between 0.40 and 0.65 mg/m^3, well below the proposed OELs (NIOSH, 2011; Morimoto et al., 2010). Most recently, Zhao et al. have investigated cardiopulmonary parameters among workers who were exposed to NP of TiO_2 to identify the related biomarkers. The total mass concentration of particles was 3.17 mg/m^3, 39% of which were NPs. Changes in several markers of lung damage, cardiovascular disease, oxidative stress and inflammation were found associated with occupational exposure, but only the surfactant protein SP-D showed a time (dose)−response pattern in the exposed workers, while other markers did not show any change (Zhao et al., 2018).

Exposure routes other than respiratory, such as the gastrointestinal system and the skin, are less important for workers than for final consumers (see below, impact of ENM on consumers). However, if ENM do not penetrate in the airway wall and are not persistent in the lung tissue, they are trapped in mucus, moved up to the pharynx by ciliated cells, and eliminated through swallowing, thus reaching the gastrointestinal system (Kreyling et al., 2013). Although cutaneous exposure to manufactured ENMs in the workplace has been tentatively quantified (Van Duuren-Stuurman et al., 2010), effective dermal penetration is, at best, uncertain (see later, impact of ENM on consumers). On the other hand, secondary exposures may also follow skin exposure, through unwanted airway contamination.

10.3 Safety by design

The strategy of SbD {prevention through design (PtD) in the United States [National Institute for Occupational Safety and Health (NIOSH)]} is based on methods to minimize occupational hazards working on the early steps of the production process. In practice, hazards should be anticipated and eliminated, or at least reduced, through modifications of design, production, exploitation, storage, and disposal.

SbD strategy was not conceived specifically for ENM, but, since several years, the NIOSH has devoted much attention to SbD/PtD strategies for ENM workers. NIOSH launched a specific governmental supported initiative (NIOSH) and, more recently, has released a publication on "General Safe Practices for Working with Engineered Nanomaterials in Research Laboratories" (NIOSH, 2012), where Sbd/PtD strategies are highlighted. When applied to ENM, as for other materials, SbD approaches must take into account the whole life cycle and can target either exposure or hazard reduction (Geraci et al., 2015), thus minimizing risks to workers throughout the process (Schulte et al., 2008). In the hierarchy of controlling workplace hazards through (1) eliminating, substituting, or modifying the nanomaterials; (2) engineering the process to minimize or eliminate exposure to the nanomaterials; (3) implementing administrative controls that limit the quantity or duration of exposure to the nanomaterials; (4) providing for use of personal protective equipment (NIOSH, 2012), SbD strategies are thus in a higher position than the use of individual protection devices.

Several projects funded by the European Commission deal with the strategy of SbD. For instance, summing up the conclusions of the NanoImpactNet project, Hunt et al. (2013) state "Regulatory bodies should encourage an industrial and innovation approach by which midlife and end-of-life information must be fed into start-of-life (design) information, closing the life cycle of NMs."

For their successful exploitation, SbD strategies require an in-depth study of the structural determinants that underlie hazard or facilitate exposure. Indeed, several early attempts to apply a SbD strategy relied on the "fiber paradigm," the attribution to the fiber-like shape of an important portion of the hazard associated with the exposure to high-aspect ratio NPs (HARN), such as MWCNT (Donaldson et al., 2011; Czarny et al., 2014; Tsuruoka et al., 2015). These approaches may be applied to other HARNs (Allegri et al., 2016a). Mitigation of MWCNT-associated hazards has been also pursued through functionalization (Hussain et al., 2016; Fanizza et al., 2015; Deligianni, 2014; Chatterjee et al., 2014), which can modify several properties of the material, such as the tendency to agglomerate or protein adsorption (Allegri et al., 2016b). Examples of other potential applications of SbD strategies to ENM have been proposed for silica-based ENM (Lehman et al., 2016) and NPs of CeO (Davidson et al., 2016), while a critical appraisal of the state of the art has recently been provided by Hjorth et al. (2017).

10.4 Impact of engineered nanomaterial on consumers

The gastrointestinal system and the skin represent the most important routes for nonprofessional ENM exposure (Pietroiusti, 2012). As far as the oral route is concerned, a percentage of nanosized particles are present in common food additives, such as TiO_2 (E171) and amorphous silica (E551). Less frequently, gastrointestinal system can also be exposed to silver (used as colorant, E174) and zinc oxide NP. Another way to introduce ENM through the oral route is due to food contamination by nanomaterials shed from packaging. Fillers with at least one nanoscale dimension are present in many types of modern biopolymer-based packaging materials to improve the mechanical and barrier properties of the material, thus leading to the formation of nanocomposites (Ghanbarzadeh et al., 2015). Some NPs used as fillers for packaging materials, such as silver, TiO_2, or ZnO NP, have also antimicrobial properties (Ghanbarzadeh et al., 2015). Lastly, as recalled earlier, inhaled ENM, cleared from the airway through the mucociliar mechanism, is swallowed and, hence, reach the gastrointestinal system.

Exposure estimates have been performed for the most widely used foodborne ENM and have yielded values of 0.036 and 1.8 mg/kg of body weight for a heavy-consumer's daily intake of, respectively, TiO_2 (Lomer et al., 2000) and SiO_2 NP (Dekkers et al., 2011). How much of the ENM intake is absorbed? In a recent study, Kreyling et al. radiolabeled TiO_2 NP and followed their fate after a single ingestion in rats. More than 99% of the ingested amount was rapidly eliminated in feces. However, 0.6% of the administered dose passed the gastrointestinal-barrier and about 0.05% was still detected after 7 days, with NP identified and quantified in several organs, such as the skeleton, uterus, spleen, brain, kidneys, lungs, and the liver (Kreyling et al., 2017b). These data should be considered at the light of Ti levels measured postmortem in human liver and spleen, which indicate that TiO_2 particles, including a significant fraction of NP, are present in human liver and spleen. Authors conclude that "The levels are below the doses regarded as safe in animals, but half are above the dose that is deemed safe for liver damage in humans when taking into account several commonly applied uncertainty factors" and that "health risks due to oral exposure to TiO_2 cannot be excluded" (Heringa et al., 2018).

Direct investigations on possible adverse effects of the foodborne ENM are still in limited number. Moreover, many of these studies have been performed with laboratory-grade materials, consisting of pure ENM preparations, rather than with food-grade materials, corresponding to those effectively used as food additives in real life, which contain a variable nanosized fraction. This discrepancy can be conspicuous. Thus, at least for TiO_2 NP, it has been proposed that gastrointestinal effects should be studied with materials endowed with characteristics resembling those of food-grade TiO_2: (1) crystalline-phase: anatase, (2) isoelectric point: very close to 4.1, (3) fraction of NPs comprised between 15% and 45%, and (4) a low specific surface area (around 10 m^2/g) (Dudefoi et al., 2017).

In vitro experiments on cells of gastrointestinal origin have indicated that, while the acute toxicity of silica and TiO_2 NP is generally low, zinc oxide NP exhibits a moderate toxicity, likely attributable to the dissolution of NP in simulated gastrointestinal fluids (McCracken et al., 2013; Setyawati et al., 2015), and other ENM, such as silver NP, have a significant acute toxicity (Schneider et al., 2017). These effects are usually attributed to the release of a sizable amount of metal ions leading to ROS production and oxidative stress (Setyawati et al., 2015). Also animal studies indicate a low acute toxicity of TiO_2 NP (Jovanovic et al., 2016), with very high values of NOAEL [no observed adverse effect level (Warheit et al., 2015)]. However, for longer exposures, significant biological effects on intestinal cell function are observed. For instance, microvilli loss, impaired barrier function, and decreased nutrient transport are observed after "chronic" (5 days) but not acute (4 hours) exposure to TiO_2 NP of a Caco-2/HT29-MTX coculture model (Guo et al., 2017). This model is particularly interesting since the coculture is covered by a mucus layer, thus mimicking in this regard healthy intestinal mucosa. In the same model impaired glucose transport caused by exposure to TiO_2 NP has been recently described (Richter et al., 2018). In vivo experiments indicate that chronic oral exposure to TiO_2 NP produces also extraintestinal changes, such as hyperglycemia (Hu et al., 2016) and cardiac dysfunction (Hong et al., 2016), along with increased levels of Ti in several organs (Hu et al., 2016).

Besides the expected local effects, such as oxidative stress and inflammation, experimental evidence, reviewed by Bergin and Witzmann (2013), Pietroiusti et al. (2016), and Mercier-Bonin et al. (2018), suggests that ENM may exert important effects through their interaction with gut microbiota (the community of organisms living within the gastrointestinal tract). Since the concept of microbiota is by no way limited to the gastrointestinal system, although most available information concerns this district, it is possible that ENM interactions with the microbiota of other districts may also have biological relevance. An interesting example of interaction among intestinal epithelial cells, ENM, and microbiota has been recently offered by Richter et al. (2018). Glucose uptake was measured in the Caco-2/HT29-MTX coculture model cited above in the presence of TiO_2 NP, using control monolayers and monolayers cultured with the commensal *Lactobacillus rhamnosus* GG (*L. rhamnosus*). Reduction in glucose transport was observed along with microvilli damage in the absence but not in the presence of the beneficial bacteria.

Gastrointestinal tract is bathed by a variety of diverse secretions, from the mouth to the colon-rectum. The characteristics of these fluids, which are highly divergent, can affect the physicochemical features of ingested ENM, such as surface chemistry, dissolution (in case of metal or metal oxide NPs, such as Ag, CuO, or ZnO), agglomeration and, hence, the interaction of the material with the mucosa surface and its absorption (Jo et al., 2016).

Even more importantly, gastrointestinal fluids may affect the biological outcome of exposure to ENM in an additional way. Indeed, biological fluids are complex solutions of low- and high-molecular weight compounds. When ENM are suspended in these matrices, they adsorb proteins and other components, due to their high surface/volume ratio, a typical "nano" property. Thus, the ENM surface, which actually interacts with cells and extracellular structures, is not the bare NP but, rather, a complex array of biological molecules derived from the biological fluid in which the NPs, or their agglomerates, are suspended. The characteristics and dynamics of this corona of molecules have been mainly investigated in reconstituted systems in which the ENM interacts with one or more proteins, hence the denomination of "protein corona." But, especially in the complex environment of the gastrointestinal tract, which is also rich of lipids and detergent molecules, it is highly likely that the ENM corona has a more complex, likely tract-specific, composition. Actually, one of the first studies on oral exposure to ENM documented the adsorption of bile salts to the nanomaterial and attributed to this interaction some biological effects (McCracken et al., 2013). The interaction between ENM and the body surface (in this case, the gastrointestinal mucosa and the mucus layer that covers wide portions of the mucosal surface) is profoundly affected by the composition of the corona. Conversely, the interaction with the mucus layer markedly limits the contact of the material with mucosal cells, its absorption, and systemic delivery (Mercier-Bonin et al., 2018). Additional complexity is given by the possibility that ingested ENM also interact with food matrices (Go et al., 2017).

In conclusion, the structural identity of the ENM (i.e., the structural determinants that influence its biological effects and should be thoroughly characterized to identify proper structure–activity relationships) should be considered together with ENM biological identity, that is the surface characteristics resulting from the interaction of the NP with the components of the biological fluid in which they are dispersed (see Section 10.5).

As far as the skin is concerned, many cosmetic products contain nanostructured components. In particular, NPs of TiO_2, SiO_2, and ZnO are present in many sunscreens. SiO_2 NP can be also present in toothpastes, antiwrinkling products or polishing creams, due to their high absorbing properties. Evidence for skin penetration of the most common ENM has been thus far negative (Krug, 2014). For example, experiments performed on animal models indicated that, as anticipated for occupational exposures, TiO_2 NPs contained in topical products are not absorbed by intact skin (Sadrieh et al., 2010). However, no systematic investigations have been performed on the possibility that ENM, contained in cosmetics, may reach basal epidermis cells, or even distribute to other organs through dermal vessels, upon exposure of damaged skin. This issue is not trivial, since, in populations characterized by high risk for skin cancer, sunscreens are used everyday, sometimes over large areas of body surface.

10.5 Structural identity versus biological identity(-ies): the role of biocorona

As stated above, a layer of adsorbed molecules covers the surface of ENM, once they are dispersed in a biological fluid. This layer is composed by a stable portion, called the "hard" corona, which directly interacts with ENM surface, and a more superficial portion, called the "soft" corona. Protein corona is a highly dynamic structure and undergoes changes depending on the prolongation of the incubation or on other environmental conditions present during the interaction (Tenzer et al., 2013; Vilanova et al., 2016; Feiner-Gracia et al., 2017; Weiss et al., 2018). Given the high adsorption capability of ENM, the composition of this corona may be very complex. This has been demonstrated also for ENM widely present in productive processes and on the market. For instance, 115 and 48 proteins were identified through liquid chromatography-tandem mass spectrometry in the corona of, respectively, negatively charged SiO_2 or TiO_2 NP dispersed in rat plasma (Shim et al., 2014). Protein variety was lowered when adsorption was investigated with arginine-coated SiO_2 NP (Shim et al., 2014). Improved analytical methods have led to the discovery of a larger variety of corona proteins. When dispersed in respiratory tract lining fluid, ENM [specifically SiO_2 and poly(vinyl) acetate NPs] adsorb hundreds of different proteins [429 vs 698 proteins identified (Kumar et al., 2016)]. In this complex structures interactions occur not only between proteins and the ENM, but also among proteins themselves, giving rise to a "corona interactome" (Pisani et al., 2017a).

As expected, corona formation is strongly influenced by structural ENM properties, such size, shape, porosity and surface charge, composition, topography, or reactive groups (Sakulkhu et al., 2015; Tenzer et al., 2011; Ma et al., 2014; Paula et al., 2014; Clemments et al., 2015; Di Cristo et al., 2016), as well as by physicochemical characteristics of the dispersion fluid, such as pH (Titma, 2018).

The interaction between ENM and biomolecules, present in biological fluids, has complex, still incompletely characterized, effects on the biological reactivity of the NP. First, biocorona adsorption changes the surface properties of ENM and, hence, surface-dependent biological effects. For example, TiO_2 NPs, which are usually considered a low-toxicity ENM, catalyze photogenerated radical production and, hence, are endowed with phototoxicity. When the NPs are suspended in a protein-rich fluid, these effects are lowered in proportion of the protein-coated NP surface and, hence, of the protein concentration of the fluid (Garvas et al., 2015). However, the ability of TiO_2 NPs to produce oxidative damage is not limited to phototoxicity. Indeed, exposure to this ENM leads to oxidation of cell membrane lipids, an effect also mitigated by the protein corona (Runa et al., 2017). While ENM surface chemistry obviously affects protein corona formation and composition, different protein coronae may, conversely, affect other physicochemical properties of the ENM, such as agglomeration tendency and, hence, toxicity (Mortensen et al., 2013; Allegri et al., 2016b).

Moreover, the protein corona may directly determine the interaction between ENM and cells or tissues. For instance, the presence of a protein corona markedly modifies ENM uptake by cells (or penetration in tissues) (Lesniak et al., 2012; Caracciolo et al., 2015; Shahabi et al., 2015; Aoyama et al., 2016; Binnemars-Postma et al., 2016; Mirshafiee et al., 2016; Saikia et al., 2016; Tavano et al., 2018) and toxicity in vitro (Panas et al., 2013; Docter et al., 2014; Fedeli et al., 2014; Liu et al. 2015; Orlando et al., 2017) or in vivo (Yoshida et al., 2015; Saikia et al., 2016). Different protein coronae explain why the biological effects of a given ENM change depending on the serum (human vs bovine) used for the dispersion of the material (Izak-Nau et al., 2013) or, simply, for cell culture (Pisani et al., 2017b). If the adsorbed protein has a specific, biologically relevant, role in that cell/tissue system, the interaction may also confer novel biological activities to the ENM. For instance, fibrinogen, a major component of plasma proteins, significantly enhances cytotoxicity and proinflammatory activities of SiO_2, carbon soot, and TiO_2 NP on murine alveolar macrophages in a dose-dependent manner (Marucco et al., 2016).

The interaction with ENM may also cause conformational alterations and/or oxidative damage (Jayaram et al., 2017) of adsorbed proteins in both hard and soft coronae (Wang et al., 2011). For example, adsorption to TiO_2 or SiO_2 NP causes characteristic, pH-dependent distortions of adsorbed bovine serum albumin (Ranjan et al., 2016; Givens et al., 2017). Indeed, at pH 2 but not at pH 4, the protein is completely unfolded when adsorbed on TiO_2 NP and markedly stretched when adsorbed to SiO_2 NP. Interestingly, structural anomalies or altered conformation of adsorbed proteins elicit specific cell responses (Jayaram et al., 2017; Borgognoni et al., 2015). If the protein is an enzyme, modification of its conformation may cause inhibition, stimulation, or more complex changes in its activity, depending on the adsorbing ENM (Deng et al., 2014). Furthermore, enzyme activation by ENM may have relevant pathophysiological consequences, even in the absence of cells. For example, very low doses of TiO_2 NP (50 ng/mL) trigger the blood contact system through FXII adsorption and activation, leading to the stimulation of kinin, complement, and coagulation cascades (Ekstrand-Hammarstrom et al., 2015). Interestingly, among plasma proteins, SiO_2 NPs exhibit a peculiar adsorption capability toward components of complement and coagulation pathways, together with lipoproteins (Tenzer et al., 2011). Proteins involved in innate immunity present in other biological fluids also exhibit a peculiar tendency to be adsorbed to ENM (Kumar et al., 2016).

Lastly, ENM corona is composed not only by proteins, but also by other components of the biological fluid in which they are dispersed. Indeed, in bronchoalveolar fluid or gastrointestinal secretion, lipids (Whitwell et al., 2016) or other organic molecules may constitute a significant portion of the surface layer. For this reason, the term biocorona seems more appropriate than protein corona. An important bioactive molecule which may enter ENM biocorona is the bacterial lipopolysaccharide (LPS), a common environmental contaminant and a powerful macrophage activator. Adsorption of LPS to various ENM, such as SiO_2 (Di Cristo et al., 2016), TiO_2 (Bianchi et al., 2015), Ag (Galbiati et al., 2018), and Au (Li et al., 2017) NP, powerfully enhances their proinflammatory effects, at least in vitro. What is more important, also the activities of the adsorbed molecule are quantitatively and qualitatively changed (Bianchi et al., 2015, 2017; Li et al., 2017), thus adding an additional layer of complexity to the mechanisms underlying ENM effects in biological systems.

Overall, even for widely used and deeply characterized ENM, the formation of biocorona provides the nanomaterial with a novel biological identity (Monopoli et al., 2012), directly responsible for ENM biological activities, the formation of which is influenced by ENM physicochemical features. This fact may have profound repercussions in toxicological studies, either in vitro or in vivo (Vranic et al., 2017; Monopoli et al., 2011; Tenzer et al., 2013; Wohlleben et al., 2016). More importantly, the composition of the biological fluids is obviously modified in several conditions of pathophysiological relevance. Therefore the biological identity of a given ENM and, hence, its effects on cells and tissues are expected to vary accordingly. These multiple identities greatly expand the complexity of a preventive estimation of possible health effects of a single ENM.

For example, after having characterized the interaction between polystyrene and TiO_2 NPs with human bronchoalveolar fluid, Whitwell et al. suggest that ENM may interact, together with surfactant-associated proteins (SP-A, -B, and -D), also with lipids and pulmonary surfactant, implying potential health effects for "people with chronic airway diseases such as asthma and chronic obstructive pulmonary disease (COPD), or those who have increased susceptibility toward other respiratory diseases" (Whitwell et al., 2016). Thus, a more thorough investigation of ENM biological identities is needed to identify possible markers of enhanced susceptibility to adverse effects. Interestingly, a different composition of protein corona of SiO_2 NP dispersed in respiratory tract lining fluids derived from asthmatic or control subjects has been recently demonstrated (Kumar et al., 2017).

Finally, it should be remembered that ENM interaction with proteins or other organic molecules is by no means limited to extracellular organic fluids. For instance, silica NPs bind also intracellular proteins, in particular those with large unstructured regions, such as RNA-binding proteins and translation initiation factors (Klein et al., 2016; Vitali et al., 2018). Interestingly, ENM adsorption of intracellular proteins has been recently exploited for biotechnological applications (Fogli et al., 2017).

10.6 The adverse outcome pathway approach

An AOP is a conceptual construction of a sequential chain of events, causally related, experimentally validated, and affecting increasingly complex levels of biological organization, which results in an AO. The AOP concept has been applied to AOs, considered relevant for risk assessment, which affect either health or environment (OECD, 2017). Between the molecular initiating event (MIE) and the AO, the pathway is described by several key events (KEs), linked by KE relationships (OECD, 2017).

Besides its importance in defining a frame to describe mechanistic relationships to explain toxic effects at organism or population levels, AOPs provide an approach to use relatively simple biological models, or even in silico modeling, to study KE linked to outcomes relevant for risk assessment. Originally developed for ecotoxicology (Ankley et al., 2010), AOPs are increasingly exploited in chemical toxicology (OECD, 2018). In particular, in 2012 OECD started a program on the development of AOPs, defined them as "the central element of a toxicological knowledge framework being built to support chemical risk assessment based on mechanistic reasoning" (OECD).

The AOP approach has been proposed as a powerful tool for linking predictive toxicology to ENM risk assessment (Schulte et al., 2018; Lai et al., 2018; Mirshafiee et al., 2017). Indeed, as underlined recently by Vinken (2018), besides helping in the identification of data gaps or the logical organization of available experimental data, AOPs may also favor categorization and grouping of toxicants, linking shared KEs and MIEs to common physicochemical features. This obviously requires an in-depth characterization of the ENM as an integral component of the AOP formulation. AOP properties are particularly important for ENM grouping, since the fast introduction into the productive cycle and, subsequently, on the market of new nanomaterials makes practically unfeasible a one-by-one preventive toxicological analysis.

However, a correct adoption of AOP approach implies the satisfaction of several requirements. The most important is that MIE identification must be based on the assumption that sufficient experimental evidence exists to support the AOP and to document the ENM role. Therefore, the definition of a robust AOP requires a series of experimental studies aimed at defining not only the MIE but also the KEs.

For this reason, examples of fully defined AOP involving ENM are still lacking, and none of the AOPs (endorsed, under review, or under progress) enlisted in OECD AOP knowledge base (https://aopkb.oecd.org/) involves specifically ENM. Examples of proposed ENM-specific AOP concern model organisms such as *Danio rerio* [inhibition of egg hatching by CuO NPs (Muller et al., 2015)] and *Caenorhabditis elegans* (with effects at whole-genome level of MWCNT, combining system biology with possible identification of AOPs).

As far as possible impacts on human pathology are concerned, the development of an ENM-related AOP seems more advanced for carbonaceous nanomaterials and, in particular, for carbon nanotubes. Wang et al. (2015) explored the possibility that a common AOP leads to fibrosis starting from respiratory exposure to different carbonaceous ENM (three types of SWCNT, graphene, and two types of graphene oxide). They first performed in vitro studies on the production of inflammatory (IL-1β) and fibrogenic (TGF-β) cytokines by macrophages and airway epithelial cells, and then correlated the results with data from experiments on animal models, where, besides cytokine production, the development of lung fibrosis was assessed (Wang et al., 2015). Authors concluded that the dispersal state and surface reactivity of ENM were important determinants of the profibrogenic AOP. The correlation of changes at gene expression level in vivo (proposed as KE) with apical endpoints (fibrosis and septal thickness) has indicated that the doses of MWCNT responsible for the KE are comparable to those validated for the final outcomes by the NIOSH (Labib et al., 2016). The AOP leading to inflammation and fibrosis upon exposure to MWCNT has been further dissected in mice, demonstrating that Stat-6 activation, rather than IL-1β production, is correlated with fibrotic changes (Nikota et al., 2017). Finally, it is known that in experimental animals the exposure to MWCNT leads to the development of mesothelioma [a well-documented effect, at least for fiber-shaped MWCNT

(Suzui et al., 2016; Rittinghausen et al., 2014; Takagi et al., 2012; Sakamoto et al., 2009; Takagi et al., 2008; Nagai et al., 2011)]. However, the pathway involved is not completely defined (Kuempel et al., 2017) since the implied KE must be not only those associated with inflammation and fibrosis, although, undoubtedly, these mechanisms are very important (Poland et al., 2008). Recent evidence suggests that another KE, consistent with the immunopathologic findings detected in mesothelioma patients, is the capability of MWCNT to induce a local immunosuppressive state, increasing the accumulation of monocytic myeloid-derived suppressor cells and thus contrasting the T-cell—dependent immune surveillance on tumor cells (Huaux et al., 2016).

An additional, important advantage of the AOP approach is that, as an example cited earlier (Eom et al., 2015) demonstrates, it would easily include data from system biology techniques, which are increasingly exploited in nanotoxicological studies (Costa and Fadeel, 2016).

10.7 Conclusions

The knowledge-based body of data generated by systems toxicology (ST) and AOP analysis approaches will certainly improve the safety assessment of ENM, but the complex and multifaceted nature of events occurring at the nanobiointerfaces (at the cell, tissue, organ, and system levels) also implies that the full replacement of in vivo assessment is not yet possible (Bergamaschi et al., 2015). To serve Risk Analysis, the data generated by ST should be validated in real exposure scenarios where more complex and unpredictable interactions can occur. Epidemiological data from human populations specifically exposed to ENM are currently very limited for many reasons (Guseva Canu et al., 2018). Epidemiological research and interventional studies, which are a necessary prerequisite for health programs and prevention, should lead to identification, selection, and validation of candidate biomarkers for generalized exposure and health effects surveillance (Bergamaschi et al., 2015). In such studies biomarkers could help to circumvent the issues of the heterogeneity of ENM, making difficult to identify and recruit enough workers with the same exposure pattern, considering that exposures to different NPs may lead to the same pathway for disease, or share common mechanisms (Schulte and Hauser, 2012; Bergamaschi et al., 2017).

In conclusion, while we share the opinion that health decisions on a variety of nanomaterial types still await better scientific bases (Warheit, 2018), we wish to stress that, until now, diseases directly linked to ENM exposure "belong to the realm of possible risk (i.e. cannot be excluded, but are unlikely)" (Pietroiusti, 2012). However, it should also be stressed that our knowledge of the consequences of ENM interactions with specific bioactive molecules present in biological fluids or with the microbial populations resident in our body compartments (first of all in the gut) is still very incomplete. The elucidation of these interactions will likely represent the most important objectives of future research on the possible health impact of ENM.

Highlights

- Although the occupational setting is the most likely situation in which low-dose, chronic exposure to Engineered Nanomaterials (ENM) occurs, usually through inhalation, no occupational ENM-related disease has been reported yet, reliable exposure biomarkers have yet to be identified, and robust methodologies to define exposure have yet to be implemented.
- Safety-by-design (SbD) approaches are considered important tools of risk mitigation and prevention for workers, and, potentially, also for consumers exposed to ENM.
- Consumers are being increasingly exposed to ENM, especially through ingestion, due to the presence of ENM in food, or cutaneous exposure, given the presence of ENM in several cosmetics.
- Due to the increasing variety of ENM produced and put on the market, categorization based on adverse outcome pathway (AOP) approaches seems a promising strategy for a biologically relevant grouping.
- ENM high adsorption capability, a typical "nano" property, leads to the formation on the ENM surface of a biocorona, including proteins and other molecules present in the biological fluid in which the ENM are dispersed. The formation of the biocorona is a complex and dynamic process. The interaction of the ENM with cells and tissues, and hence their potential health effects, is strongly dependent on the biocorona that confers a new, evolving biological identity to the nanomaterial.

Key points

1. No ENM-specific toxic effect on humans has been demonstrated yet.
2. Nanoscale particles are common in workplaces, and are not necessarily associated with ENM production.
3. Workplace exposure to ENM mainly occurs through inhalation, thus rendering airways and the lungs important, yet not exclusive, target organs.
4. Robust strategies to estimate exposure specifically related to ENM are yet to be identified thus justifying protective measures based on precautionary approaches.
5. SbD strategies are in high position in the hierarchy of risk mitigation measures. These are based on methods to minimize ENM-related occupational hazards working on the early steps of the production process

through modifications of the design, production, exploitation, storage, and disposal.

6. Consumers are exposed to ENM mainly through ingestion or cutaneous application of widely marketed products, such as food additives and cosmetics, several of which contain nanosized components.
7. Although minor, a sizable fraction of ENM present in food additives is absorbed and distributed to other body compartments.
8. Conversely, no evidence of ENM absorption by intact skin has been obtained so far.
9. When interacting with body tissues, ENM adsorb bioactive molecules present in the biological fluids. As a result, the biological activities of both the ENM and the adsorbed molecule may change, with the complex acquiring a novel, tissue-dependent biological identity.
10. At the light of the ever increasing number and variety of ENM produced and marketed, preventive assessment of potential hazards requires grouping and categorization approaches, which should be based on both ENM structural features and expected biological identity.
11. Research activity should pursue the AOP strategy, thus favoring a preventive assessment of ENM-related hazard.

Acknowledgments

Supported, in part, by Grant Agreements LIFE 17 ENV/GR/000285 (LIFE NanoEXPLORE) and 760928 (Horizon2020 project BIORIMA) to E.B. M.C. is supported by a fellowship of "Associazione Italiana per la ricerca sul cancro" (AIRC, no. 19272).

References

Allegri, M., Bianchi, M.G., Chiu, M., Varet, J., Costa, A.L., Ortelli, S., et al., 2016a. Shape-related toxicity of titanium dioxide nanofibres. PLoS One 11, e0151365.

Allegri, M., Perivoliotis, D.K., Bianchi, M.G., Chiu, M., Pagliaro, A., Koklioti, M.A., et al., 2016b. Toxicity determinants of multi-walled carbon nanotubes: the relationship between functionalization and agglomeration. Toxicol. Rep. 3, 230–243.

Ankley, G.T., Bennett, R.S., Erickson, R.J., Hoff, D.J., Hornung, M.W., Johnson, R.D., et al., 2010. Adverse outcome pathways: a conceptual framework to support ecotoxicology research and risk assessment. Environ. Toxicol. Chem. 29, 730–741.

Aoyama, M., Hata, K., Higashisaka, K., Nagano, K., Yoshioka, Y., Tsutsumi, Y., 2016. Clusterin in the protein corona plays a key role in the stealth effect of nanoparticles against phagocytes. Biochem. Biophys. Res. Commun. 480, 690–695.

Bergamaschi, E., Murphy, F., Poland, C.A., Mullins, M., Costa, A.L., McAlea, E., et al., 2015. Impact and effectiveness of risk mitigation strategies on the insurability of nanomaterial production: evidences from industrial case studies. Wiley Interdiscip. Rev. Nanomed. Nanobiotechnol. 7, 839–855.

Bergamaschi, E., Guseva-Canu, I., Prina-Mello, A., Magrini, A., 2017. Biomonitoring. In: Fadeel, B., Pietroiusti, A., Shvedova, A. (Eds.), Adverse Effects of Engineered Nanomaterials, second ed. Academic Press, London.

Bergin, I.L., Witzmann, F.A., 2013. Nanoparticle toxicity by the gastrointestinal route: evidence and knowledge gaps. Int. J. Biomed. Nanosci. Nanotechnol. 3, 163–210.

Bianchi, M.G., Allegri, M., Costa, A.L., Blosi, M., Gardini, D., Del Pivo, C., et al., 2015. Titanium dioxide nanoparticles enhance macrophage activation by LPS through a TLR4-dependent intracellular pathway. Toxicol. Res. 4, 385–398.

Bianchi, M.G., Allegri, M., Chiu, M., Costa, A.L., Blosi, M., Ortelli, S., et al., 2017. Lipopolysaccharide adsorbed to the bio-corona of TiO_2 nanoparticles powerfully activates selected pro-inflammatory transduction pathways. Front. Immunol. 8, 866.

Binnemars-Postma, K.A., Ten Hoopen, H.W., Storm, G., Prakash, J., 2016. Differential uptake of nanoparticles by human M1 and M2 polarized macrophages: protein corona as a critical determinant. Nanomed. (Lond.) 11, 2889–2902.

Borgognoni, C.F., Mormann, M., Qu, Y., Schafer, M., Langer, K., Ozturk, C., et al., 2015. Reaction of human macrophages on protein corona covered TiO(2) nanoparticles. Nanomedicine 11, 275–282.

Borm, P.J., Robbins, D., Haubold, S., Kuhlbusch, T., Fissan, H., Donaldson, K., et al., 2006. The potential risks of nanomaterials: a review carried out for ECETOC. Part. Fibre Toxicol. 3, 11.

Brouwer, D., 2010. Exposure to manufactured nanoparticles in different workplaces. Toxicology 269, 120–127.

Caracciolo, G., Palchetti, S., Colapicchioni, V., Digiacomo, L., Pozzi, D., Capriotti, A.L., et al., 2015. Stealth effect of biomolecular corona on nanoparticle uptake by immune cells. Langmuir 31, 10764–10773.

Chatterjee, N., Yang, J., Kim, H.M., Jo, E., Kim, P.J., Choi, K., et al., 2014. Potential toxicity of differential functionalized multiwalled carbon nanotubes (MWCNT) in human cell line (BEAS2B) and *Caenorhabditis elegans*. J. Toxicol. Environ. Health A 77, 1399–1408.

Chou, C.C., Hsiao, H.Y., Hong, Q.S., Chen, C.H., Peng, Y.W., Chen, H.W., et al., 2008. Single-walled carbon nanotubes can induce pulmonary injury in mouse model. Nano Lett. 8, 437–445.

Clemments, A.M., Botella, P., Landry, C.C., 2015. Protein adsorption from biofluids on silica nanoparticles: corona analysis as a function of particle diameter and porosity. ACS Appl. Mater. Interfaces 7, 21682–21689.

Costa, P.M., Fadeel, B., 2016. Emerging systems biology approaches in nanotoxicology: towards a mechanism-based understanding of nanomaterial hazard and risk. Toxicol. Appl. Pharmacol. 299, 101–111.

Czarny, B., Georgin, D., Berthon, F., Plastow, G., Pinault, M., Patriarche, G., et al., 2014. Carbon nanotube translocation to distant organs after pulmonary exposure: insights from in situ (14)C-radiolabeling and tissue radioimaging. ACS Nano 8, 5715–5724.

Davidson, D.C., Derk, R., He, X., Stueckle, T.A., Cohen, J., Pirela, S.V., et al., 2016. Direct stimulation of human fibroblasts by $nCeO_2$ in vitro is attenuated with an amorphous silica coating. Part. Fibre Toxicol. 13, 23.

Dekkers, S., Krystek, P., Peters, R.J., Lankveld, D.P., Bokkers, B.G., van Hoeven-Arentzen, P.H., et al., 2011. Presence and risks of nanosilica in food products. Nanotoxicology 5, 393–405.

Deligianni, D.D., 2014. Multiwalled carbon nanotubes enhance human bone marrow mesenchymal stem cells' spreading but delay their proliferation in the direction of differentiation acceleration. Cell Adh. Migr. 8, 558–562.

Deng, Z.J., Butcher, N.J., Mortimer, G.M., Jia, Z., Monteiro, M.J., Martin, D.J., et al., 2014. Interaction of human arylamine *N*-acetyltransferase 1 with different nanomaterials. Drug Metab. Dispos. 42, 377–383.

Di Cristo, L., Movia, D., Bianchi, M.G., Allegri, M., Mohamed, B.M., Bell, A.P., et al., 2016. Proinflammatory effects of pyrogenic and precipitated amorphous silica nanoparticles in innate immunity cells. Toxicol. Sci. 150, 40–53.

Ding, Y., Kuhlbusch, T.A.J., Van Tongeren, M., Jimenez, A.S., Tuinman, I., Chen, R., et al., 2017. Airborne engineered nanomaterials in the workplace-a review of release and worker exposure during nanomaterial production and handling processes. J. Hazard. Mater. 322, 17–28.

Docter, D., Bantz, C., Westmeier, D., Galla, H.J., Wang, Q., Kirkpatrick, J.C., et al., 2014. The protein corona protects against size- and dose-dependent toxicity of amorphous silica nanoparticles. Beilstein J. Nanotechnol. 5, 1380–1392.

Donaldson, K., Murphy, F., Schinwald, A., Duffin, R., Poland, C.A., 2011. Identifying the pulmonary hazard of high aspect ratio nanoparticles to enable their safety-by-design. Nanomed. (Lond.) 6, 143–156.

Dudefoi, W., Terrisse, H., Richard-Plouet, M., Gautron, E., Popa, F., Humbert, B., et al., 2017. Criteria to define a more relevant reference sample of titanium dioxide in the context of food: a multiscale approach. Food Addit. Contam. Part A Chem. Anal. Control Expo. Risk Assess. 34, 653–665.

Ekstrand-Hammarstrom, B., Hong, J., Davoodpour, P., Sandholm, K., Ekdahl, K.N., Bucht, A., et al., 2015. TiO_2 nanoparticles tested in a novel screening whole human blood model of toxicity trigger adverse activation of the kallikrein system at low concentrations. Biomaterials 51, 58–68.

Eom, H.J., Roca, C.P., Roh, J.Y., Chatterjee, N., Jeong, J.S., Shim, I., et al., 2015. A systems toxicology approach on the mechanism of uptake and toxicity of MWCNT in *Caenorhabditis elegans*. Chem. Biol. Interact. 239, 153–163.

Fanizza, C., Casciardi, S., Incoronato, F., Cavallo, D., Ursini, C.L., Ciervo, A., et al., 2015. Human epithelial cells exposed to functionalized multiwalled carbon nanotubes: interactions and cell surface modifications. J. Microsc. 259, 173–184.

Fedeli, C., Segat, D., Tavano, R., De Franceschi, G., de Laureto, P.P., Lubian, E., et al., 2014. Variations of the corona HDL:albumin ratio determine distinct effects of amorphous SiO_2 nanoparticles on monocytes and macrophages in serum. Nanomed. (Lond.) 9, 2481–2497.

Feiner-Gracia, N., Beck, M., Pujals, S., Tosi, S., Mandal, T., Buske, C., et al., 2017. Super-resolution microscopy unveils dynamic heterogeneities in nanoparticle protein corona. Small 13, 1701631.

Fogli, S., Montis, C., Paccosi, S., Silvano, A., Michelucci, E., Berti, D., et al., 2017. Inorganic nanoparticles as potential regulators of immune response in dendritic cells. Nanomed. (Lond.) 12, 1647–1660.

Galbiati, V., Cornaghi, L., Gianazza, E., Potenza, M.A., Donetti, E., Marinovich, M., et al., 2018. In vitro assessment of silver nanoparticles immunotoxicity. Food Chem. Toxicol. 112, 363–374.

Ganguly, K., Ettehadieh, D., Upadhyay, S., Takenaka, S., Adler, T., Karg, E., et al., 2017. Early pulmonary response is critical for extra-pulmonary carbon nanoparticle mediated effects: comparison of inhalation versus intra-arterial infusion exposures in mice. Part. Fibre Toxicol. 14, 19.

Garvas, M., Testen, A., Umek, P., Gloter, A., Koklic, T., Strancar, J., 2015. Protein corona prevents TiO_2 phototoxicity. PLoS One 10, e0129577.

Geraci, C., Heidel, D., Sayes, C., Hodson, L., Schulte, P., Eastlake, A., et al., 2015. Perspectives on the design of safer nanomaterials and manufacturing processes. J. Nanopart. Res. 17, 366.

Ghanbarzadeh, B., Oleyaei, S.A., Almasi, H., 2015. Nanostructured materials utilized in biopolymer-based plastics for food packaging applications. Crit. Rev. Food Sci. Nutr. 55, 1699–1723.

Givens, B.E., Xu, Z., Fiegel, J., Grassian, V.H., 2017. Bovine serum albumin adsorption on SiO_2 and TiO_2 nanoparticle surfaces at circumneutral and acidic pH: a tale of two nano-bio surface interactions. J. Colloid Interface Sci. 493, 334–341.

Go, M.R., Bae, S.H., Kim, H.J., Yu, J., Choi, S.J., 2017. Interactions between food additive silica nanoparticles and food matrices. Front. Microbiol. 8, 1013.

Guo, Z., Martucci, N.J., Moreno-Olivas, F., Tako, E., Mahler, G.J., 2017. Titanium dioxide nanoparticle ingestion alters nutrient absorption in an in vitro model of the small intestine. NanoImpact 5, 70–82.

Guseva Canu, I., Schulte, P.A., Riediker, M., Fatkhutdinova, L., Bergamaschi, E., 2018. Methodological, political and legal issues in the assessment of the effects of nanotechnology on human health. J. Epidemiol. Community Health 72, 148–153.

Heringa, M.B., Peters, R.J.B., Bleys, R., van der Lee, M.K., Tromp, P.C., van Kesteren, P. C.E., et al., 2018. Detection of titanium particles in human liver and spleen and possible health implications. Part. Fibre Toxicol. 15, 15.

Hjorth, R., van Hove, L., Wickson, F., 2017. What can nanosafety learn from drug development? The feasibility of "safety by design". Nanotoxicology 11, 305–312.

Hong, F., Wu, N., Zhao, X., Tian, Y., Zhou, Y., Chen, T., et al., 2016. Titanium dioxide nanoparticle-induced dysfunction of cardiac hemodynamics is involved in cardiac inflammation in mice. J. Biomed. Mater. Res. A 104, 2917–2927.

Hu, H., Li, L., Guo, Q., Jin, S., Zhou, Y., Oh, Y., et al., 2016. A mechanistic study to increase understanding of titanium dioxide nanoparticles-increased plasma glucose in mice. Food Chem. Toxicol. 95, 175–187.

Huaux, F., d'Ursel de Bousies, V., Parent, M.A., Orsi, M., Uwambayinema, F., Devosse, R., et al., 2016. Mesothelioma response to carbon nanotubes is associated with an early and selective accumulation of immunosuppressive monocytic cells. Part. Fibre Toxicol. 13, 46.

Hunt, G., Lynch, I., Cassee, F., Handy, R.D., Fernandes, T.F., Berges, M., et al., 2013. Towards a consensus view on understanding nanomaterials hazards and managing exposure: knowledge gaps and recommendations. Materials (Basel) 6, 1090–1117.

Hussain, S., Ji, Z., Taylor, A.J., DeGraff, L.M., George, M., Tucker, C.J., et al., 2016. Multiwalled carbon nanotube functionalization with high molecular weight hyaluronan significantly reduces pulmonary injury. ACS Nano 10, 7675–7688.

IARC, 2010. Carbon Black, Titanium Dioxide, and Talc. IARC, Lyon.

IARC, 2017. Some Nanomaterials and Some Fibres. IARC, Lyon.

Izak-Nau, E., Voetz, M., Eiden, S., Duschl, A., Puntes, V.F., 2013. Altered characteristics of silica nanoparticles in bovine and human serum: the importance of nanomaterial characterization prior to its toxicological evaluation. Part. Fibre Toxicol. 10, 56.

Jayaram, D.T., Runa, S., Kemp, M.L., Payne, C.K., 2017. Nanoparticle-induced oxidation of corona proteins initiates an oxidative stress response in cells. Nanoscale 9, 7595–7601.

Jo, M.R., Yu, J., Kim, H.J., Song, J.H., Kim, K.M., Oh, J.M., et al., 2016. Titanium dioxide nanoparticle-biomolecule interactions influence oral absorption. Nanomater. (Basel) 6, 125.

Jovanovic, B., Cvetkovic, V.J., Mitrovic, T., 2016. Effects of human food grade titanium dioxide nanoparticle dietary exposure on *Drosophila melanogaster* survival, fecundity, pupation and expression of antioxidant genes. Chemosphere 144, 43–49.

Klein, G., Mathe, C., Biola-Clier, M., Devineau, S., Drouineau, E., Hatem, E., et al., 2016. RNA-binding proteins are a major target of silica nanoparticles in cell extracts. Nanotoxicology 10, 1555–1564.

Kreyling, W.G., Semmler-Behnke, M., Takenaka, S., Moller, W., 2013. Differences in the biokinetics of inhaled nano- versus micrometer-sized particles. Acc. Chem. Res. 46, 714–722.

Kreyling, W.G., Holzwarth, U., Haberl, N., Kozempel, J., Wenk, A., Hirn, S., et al., 2017a. Quantitative biokinetics of titanium dioxide nanoparticles after intratracheal instillation in rats: Part 3. Nanotoxicology 11, 454–464.

Kreyling, W.G., Holzwarth, U., Schleh, C., Kozempel, J., Wenk, A., Haberl, N., et al., 2017b. Quantitative biokinetics of titanium dioxide nanoparticles after intravenous injection in rats: Part 2. Nanotoxicology 11, 443–453.

Krug, H.F., 2014. Nanosafety research—are we on the right track? Angew. Chem. Int. Ed. Engl. 53, 12304–12319.

Kuempel, E.D., Jaurand, M.C., Moller, P., Morimoto, Y., Kobayashi, N., Pinkerton, K.E., et al., 2017. Evaluating the mechanistic evidence and key data gaps in assessing the potential carcinogenicity of carbon nanotubes and nanofibers in humans. Crit. Rev. Toxicol. 47, 1–58.

Kuhlbusch, T.A., Asbach, C., Fissan, H., Gohler, D., Stintz, M., 2011. Nanoparticle exposure at nanotechnology workplaces: a review. Part. Fibre Toxicol. 8, 22.

Kumar, A., Bicer, E.M., Morgan, A.B., Pfeffer, P.E., Monopoli, M., Dawson, K.A., et al., 2016. Enrichment of immunoregulatory proteins in the biomolecular corona of nanoparticles within human respiratory tract lining fluid. Nanomedicine 12, 1033–1043.

Kumar, A., Bicer, E.M., Pfeffer, P., Monopoli, M.P., Dawson, K.A., Eriksson, J., et al., 2017. Differences in the coronal proteome acquired by particles depositing in the lungs of asthmatic versus healthy humans. Nanomedicine 13, 2517–2521.

Labib, S., Williams, A., Yauk, C.L., Nikota, J.K., Wallin, H., Vogel, U., et al., 2016. Nano-risk science: application of toxicogenomics in an adverse outcome pathway framework for risk assessment of multi-walled carbon nanotubes. Part. Fibre Toxicol. 13, 15.

Lai, R.W.S., Yeung, K.W.Y., Yung, M.M.N., Djurisic, A.B., Giesy, J.P., Leung, K.M.Y., 2018. Regulation of engineered nanomaterials: current challenges, insights and future directions. Environ. Sci. Pollut. Res. Int. 25, 3060–3077.

Lehman, S.E., Morris, A.S., Mueller, P.S., Salem, A.K., Grassian, V.H., Larsen, S.C., 2016. Silica nanoparticle-generated ROS as a predictor of cellular toxicity: mechanistic insights and safety by design. Environ. Sci. Nano 3, 56–66.

Lesniak, A., Fenaroli, F., Monopoli, M.P., Aberg, C., Dawson, K.A., Salvati, A., 2012. Effects of the presence or absence of a protein corona on silica nanoparticle uptake and impact on cells. ACS Nano 6, 5845–5857.

Li, Y., Shi, Z., Radauer-Preiml, I., Andosch, A., Casals, E., Luetz-Meindl, U., et al., 2017. Bacterial endotoxin (lipopolysaccharide) binds to the surface of gold nanoparticles, interferes with biocorona formation and induces human monocyte inflammatory activation. Nanotoxicology 11, 1157–1175.

Liou, S.H., Tsai, C.S., Pelclova, D., Schubauer-Berigan, M.K., Schulte, P.A., 2015. Assessing the first wave of epidemiological studies of nanomaterial workers. J. Nanopart. Res. 17, 413.

Liu, T.P., Wu, S.H., Chen, Y.P., Chou, C.M., Chen, C.T., 2015. Biosafety evaluations of well-dispersed mesoporous silica nanoparticles: towards in vivo-relevant conditions. Nanoscale 7, 6471–6480.

Lomer, M.C., Thompson, R.P., Commisso, J., Keen, C.L., Powell, J.J., 2000. Determination of titanium dioxide in foods using inductively coupled plasma optical emission spectrometry. Analyst 125, 2339–2343.

Ma, Z., Bai, J., Wang, Y., Jiang, X., 2014. Impact of shape and pore size of mesoporous silica nanoparticles on serum protein adsorption and RBCs hemolysis. ACS Appl. Mater. Interfaces 6, 2431–2438.

Marucco, A., Gazzano, E., Ghigo, D., Enrico, E., Fenoglio, I., 2016. Fibrinogen enhances the inflammatory response of alveolar macrophages to TiO_2, SiO_2 and carbon nanomaterials. Nanotoxicology 10, 1–9.

McCracken, C., Zane, A., Knight, D.A., Dutta, P.K., Waldman, W.J., 2013. Minimal intestinal epithelial cell toxicity in response to short- and long-term food-relevant inorganic nanoparticle exposure. Chem. Res. Toxicol. 26, 1514–1525.

Mercier-Bonin, M., Despax, B., Raynaud, P., Houdeau, E., Thomas, M., 2018. Mucus and microbiota as emerging players in gut nanotoxicology: the example of dietary silver and titanium dioxide nanoparticles. Crit. Rev. Food. Sci. Nutr. 58, 1023–1032.

Mirshafiee, V., Kim, R., Park, S., Mahmoudi, M., Kraft, M.L., 2016. Impact of protein precoating on the protein corona composition and nanoparticle cellular uptake. Biomaterials 75, 295–304.

Mirshafiee, V., Jiang, W., Sun, B., Wang, X., Xia, T., 2017. Facilitating translational nanomedicine via predictive safety assessment. Mol. Ther. 25, 1522–1530.

Monopoli, M.P., Walczyk, D., Campbell, A., Elia, G., Lynch, I., Bombelli, F.B., et al., 2011. Physical-chemical aspects of protein corona: relevance to in vitro and in vivo biological impacts of nanoparticles. J. Am. Chem. Soc. 133, 2525–2534.

Monopoli, M.P., Aberg, C., Salvati, A., Dawson, K.A., 2012. Biomolecular coronas provide the biological identity of nanosized materials. Nat. Nanotechnol. 7, 779–786.

Morimoto, Y., Kobayashi, N., Shinohara, N., Myojo, T., Tanaka, I., Nakanishi, J., 2010. Hazard assessments of manufactured nanomaterials. J. Occup. Health 52, 325–334.

Mortensen, N.P., Hurst, G.B., Wang, W., Foster, C.M., Nallathamby, P.D., Retterer, S.T., 2013. Dynamic development of the protein corona on silica nanoparticles: composition and role in toxicity. Nanoscale 5, 6372–6380.

Muller, E.B., Lin, S., Nisbet, R.M., 2015. Quantitative adverse outcome pathway analysis of hatching in zebrafish with CuO nanoparticles. Environ. Sci. Technol. 49, 11817–11824.

NIOSH, 2011. NIOSH Current Intelligence Bulletin 63: Occupational Exposure to Titanium Dioxide. NIOSH, Cincinnati, OH.

NIOSH, 2012. General safe practices for working with engineered nanomaterials in research laboratories. <https://www.cdc.gov/niosh/docs/2012-147/default.html> (accessed 14.05.18.).

Nagai, H., Okazaki, Y., Chew, S.H., Misawa, N., Yamashita, Y., Akatsuka, S., et al., 2011. Diameter and rigidity of multiwalled carbon nanotubes are critical factors in mesothelial injury and carcinogenesis. Proc. Natl. Acad. Sci. U.S.A. 108, E1330–E1338.

Nikota, J., Banville, A., Goodwin, L.R., Wu, D., Williams, A., Yauk, C.L., et al., 2017. Stat-6 signaling pathway and not interleukin-1 mediates multi-walled carbon nanotube-induced lung fibrosis in mice: insights from an adverse outcome pathway framework. Part. Fibre Toxicol. 14, 37.

OECD, 2017. Adverse outcome pathways, molecular screening and toxicogenomics. <http://www.oecd.org/chemicalsafety/testing/adverse-outcome-pathways-molecular-screening-and-toxicogenomics.htm> (accessed 19.04.18.).

OECD, 2018. Users' Handbook Supplement to the Guidance Document for Developing and Assessing Adverse Outcome Pathways. OECD Series on Adverse Outcome Pathways, No. 1. OECD Publishing, Paris. https://doi.org/10.1787/5jlv1m9d1g32-en.

Orlando, A., Cazzaniga, E., Tringali, M., Gullo, F., Becchetti, A., Minniti, S., et al., 2017. Mesoporous silica nanoparticles trigger mitophagy in endothelial cells and perturb neuronal network activity in a size- and time-dependent manner. Int. J. Nanomed. 12, 3547–3559.

Panas, A., Marquardt, C., Nalcaci, O., Bockhorn, H., Baumann, W., Paur, H.R., et al., 2013. Screening of different metal oxide nanoparticles reveals selective toxicity and inflammatory potential of silica nanoparticles in lung epithelial cells and macrophages. Nanotoxicology 7, 259–273.

Paula, A.J., Silveira, C.P., Martinez, D.S., Souza Filho, A.G., Romero, F.V., Fonseca, L.C., et al., 2014. Topography-driven bionano-interactions on colloidal silica nanoparticles. ACS Appl. Mater. Interfaces 6, 3437–3447.

Pelclova, D., Zdimal, V., Fenclova, Z., Vlckova, S., Turci, F., Corazzari, I., et al., 2016a. Markers of oxidative damage of nucleic acids and proteins among workers exposed to TiO_2 (nano) particles. Occup. Environ. Med. 73, 110–118.

Pelclova, D., Zdimal, V., Kacer, P., Fenclova, Z., Vlckova, S., Komarc, M., et al., 2016b. Leukotrienes in exhaled breath condensate and fractional exhaled nitric oxide in workers exposed to TiO_2 nanoparticles. J. Breath Res. 10, 036004.

Pelclova, D., Zdimal, V., Kacer, P., Fenclova, Z., Vlckova, S., Syslova, K., et al., 2016c. Oxidative stress markers are elevated in exhaled breath condensate of workers exposed to nanoparticles during iron oxide pigment production. J. Breath Res. 10, 016004.

Pelclova, D., Zdimal, V., Kacer, P., Zikova, N., Komarc, M., Fenclova, Z., et al., 2017. Markers of lipid oxidative damage in the exhaled breath condensate of nano TiO_2 production workers. Nanotoxicology 11, 52–63.

Pietroiusti, A., 2012. Health implications of engineered nanomaterials. Nanoscale 4, 1231–1247.

Pietroiusti, A., Magrini, A., Campagnolo, L., 2016. New frontiers in nanotoxicology: gut microbiota/microbiome-mediated effects of engineered nanomaterials. Toxicol. Appl. Pharmacol. 299, 90–95.

Pisani, C., Gaillard, J.C., Odorico, M., Nyalosaso, J.L., Charnay, C., Guari, Y., et al., 2017a. The timeline of corona formation around silica nanocarriers highlights the role of the protein interactome. Nanoscale 9, 1840–1851.

Pisani, C., Rascol, E., Dorandeu, C., Gaillard, J.C., Charnay, C., Guari, Y., et al., 2017b. The species origin of the serum in the culture medium influences the in vitro toxicity of silica nanoparticles to HepG2 cells. PLoS One 12, e0182906.

Poland, C.A., Duffin, R., Kinloch, I., Maynard, A., Wallace, W.A., Seaton, A., et al., 2008. Carbon nanotubes introduced into the abdominal cavity of mice show asbestos-like pathogenicity in a pilot study. Nat. Nanotechnol. 3, 423–428.

Ranjan, S., Dasgupta, N., Srivastava, P., Ramalingam, C., 2016. A spectroscopic study on interaction between bovine serum albumin and titanium dioxide nanoparticle synthesized from microwave-assisted hybrid chemical approach. J. Photochem. Photobiol. B 161, 472–481.

Richter, J.W., Shull, G.M., Fountain, J.H., Guo, Z., Musselman, L.P., Fiumera, A.C., et al., 2018. Titanium dioxide nanoparticle exposure alters metabolic homeostasis in a cell culture model of the intestinal epithelium and *Drosophila melanogaster*. Nanotoxicology 12, 390–406.

Rittinghausen, S., Hackbarth, A., Creutzenberg, O., Ernst, H., Heinrich, U., Leonhardt, A., et al., 2014. The carcinogenic effect of various multi-walled carbon nanotubes (MWCNTs) after intraperitoneal injection in rats. Part. Fibre Toxicol. 11, 59.

Rotoli, B.M., Gatti, R., Movia, D., Bianchi, M.G., Di Cristo, L., Fenoglio, I., et al., 2015. Identifying contact-mediated, localized toxic effects of MWCNT aggregates on epithelial monolayers: a single-cell monitoring toxicity assay. Nanotoxicology 9, 230–241.

Runa, S., Lakadamyali, M., Kemp, M.L., Payne, C.K., 2017. TiO_2 nanoparticle-induced oxidation of the plasma membrane: importance of the protein corona. J. Phys. Chem. B 121, 8619–8625.

Sadrieh, N., Wokovich, A.M., Gopee, N.V., Zheng, J., Haines, D., Parmiter, D., et al., 2010. Lack of significant dermal penetration of titanium dioxide from sunscreen formulations containing nano- and submicron-size TiO_2 particles. Toxicol. Sci. 115, 156–166.

Saikia, J., Yazdimamaghani, M., Hadipour Moghaddam, S.P., Ghandehari, H., 2016. Differential protein adsorption and cellular uptake of silica nanoparticles based on size and porosity. ACS Appl. Mater. Interfaces 8, 34820–34832.

Sakamoto, Y., Nakae, D., Fukumori, N., Tayama, K., Maekawa, A., Imai, K., et al., 2009. Induction of mesothelioma by a single intrascrotal administration of multi-wall carbon nanotube in intact male Fischer 344 rats. J. Toxicol. Sci. 34, 65–76.

Sakulkhu, U., Mahmoudi, M., Maurizi, L., Coullerez, G., Hofmann-Amtenbrink, M., Vries, M., et al., 2015. Significance of surface charge and shell material of superparamagnetic iron oxide nanoparticle (SPION) based core/shell nanoparticles on the composition of the protein corona. Biomater. Sci. 3, 265–278.

Sanchez, V.C., Pietruska, J.R., Miselis, N.R., Hurt, R.H., Kane, A.B., 2009. Biopersistence and potential adverse health impacts of fibrous nanomaterials: what have we learned from asbestos? Wiley Interdiscip. Rev. Nanomed. Nanobiotechnol. 1, 511–529.

Sargent, L.M., Porter, D.W., Staska, L.M., Hubbs, A.F., Lowry, D.T., Battelli, L., et al., 2014. Promotion of lung adenocarcinoma following inhalation exposure to multi-walled carbon nanotubes. Part. Fibre Toxicol. 11, 3.

Savolainen, K., Pietroiusti, A., 2017. Exposure assessment. In: Fadeel, B., Pietroiusti, A., Shvedova, A. (Eds.), Adverse Effects of Engineered Nanomaterial Exposure, Toxicology, and Impact on Human Health, second ed. Elsevier—Academic Press, London.

Schneider, T., Westermann, M., Glei, M., 2017. In vitro uptake and toxicity studies of metal nanoparticles and metal oxide nanoparticles in human HT29 cells. Arch. Toxicol. 91, 3517–3527.

Schulte, P.A., Hauser, J.E., 2012. The use of biomarkers in occupational health research, practice, and policy. Toxicol. Lett. 213, 91–99.

Schulte, P.A., Rinehart, R., Okun, A., Geraci, C.L., Heidel, D.S., 2008. National prevention through design (PtD) initiative. J. Safety Res. 39, 115–121.

Schulte, P.A., Kuempel, E.D., Drew, N.M., 2018. Characterizing risk assessments for the development of occupational exposure limits for engineered nanomaterials. Regul. Toxicol. Pharmacol. 95, 207–219.

Setyawati, M.I., Tay, C.Y., Leong, D.T., 2015. Mechanistic investigation of the biological effects of SiO(2), TiO(2), and ZnO nanoparticles on intestinal cells. Small 11, 3458–3468.

Shahabi, S., Treccani, L., Dringen, R., Rezwan, K., 2015. Modulation of silica nanoparticle uptake into human osteoblast cells by variation of the ratio of amino and sulfonate surface groups: effects of serum. ACS Appl. Mater. Interfaces 7, 13821–13833.

Shim, K.H., Hulme, J., Maeng, E.H., Kim, M.K., An, S.S., 2014. Analysis of SiO2 nanoparticles binding proteins in rat blood and brain homogenate. Int. J. Nanomed. 9 (Suppl 2), 207–215.

Suzui, M., Futakuchi, M., Fukamachi, K., Numano, T., Abdelgied, M., Takahashi, S., et al., 2016. Multiwalled carbon nanotubes intratracheally instilled into the rat lung induce development of pleural malignant mesothelioma and lung tumors. Cancer Sci. 107, 924–935.

Takagi, A., Hirose, A., Nishimura, T., Fukumori, N., Ogata, A., Ohashi, N., et al., 2008. Induction of mesothelioma in p53$^{+/-}$ mouse by intraperitoneal application of multi-wall carbon nanotube. J. Toxicol. Sci. 33, 105–116.

Takagi, A., Hirose, A., Futakuchi, M., Tsuda, H., Kanno, J., 2012. Dose-dependent mesothelioma induction by intraperitoneal administration of multi-wall carbon nanotubes in p53 heterozygous mice. Cancer Sci. 103, 1440–1444.

Takanashi, S., Hara, K., Aoki, K., Usui, Y., Shimizu, M., Haniu, H., et al., 2012. Carcinogenicity evaluation for the application of carbon nanotubes as biomaterials in rasH2 mice. Sci. Rep. 2, 498.

Tavano, R., Gabrielli, L., Lubian, E., Fedeli, C., Visentin, S., Polverino de Laureto, P., et al., 2018. C1q-mediated complement activation and C3 opsonization trigger recognition of stealth poly(2-methyl-2-oxazoline)-coated silica nanoparticles by human phagocytes. ACS Nano. 12, 5834–5847.

Tenzer, S., Docter, D., Rosfa, S., Wlodarski, A., Kuharev, J., Rekik, A., et al., 2011. Nanoparticle size is a critical physicochemical determinant of the human blood plasma corona: a comprehensive quantitative proteomic analysis. ACS Nano 5, 7155–7167.

Tenzer, S., Docter, D., Kuharev, J., Musyanovych, A., Fetz, V., Hecht, R., et al., 2013. Rapid formation of plasma protein corona critically affects nanoparticle pathophysiology. Nat. Nanotechnol. 8, 772–781.

Titma, T., 2018. The effect of surface charge and pH on the physiological behaviour of cobalt, copper, manganese, antimony, zinc and titanium oxide nanoparticles in vitro. Toxicol. In Vitro 50, 11–21.

Tsuruoka, S., Matsumoto, H., Koyama, K., Akiba, E., Yanagisawa, T., Cassee, F.R., et al., 2015. Radical scavenging reaction kinetics with multiwalled carbon nanotubes. Carbon N.Y. 83, 232–239.

Van Duuren-Stuurman, B., Pelzer, J., Moehlmann, C., Berges, M., Bard, D., Wake, D., et al., 2010. A structured observational method to assess dermal exposure to manufactured nanoparticles DREAM as an initial assessment tool. Int. J. Occup. Environ. Health 16, 399–405.

Vilanova, O., Mittag, J.J., Kelly, P.M., Milani, S., Dawson, K.A., Radler, J.O., et al., 2016. Understanding the kinetics of protein-nanoparticle corona formation. ACS Nano 10 10842–10850.

Vinken, M., 2018. Taking adverse outcome pathways to the next level. Toxicol. In Vitro 50, A1–A2.

Vitali, M., Rigamonti, V., Natalello, A., Colzani, B., Avvakumova, S., Brocca, S., et al., 2018. Conformational properties of intrinsically disordered proteins bound to the surface of silica nanoparticles. Biochim. Biophys. Acta 1862, 1556–1564.

Vranic, S., Gosens, I., Jacobsen, N.R., Jensen, K.A., Bokkers, B., Kermanizadeh, A., et al., 2017. Impact of serum as a dispersion agent for in vitro and in vivo toxicological assessments of TiO_2 nanoparticles. Arch Toxicol. 91, 353–363.

Wang, J., Jensen, U.B., Jensen, G.V., Shipovskov, S., Balakrishnan, V.S., Otzen, D., et al., 2011. Soft interactions at nanoparticles alter protein function and conformation in a size dependent manner. Nano Lett. 11, 4985–4991.

Wang, X., Duch, M.C., Mansukhani, N., Ji, Z., Liao, Y.P., Wang, M., et al., 2015. Use of a pro-fibrogenic mechanism-based predictive toxicological approach for tiered testing and decision analysis of carbonaceous nanomaterials. ACS Nano 9, 3032–3043.

Wang, X., Sun, B., Liu, S., Xia, T., 2017. Structure activity relationships of engineered nanomaterials in inducing NLRP3 inflammasome activation and chronic lung fibrosis. NanoImpact 6, 99–108.

Warheit, D.B., 2018. Hazard and risk assessment strategies for nanoparticle exposures: how far have we come in the past 10 years? F1000Res. 7, 376.

Warheit, D.B., Boatman, R., Brown, S.C., 2015. Developmental toxicity studies with 6 forms of titanium dioxide test materials (3 pigment-different grade & 3 nanoscale) demonstrate an absence of effects in orally-exposed rats. Regul. Toxicol. Pharmacol. 73, 887–896.

Weiss, A.C.G., Kempe, K., Forster, S., Caruso, F., 2018. Microfluidic examination of the "hard" biomolecular corona formed on engineered particles in different biological milieu. Biomacromolecules 19, 2580–2594.

Whitwell, H., Mackay, R.M., Elgy, C., Morgan, C., Griffiths, M., Clark, H., et al., 2016. Nanoparticles in the lung and their protein corona: the few proteins that count. Nanotoxicology 10, 1385–1394.

Wohlleben, W., Driessen, M.D., Raesch, S., Schaefer, U.F., Schulze, C., Vacano, B., et al., 2016. Influence of agglomeration and specific lung lining lipid/protein interaction on short-term inhalation toxicity. Nanotoxicology 10, 970–980.

Xia, T., Kovochich, M., Brant, J., Hotze, M., Sempf, J., Oberley, T., et al., 2006. Comparison of the abilities of ambient and manufactured nanoparticles to induce cellular toxicity according to an oxidative stress paradigm. Nano Lett. 6, 1794–1807.

Yoshida, T., Yoshioka, Y., Morishita, Y., Aoyama, M., Tochigi, S., Hirai, T., et al., 2015. Protein corona changes mediated by surface modification of amorphous silica nanoparticles suppress acute toxicity and activation of intrinsic coagulation cascade in mice. Nanotechnology 26, 245101.

Zhao, L., Zhu, Y., Chen, Z., Xu, H., Zhou, J., Tang, S., et al., 2018. Cardiopulmonary effects induced by occupational exposure to titanium dioxide nanoparticles. Nanotoxicology 12, 169–184.

Further reading

NIOSH, 2018. Prevention through design. <https://www.cdc.gov/niosh/topics/PtD/> (accessed 18.04.18.).

SECTION

IV

Social and regulatory issues in application of engineered nanomaterials

CHAPTER

Using "nano tools" as the basis for a hands-on experiential course in nanotechnology

11

Geoffrey D. Bothun[1], Vinka Oyanedel-Craver[2] and Keunhan Park[3]

[1]*Department of Chemical Engineering, University of Rhode Island, Kingston, RI, United States*

[2]*Department of Civil and Environmental Engineering, University of Rhode Island, Kingston, RI, United States*

[3]*Department Mechanical Engineering, University of Utah, Salt Lake City, UT, United States*

11.1 Introduction

In order to ensure that nanomaterials and nano-enabled products are beneficial to our society, it is essential to educate a workforce capable of performing research and development activities that translate into manufacturing opportunities. As such, an emerging "nanotechnology workforce" requires training not only in the fundamental aspects of nanoscale properties and processes but also in the use of advance characterization techniques and metrology for nanomaterials, as well as the application and implication of nano-enabled products (Jackman et al., 2016). These training components rely on the convergence of life sciences, natural sciences, and engineering to enable a collaborative approach for the development of nanotechnology workforce across disciplines.

Training and educating the nanotechnology workforce is a priority of several agencies worldwide, notably the US National Science Foundation (NSF), which has awarded more than 200 grants in this area between 1996 and 2015 (Bainbridge and Roco, 2016). This funding has supported the development of nanotechnology and nanoscience programs across the United States, as well as collaborative efforts with other countries. Many training and education efforts have been focused on developing curriculum to preparing students in technical and fundamental skills. However, due to the broader implications of the application of nanomaterials and nanoproducts, the nanotechnology workforce needs additional knowledge and skills addressing issues relating to environmental health and safety (EHS), and ethical, legal, and societal implications (ELSI). This broad training is often at odds with typical discipline-specific training and education

Exposure to Engineered Nanomaterials in the Environment. DOI: https://doi.org/10.1016/B978-0-12-814835-8.00011-X

efforts but is necessary to promote sustainable nanomanufacturing through workforce development (Rocabert et al., 2017).

This chapter discusses the development of an interdisciplinary undergraduate-level nanotechnology course for science and engineering students that is rooted in hands-on, experiential learning. The approach builds on student knowledge gained through introductory chemistry and physics courses, providing an experience that is amenable to a wide range of science and engineering majors. Through the course, students examined the physical phenomena underlying key nanotechnology instrumentation (i.e., "nano tools"), including pathways of photons and electrons, and why these phenomena enabled nanoscale characterization. With the tools at hand, students applied fundamental concepts relating to metrology and synthesis to design, create, and characterize nanomaterials through laboratories and mentor-guided team projects. Team projects, coupled with focused workshops, facilitated collaboration across disciplines, which was aided by communication training to identify fundamental nanomaterial properties and enable applications and to reveal potential impacts on EHS and associated ethical, legal and society aspects (ELSA). Course improvement was achieved using a feedback loop approach to balance breadth versus depth and to advance interdisciplinary group problem and project-based learning. A summary of assessment results and a discussion of the logistical and pedagogical challenges that were encountered over a 3-year period are provided.

11.2 Background

An interdisciplinary "nano tools" course, supported by the NSF Nanotechnology Undergraduate Education in Engineering program, was offered three times between 2014 and 2016 at the University of Rhode Island to develop nanotechnology concept competences and professional skills in undergraduate students by exposing them to state-of-the-art instruments commonly used in nanotechnology. This course was led by the Departments of Chemical Engineering and Civil & Environmental Engineering, with support from the Rhode Island Consortium for Nanoscience and Nanotechnology and the College of Engineering. The objectives of nano tools were:

1. to provide basic knowledge of the principles and operation of nanoscale instrumentations;
2. to foster problem-based, peer-to-peer learning through research-oriented group projects;
3. to enhance students' technical communication skills through presentations, journal-formatted project reports, and online learning;
4. to enhance student–faculty and faculty–faculty collaboration; and
5. to expose students to societal, ethical, economic, environmental, and entrepreneurial/commercial implications of nanotechnology.

Table 11.1 Demographics of students enrolled in the course.

	2014 (%)	2015 (%)	2016 (%)
Engineering	78	50	60
Sciences	22	50	40
Undergraduate	*56*	*100*	*70*
Graduate	*44*	*0*	*30*

The objectives were achieved using an approach that differs from many nanotechnology courses. First, focus was given to three techniques—dynamic light scattering (DLS), electron microscopy, and atomic force microscopy (AFM)—and the underlying physics behind these techniques that permit nanoscale characterization. These three techniques were selected not only because they are able to provide characterization of different aspects of nanoparticles but also because of their differences in terms of operational fundamentals (see below for more details). Second, emphasis was placed on obtaining complimentary information from these three techniques through intense hands-on training. Third, the knowledge and training gained by the students was applied to an independent problem-based research project led by a faculty mentor. This approach led to a rich educational experience and an in-depth understanding of how properties at the nanoscale relate to those at higher dimensions, and how these properties can be assessed.

The course was open to science, technology, engineering and mathematics (STEM) majors at the junior and senior level as a professional elective that addresses student outcomes used for ABET accreditation (ABET was an acronym for the Accreditation Board for Engineering and Technology; now ABET stands for confidence in technical education). The percentages of students in engineering and science, and at the undergraduate and graduate levels, are shown in Table 11.1. While undergraduate students were given priority, graduate students were also invited to participate in the course. A mix of undergraduate majors from different STEM disciplines and the combination of undergraduate and graduate students fostered peer-to-peer learning and created an undergraduate–graduate mentoring dynamic.

11.3 Course design

The nano tools course was delivered in three phases: an introductory, a nano tools (instrumentation), and a project phase (Fig. 11.1). The introductory phase familiarized students with nanotechnology concepts, terms, and processes such as length scale and surface-to-volume ratios, as well as reactivity, surface modification, and surface interactions [i.e., Derjaguin, Landau, Verwey, and Overbeek (DLVO) theory accounting for van der Waals dispersion and electrostatic interactions] as a function of material and coating composition. During

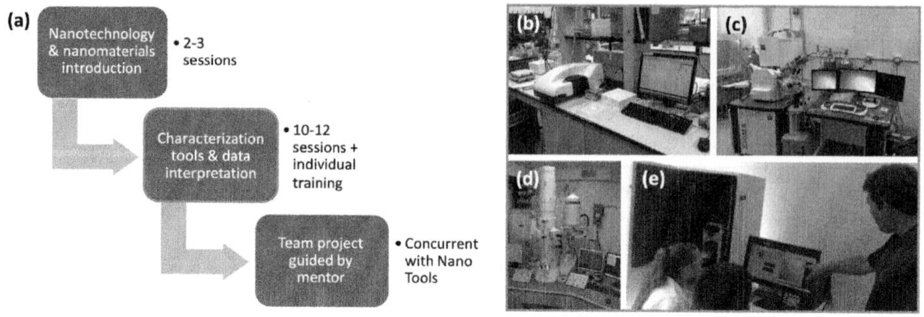

FIGURE 11.1

Nano tools course overview. (A) The three phases of the course and (B–E) the nano tools employed: (B) DLS, (C) SEM, (D) TEM, and (E) AFM. *AFM*, Atomic force microscopy; *DLS*, dynamic light scattering; *SEM*, scanning electron microscopy; *TEM*, transmission electron microscopy.

this phase, students participated in laboratory sessions through which they acquired general skills in laboratory techniques and experimental design, and safety awareness related with working with nanomaterials. This created a common knowledge base among the students from different disciplines. Students synthesized their own gold nanoparticles with different surface coatings that were used, along with a standard nanoparticle certified by the National Institutes of Science and Technology, throughout the entire course to illustrate how different instruments provided complimentary information needed to provide a more complete characterization of a given material. The nanoparticles were in colloidal dispersion in water, which also allowed the students to examine the effects of aging over the semester (e.g., aggregation) and how sample preparation influenced characterization (e.g., drying to prepare microscopy specimens). Homework assignments and laboratory reports were the principle modes of student assessment during the introductory phase.

The second phase developed competency in nano tools characterization. Between two to three sessions (weeks) were allocated for each tool through which fundamental principles of the tool, the operational protocols, and data analysis were emphasized (Fig. 11.2). Lectures connected the operational principles of a tool such as the pathway of light, electrons, or a cantilever to the ability to determine nanoscale properties. The three tools employed and the rationale for their selection are listed below. Laboratory reports were the principle mode of student assessment during the nano tool phase.

- DLS. DLS provided hydrodynamic diameters and zeta potentials of colloidal dispersions, allowing students to connect the synthesis process to particle size and surface charge density. Students measured the size and charge of the gold nanoparticles, they synthesized as a function of salt concentration and salt valency to illustrate how changes in electrostatic repulsion (according to

DLVO) led to changes in the effective particle size due to aggregation (Fig. 11.3). Results were reported and discussed based on the principles of DLS, which consisted of converting raw scattering intensity data into intensity, volume, and number-averaged histogram plots.

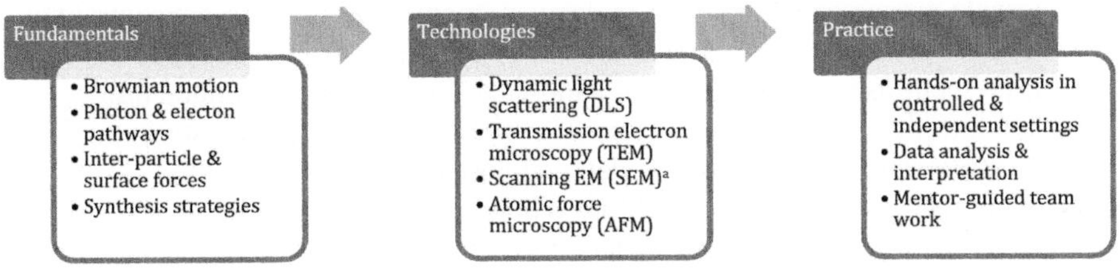

FIGURE 11.2

Nano tool competency development strategy. SEM was removed from the course after the first offering (2014) for two reasons: (1) to provide more time for deeper learning of remaining tools and (2) it was not suitable for detailed nanoparticle characterization. *SEM*, Scanning electron microscopy.

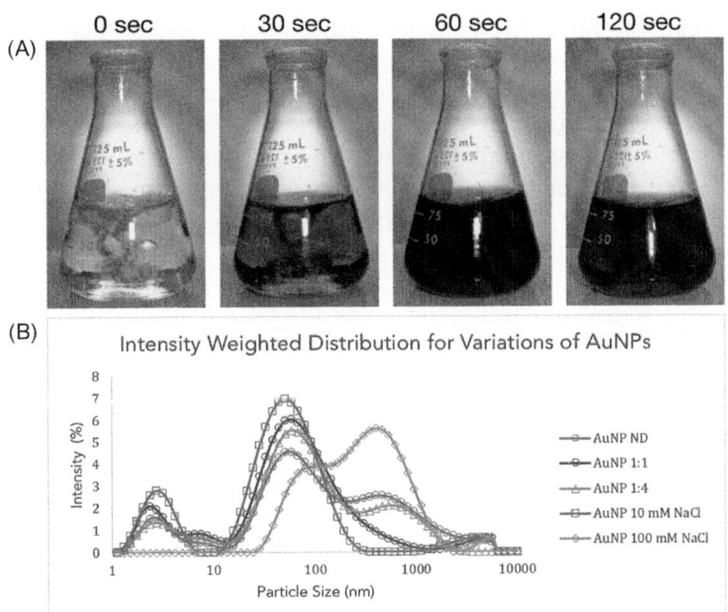

FIGURE 11.3

Example student results of gold nanoparticle synthesis and DLS analysis (abbreviated here as AuNP). (A) Nanoparticles capped with (sodium) citrate prepared by a modified Brust method. (B) DLS results as a function of dilution and salt (NaCl) concentration. Results are shown as presented by the students without modification. *DLS*, Dynamic light scattering.

- Transmission electron microscopy (TEM). TEM provided core nanoparticle diameters and allowed the students to compare these to hydrodynamic diameters from DLS to examine the effect of surface coating, charged layers (i.e., ionic Stern and double layers), and aggregation on the difference between the two diameters. These coating or solution properties further reflect nanoscale processes where local molecular orientations and ion concentrations differ from the bulk, providing an instructional basis for connecting measurement data to the effects of nanoscale dimensions and interactions. Histogram plots of nanoparticle diameter were compared to polystyrene particle standards to demonstrate data validation (Fig. 11.4).
- Scanning electron microscopy (SEM). While SEM provided nanoparticle diameters, it is not ideal for characterizing small nanoparticles such as those synthesized by the students (~10–50 nm in diameter). SEM was used to demonstrate how specimen preparation (sample drying) led to aggregates where individual particles were difficult to distinguish. Students diluted their samples, guided by particle number concentration calculations, to determine the concentration that provided individual particle resolution on the specimen. **Note that SEM was used only in 2014 and removed from the list of tools in 2015 and 2016 to provide more time for a deeper learning experience with the other tools.*
- AFM. Similar to SEM, AFM provided nanoparticle diameters that could be compared to TEM, SEM, and DLS results. Depending on the mode of operation, AFM measurements are based on the deflection of a mechanical cantilever due to topography and/or surface forces. AFM was particularly useful in relating content in introductory physics courses (deflection distance, material spring constants) to the ability to measure nanoscale properties. Students utilized line scan features with serial nanoparticle dilution to obtain an average diameter (Fig. 11.5).

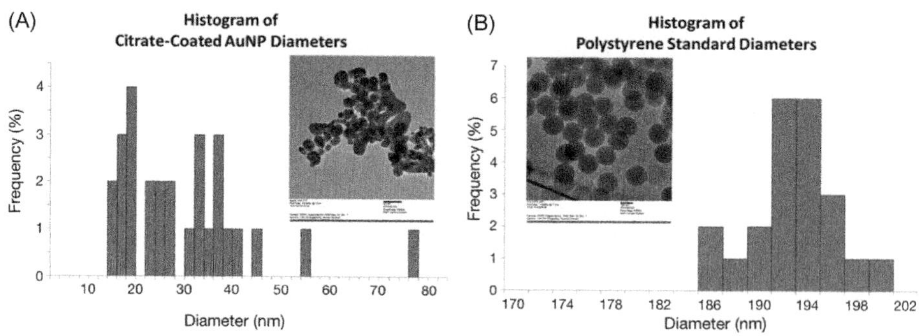

FIGURE 11.4

Example student results of TEM analysis. Histograms of (A) gold nanoparticles and (B) polystyrene particle standards are compared. Image analysis for particle diameter performed with imageJ. Results are shown as presented by the students without modification. *TEM*, Transmission electron microscopy.

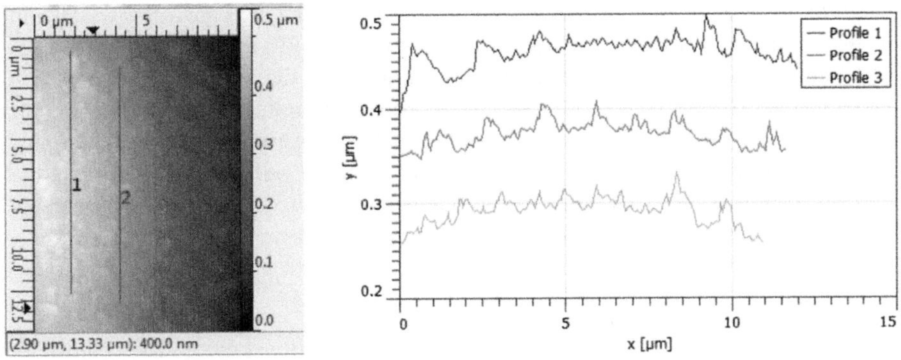

FIGURE 11.5

Example student results of AFM analysis. An AFM image is shown with the corresponding line scans. Results are shown as presented by the students without modification. *AFM*, atomic force microscopy.

During the nano tools phase, groups presented, compared, and discussed their results comprehensively, combining results for all tools as the semester progressed. Comprehensive nanomaterial analysis is critical for properly characterizing materials in the context of nanotechnology EHS. This was discussed extensively with the students, and examples were provided from the course and from scientific literature where incomplete analysis with just a single tool does not capture information needed to assess EHS (e.g., surface reactivity of a nanoparticle is a function of its diameter; DLS alone would not provide this information). Discussing the role of characterization in determining EHS complemented the first laboratory session where students learned about risk assessment and management during nanomanufacturing and how management varied based on material properties. Students were briefed about current occupational safety and health procedures as well as national, state, and institutional safety requirements for nanotechnology activities in the laboratory and at the industrial scale.

Our team recognized that the ELSI of nanotechnology is critical to build public confidence and support for innovation and commercialization of nanomaterials and nanotechnologies (N.S.A.T.C.C.O Technology, 2016). Therefore we proactively addressed ELSI within the context of nanotechnology by promoting awareness and education of these aspects to the students in two ways. First, students participated in a laboratory session comprised by a lecture followed by a role playing scenario and second, students viewed a series of videos and were asked a series of reflective questions. During the laboratory session, topics such as responsible nanomanufacturing, life cycle assessment, national and global regulations, voluntary reporting initiatives, and public opinion were covered. Then, the students participated in a role playing game called "You Decide" developed by NISE Network (2014) in which students discusses how technology and society interact with each other, as well as how personal and societal values can impact the development, use, and adoption of nanotechnologies. Finally, students viewed a series of videos developed by the Project on Emerging Nanotechnologies and the NSF called "Nanotechnology: The Power of the Small" (http://powerofsmall.org/index.php). This series covers topics related to environment and human health as well as public perception of the impacts of nanotechnology on societal needs such as food production, privacy, and healthcare. Students answered questions such as follows:

- What are the benefits and drawbacks of using nanotechnology-derived pollution remediation methods?
- Would you vote in favor of or against running the pilot remediation project?
- As a consumer, would you like to see products containing nanoparticles labeled as such?
- Twenty years from now, how do you think nanotechnology will be a part of our lives?
- What opportunities do you think nanotechnology might create?

During the third phase, students applied the knowledge and skills learned to an interdisciplinary team project where students were mentored by faculty members other than the course instructors. Engaging faculty mentors provided an additional level of assessment from an independent expert in the field of nanotechnology. Students were required to employ at least two of the tools used during the second (nano tools) phase. Team projects were presented and assigned at the beginning of the course, and students met with their mentors throughout the semester to develop a background understanding of the project and to design an experimental approach. Mentors received a "mentor packet" that guided their interaction with the students and provided the basis for a research proposal (overview, resources required, timeline) written by the teams. Examples of project topics are listed below:

- nanoparticle synthesis and impregnation into hydrogels for controlled release;
- polymer nanoparticle synthesis and characterization for inhalation therapy;
- graphene-based electrodes for lithium-ion batteries;
- self-assembled coatings on magnetic nanoparticles; and
- nanoparticle characterization in environmentally relevant conditions.

The independent nature of the project required that the students develop collaboration, project management, and communication skills. During the project phase, lecture time was devoted to project updates and class discussions, and additional user time and instruction were provided on each of the tools. Class discussions were particularly insightful and used to engage students in critical thinking about potential EHS impacts of the materials they were working with. Update and final presentations, and peer and mentor evaluations, were the principle modes of student assessment during the project phase.

11.4 Course revisions and assessment

The first course offering in 2014 was used as the basis for subsequent course revisions in 2015 and 2016. Additional course revisions were made based on annual and continuous feedback—continuous feedback in the form of student performance and informal surveys allowed for the course to be tailored "on the fly" to achieve sufficient breadth and depth on the topics. Pre- and postsurveys were given to the students to assess their initial and gained knowledge and competences proposed for the course, respectively. Table 11.2 shows an example of the comments received from the students in 2014, and the actions taken to address them and the outcomes of these actions in 2015 and 2016.

Most student comments in 2014 pertained to the laboratory experiments and the team project. With respect to the laboratories, students indicated that more time was needed on each instrument (more depth) to gain competency and the ability to analyze results (connecting in class and experiential learning). To provide a more effective in-laboratory experience a series of videos was prepared by the instructors and graduate-teaching assistants introducing students to each tool and outlining the

Table 11.2 Students feedback and actions taken.

What are your suggestions for changes that would improve this course (2014)?	Actions taken (2015–16)	Outcome (2015–16)
Use each instrument and write a status report, then at the end of the year present all data gathered	Required students to compare and analyze data cumulatively in each lab report	A better understanding of how multiple tools are needed for characterization of nanomaterials; and determine which of them is more appropriate for each material
Provide more information regarding the content being asked for the lab reports	Posed specific questions to address pertaining to analysis, data quality, and comparison to previous techniques (e.g., describe what properties DLS and TEM measure and why they are different)	Ability to produce more technically accurate and informative lab reports with improved analyses
Make project a semester long learning experience	Projects and mentors assigned at beginning of class; work ongoing throughout semester linking tools with specific project elements	A more in-depth experience where real connections were made between the tool and how it enables applied research; students were able to present results at conferences

DLS, *Dynamic light scattering;* TEM, *transmission electron microscopy.*

basic hardware and software operation. Students were required to watch the videos and prepare prelaboratory reports before attending the laboratory session. These videos allowed us to reduce the time allocated for basic instruction, increasing the time devoted by the students for sample preparation and characterization of their nanomaterials. In addition to the videos, extra time was provided to the students for individualized or team training on all the tools after the initial laboratory session(s). Students seeking more time on a specific instrument were allowed to schedule personal time with the respective equipment manager.

In addition to providing videos and expanding the students' access to the tools, laboratory report requirements were revised to guide the experimental approach and the data analysis. Specific questions were added to the laboratory assignment to focus student's attention on discussing how the operational fundamentals of the tool facilitated nanoscale characterization. In-class examples of data analysis, presentation, and reporting were provided. Finally, students were required to add a section in their report comparing the results obtained from their nanoparticles to NIST standard nanoparticles. This was done so that students could discuss the complementarity of the results from different tools as well as identify the limitations and possible artifacts of each of the tools used.

Knowledge competency assessments were evaluated using pre- and postsurveys. Three domains of knowledge were assessed:

- *Fundamentals*: At the nanoscale, factors relating to size and scale (e.g., size, scale, scaling, shape, proportionality, dimensionality) help describe matter and predict its behavior. A basic understanding of the DLVO theory and quantum mechanics, and the effect of nanoparticle composition and the physicochemical characteristics of the bulk solution, is used to determine interparticles forces and assess the stability of nanoparticle dispersions.
- *Technical knowledge*: The principles and basic operation of the different tools is applied to provide complimentary information on a common sample that can be gained from the different tools.
- *EHS and ELSI*: An understanding of the environmental, health, and safety effect of nanotechnology development/applications, possible risks, and mitigation strategies. An understanding of the broad implications of nanotechnology on social, economic, workforce, educational, ethical, and legal societal aspects.

We developed a series of questions both quantitative and qualitative in nature that the students answered. The answers provided by the students were scored using a four-tier rubric matrix developed based on a modified scientific ability rubric from Rutgers University Physics and Astronomy Education Research (Etkina et al., 2010; Wansom et al., 2009). Fig. 11.6 shows the evolution of the scores over the 3 years in which the course was offered. The figure shows how the modifications of the course based on feedback and assessment in 2014 (year 1) increased the level of competency achieved by the students in 2015 and 2016. These results also support the use of the tools as vehicle to expand the knowledge

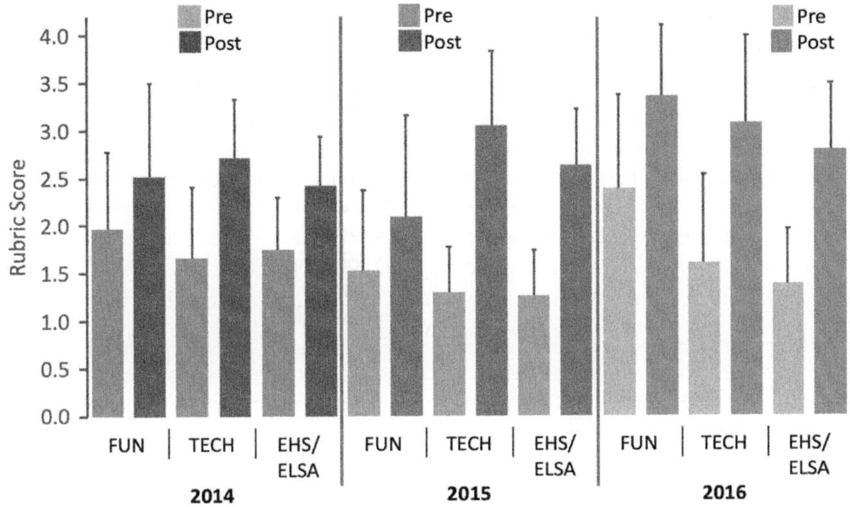

FIGURE 11.6

Rubric score for each of the knowledge domains over the 3-year period. Lighter colors show average scores for the initial presurveys and darker colors for the postsurveys. Standard deviations are shown corresponding to $n = 9$, 10, 12 for 2014, 2015, and 2016, respectively. *FUN*, fundamentals; *TECH*, technology; *EHS/ELSA*, environmental health and safety and ethical, legal, and societal issues.

of students about fundamental nanotechnology processes as it demonstrated as score increment between pre- and postsurvey with regard to the domain of fundamental knowledge. Students' perceptions about their acquired knowledge and skills during the course were collected through open-ended questions (Table 11.3). Students valued learning fundamentals and technical skills associated with each tool. Students also perceived that they not only gained knowledge regarding the operation of the tools and analysis of the results but also learned about nanoscale fundamental processes. This theoretical–practical approach supports the development of the "nanotechnology workforce" in two ways: first by developing essential nanomanufacturing and nano-characterization skills, and second by providing comprehensive knowledge about nanotechnology principles. Similar courses reported at the graduate level have also been successful (de Melo et al., 2017), reinforcing the idea that this approach could be replicated at different institutions and with undergraduate students.

Students identified the team projects as one of their most valued educational experiences in the course. Since teams were formed with students from different majors and levels, members were able to contribute complementary skill to tackle the problems they faced regarding sample preparation, tool operation, and results analysis. Finally, another positive aspect of the participation in interdisciplinary teams was to foster effective communication across fields, which allowed them to

Table 11.3 Students perceptions about knowledge and skills acquired.

What was the most effective part of this course?	What do you consider to be the most important?
Teaching the instruments in lab first, letting us try it, then using it in lab the next week, and then use it again in the research project	Being introduced to the nanoworld at all. This course made me think a lot about my future in terms of research or even a career
Very good overview of applications, as well as thorough intro of each tools	Developing skills to use the instrumentation and more importantly analyzing and interpreting the results from the instrumentation
The research project. Learned how to work better in groups, time management with respect to balancing research projects and other classes. Learn about mentors	Pro and cons of various instruments. What each measurement can tell you
The quantification of the collected data and what can be said about them	How to work effectively as group

inform their findings in the laboratory reports in which comparative analysis among the results obtained from different tools was required.

11.5 Challenges

The course was successfully offered over a 3-year period, over which several challenges were overcome. While there was a general appreciation of the course content and dynamic among students and instructors, the course proved very time consuming due to the high number of hand-on activities (laboratory sessions) and the extensive training required to progress on team projects. As a result, the course required significant human and laboratory resources. Due to the fast pace of the course and complex technical content, it was determined that the course could not support more than 12 students (or four groups of three students each) and that a dedicated teaching assistance was required to help manage laboratory training and user time on instruments. This required institutional support, which could be limiting to schools with limited resources and/or limited institutional investment.

The diversity of disciplines represented by the students was the strength of the course; however, this raised some challenges when trying to develop a common knowledge base for the course. For example, some students enrolled in the course with a strong laboratory foundation, while for others, this course was the first intensive laboratory experience.

Finally, there was a significant level of coordination needed between the instructors and the team projects mentors so that the mentors understood the objectives of the course and aligned their projects accordingly. This was addressed in part by creating a detailed mentor packet that helped guide the projects.

11.6 Conclusion

The nano tools course brought together students and faculty form different STEM fields to provide cross-training of essential technical skills as a means of developing a capable workforce that can support increasing nanomanufacturing activity.

This course demonstrated that by starting with the tool, students learn how nanoscale properties are determined through theory reinforced by experimentation. In addition the students develop the knowledge needed to discuss complimentary results from different instruments in order to reinforce the fundamental basis for each technique and empower the students to make decisions on what properties to measure, how to measure them, and why they are important from a technical and EHS aspect. Finally, it was shown that applying nano tools to mentored research projects connects the instrumentation to applications, reinforcing knowledge, and developing interdisciplinary competences among the students.

Key points

- Interdisciplinary course was designed and implemented to develop fundamental and technical nanotechnology competences through the use of a variety characterization techniques.
- Formal and informal assessment demonstrated that using the characterization tools, students learn how nanoscale properties are determined through the theory reinforced by experimentation.

Acknowledgment

This work was funded by the National Science Foundation award #EEC 1242129.

Appendix

Assessment survey

The following survey was conducted on the first day of class (preassessment) and on the last day of class (postassessment). The questions posed are as follows:

1. *Size and scale.* Place the following objects on the ruler according their approximate size (use diameter unless otherwise specified): (1) bacterium, (2) ant, (3) water molecule, (4) human hair, (5) protein, and (6) gold atom.

1 m	1 dm	1 cm	1 mm	100 μm	10 μm	1 μm	100 nm	10 nm	1 nm	1 Å	0,1 Å
1 m	10^{-1} m	10^{-2} m	10^{-3} m	10^{-4} m	10^{-5} m	10^{-6} m	10^{-7} m	10^{-8} m	10^{-9} m	10^{-10} m	10^{-11} m

2. *Forces.* List forces that exist (1) at the nanometer length scale (i.e., nanoscale) and (2) at the macroscopic length scale (i.e., macroscale). Some forces may exist at both length scales.
3. *Tools, instrumentation, characterization.* Name the governing principle for each of these nano tools. (1) DLS; (2) TEM; (3) AFM.
4. *EHS; ethical, legal, societal impacts.* In your opinion, what are the main barriers to the commercial use of products containing nanomaterials?
5. *Data analysis; tools, instrumentation, characterization.* The figures below show two different batches of gold nanoparticles. Can you identify the tool used? In words, what type of quantitative and qualitative information can you obtain from the images? What type of information cannot be obtained?

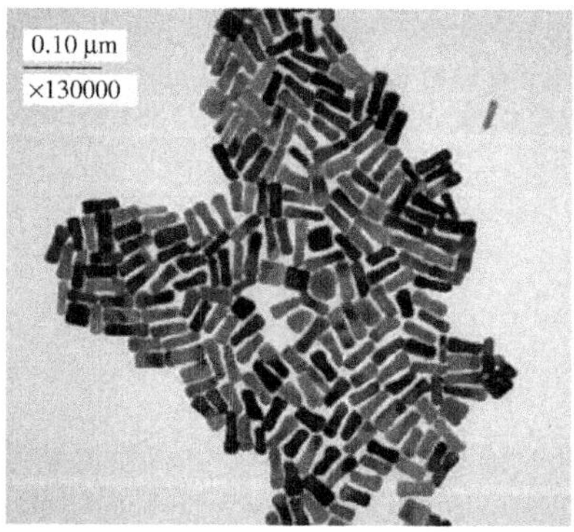

6. *Comprehensive.* When selecting courses for this semester, how did you explain (or would have explained) to your friends or family what this course was about?

References

Bainbridge, W.S., Roco, M.C., 2016. Science and technology convergence: with emphasis for nanotechnology-inspired convergence. J. Nanopart. Res. 18 (7), 211.

de Melo, N.F.S., Fraceto, L.F., Grillo, R., 2017. Heightening awareness for graduate students of the potential impacts of nanomaterials on human health and the environment using a theoretical–practical approach. J. Chem. Educ. 94 (10), 1471–1479.

Etkina, E., Karelina, A., Ruibal-Villasenor, M., Rosengrant, D., Jordan, R., Hmelo-Silver, C.E., 2010. Design and reflection help students develop scientific abilities: learning in introductory physics laboratories. J. Learn. Sci. 19 (1), 54–98.

Jackman, J.A., Cho, D.J., Lee, J., Chen, J.M., Besenbacher, F., Bonnell, D.A., et al., 2016. Nanotechnology education for the global world: training the leaders of tomorrow. ACS Nano 10 (6), 5595–5599.

NISE Network, 2014. Exploring nano & society—you decide!. Available from: <http://www.nisenet.org/catalog/programs/exploring_nano_society_-_you_decide> (accessed 25.01.14).

Rocabert, C., Knibbe, C., Consuegra, J., Schneider, D., Beslon, G., 2017. Beware batch culture: seasonality and niche construction predicted to favor bacterial adaptive diversification. PLoS Comput. Biol. 13 (3), e1005459.

N.S.A.T.C.C.O Technology, 2016. National Nanotechnology Initiative: Strategy Plan.

Wansom, S., Mason, T.O., Hersam, M.C., Drane, D., Light, G., Cormia, R., et al., 2009. A rubric for post-secondary degree programs in nanoscience and nanotechnology. Int. J. Eng. Educ. 23 (3), 615–627.

CHAPTER

Engineered nanomaterials and consumers: acceptance and rejection

12

Elena Maestri[1,2], Nelson Marmiroli[1,3], Jing Song[4] and Jason C. White[5]

[1]*Department of Chemistry, Life Sciences and Environmental Sustainability, University of Parma, Parma, Italy*

[2]*SITEIA.PARMA, University of Parma, Parma, Italy*

[3]*National Interuniversity Consortium for Environmental Sciences (CINSA), Parma, Italy*

[4]*Institute of Soil Science, Chinese Academy of Sciences, Nanjing, P.R. China*

[5]*Connecticut Agricultural Experiment Station, New Haven, CT, United States*

12.1 Introduction

Nanotechnology is a rapidly developing field, and as novel materials are synthesized, additional applications continue to be found. As discussed in previous chapters, there are still significant gaps in regulatory framework in many countries; similarly, there are uncertainties and wide variability over the acceptance of this technology by the public (Dorbeck-Jung and Shelley-Egan, 2014). For example, in fields such as the semiconductor industry, acceptance by consumers is widespread. However, applications of nanotechnology now include agriculture and the food industry, with new strategies to improve food production and processing along the entire supply chain, including enhanced agricultural practices; techniques to increase plant performance, nutrient absorption, and pest resistance; safer and more effective industrial processing; advanced analytical and sensor platforms; and techniques to promote shelf life of fresh and processed products. Importantly, consumer acceptance of nanotechnology in the food sector is uncertain. The progression from invention to innovation requires several stages of acceptance (Rogers, 1983): (1) relative advantage: it must be perceived as superior to the existing; (2) observability: the benefits must be visible; (3) trialability: it must be possible to try it; and (4) complexity: the innovation must not be difficult to learn. The application of nanotechnology could potentially fall short in any one of these steps.

12.2 Nanotechnology in consumer products

Nanotechnology-based food products and food packaging materials are already available to consumers in some countries, such as titanium dioxide in sweets or

Exposure to Engineered Nanomaterials in the Environment. DOI: https://doi.org/10.1016/B978-0-12-814835-8.00012-1

metals in supplements (as a coloring or caking agent), but many additional products and applications are currently at the research and development stage. Some of these approaches have reached a high level of technology readiness, and deployment may be imminent. In fact it is likely that a broad range of nanotechnology-derived products will be increasingly available to stakeholders, agri-food business operators, the food industry, and consumers in the near future. A survey of the available databases on nanotechnology was performed by the European Commission (2012) and select findings are reported in Table 12.1.

The issue of labeling is an area of debate in many countries and is the most obvious way to provide consumers with information, or at least the right to choose. The EU advocates labeling for nanoingredients in cosmetics, medical devices, and food (Rauscher et al., 2017). Labeling has a dual purpose: it informs consumers and allows freedom of choice for those willing to purchase nanofood and provides citizens who are against nanotechnologies with critical information (Chuah et al., 2018). Consequently, the label is simultaneously attractive for some consumers and repulsive for others. It must be also noted that labeling creates additional burdens in control and regulatory procedures (i.e., label compliance or guarantees) and possibly trade barriers for countries which do not use labels or that have different requirements. Although mandatory labeling of cosmetics in EU has not brought about rejection from consumers, labeling of food and food products is expected to have a more significant influence on consumer attitudes (Capon et al., 2015; Frewer, 2017).

12.3 Rationale for acceptance of nanotechnologies

As evident in this text, nanotechnology is an exciting field and rapidly developing field that holds great promise to improve the lives of many consumers. However, it is clear that although nanotechnologies have already been applied to a number of sectors impacting everyday life, significant uncertainty about the risks and the benefits remains (Sodano, 2018). The EC Code of Conduct for Responsible Nanosciences and Nanotechnologies Research (European Commission, 2009) states that nanotechnology research has to be meaningful, comprehensible, in the interest of the well-being of individuals and society. The benefits of research and applications must be clear and disseminated to the public.

A survey of Swiss University students highlighted that nanotechnology is strongly associated with positive images and terminology, such as "fascinating," "promising," and "something one should encourage," "versatile," "sophisticated"; however, the attribute "risky" was also identified (Ineichen et al., 2017). Quite interestingly, students from natural sciences were more positive in their evaluations as compared to students from humanistic and social sciences. Gupta et al. (2015) conducted a large study and showed that acceptance can be very high for nanotechnologies applied to sports, remediation, water filtration, medical

Table 12.1 Databases and websites with information on nanomaterials and nanotechnologies.

Database	Website	Organization	Content	Status in 2018
Nanowatch.de	http://archiv.bund.net/nc/themen_und_projekte/nanotechnologie/nanoproduktdatenbank/produktsuche/	BUND, Friends of the Earth Germany	Products containing nanomaterials	Active
ANEC/BEUC[a] 2010 inventory	www.beuc.org www.anec.eu	ANEC-BEUC	Consumer products with nanosilver	Still available, not updated
Consumer Products Inventory	http://www.nanotechproject.org/cpi/	The Project on Emerging Nanotechnologies	Nanotechnology-based consumer products	Active
DaNa Knowledge Base Nanomaterials	https://nanopartikel.info/en/nanoinfo/knowledge-base	Project DaNa2.0	Applications of nanomaterials	Active
The Database of Nanotechnologies for the Greater Region	http://www.nanodaten.de/site/page_de_garde.html	Interreg Project	Information about existing products	Active
EUON	https://euon.echa.europa.eu/home	ECHA	EU reference website	Active
InfoNano	https://infonano.agirpourlenvironnement.org/	Agir pour l'Environnement	Food products	Active
JRC Web Platform on Nanomaterials	https://ihcp.jrc.ec.europa.eu/scientific-tools/web-platform-on-nanomaterials	JRC	All nanotechnologies	Stopped in 2016
The Nanodatabase	http://nanodb.dk/	DTU Environment, the Danish Ecological Council and Danish Consumer Council	Products containing nanomaterials	Active
NANO supermarket	http://www.nanosupermarket.org/	Next Nature Network	Speculative nanotech products	Active
Nanotechnology in food	https://www.centerforfoodsafety.org/nanotechnology-in-food	Center for Food Safety	Food-related products	Active
Nanotechnology Products and Applications	https://www.nanowerk.com/products/products.php	Nanowerk	Industrial and commercial applications	Active
Nanotechnology Products Database	http://product.statnano.com/	StatNano	Products	Active
ObservatoryNano	http://www.safenano.org/research/observatorynano/	Institute of Nanotechnology, United Kingdom	Overview of nanomaterials	Finished in 2012
Progetto RInnovaReNano	https://nanotecnologie.iss.it/?page_id = 881	Istituto Superiore di Sanità	Applications of nanomaterials	Active
Registered substances	https://echa.europa.eu/web/guest/information-on-chemicals/registered-substances	ECHA	Registration dossiers	Active

[a]ANEC, *European Association for the Co-ordination of Consumer; Representation in standardization*, BEUC, *Bureau Européen des Unions de Consommateurs.*

Table 12.2 Classification of cultural worldviews.

	Dismissing environmental and technological risks	**Giving credence to environmental and technological risks**
Vertical dimension	*Hierarchy*	*Egalitarianism*
Attitude toward social ordering	Social ordering based on individual characteristics (race, gender, etc.)	Social ordering should not be based on individual characteristics
Horizontal dimension	*Individualism*	*Communitarianism*
Attitude toward individual/collective interests	Society is competitive and individual interests should prevail	Value of solidarity, social welfare should prevail over individual interests

Source: *Based on Yang, Y., 2018. Deconstructing Public Perceptions of Novel Food Technologies: Human Values and Information Communication Strategies (Ph.D. thesis). University of Saskatchewan, Canada. Available from: <http://hdl.handle.net/10388/8506>.*

applications; much less acceptance was evident for food packaging, sensors, cosmetics, and nutrients in food. Positive factors were perceived as general benefits for people, daily use, environmental aspects, and health; negative attitudes were focused on possible abuse, fear, minimal knowledge, and privacy concerns.

According to the theory of cultural cognition, acceptance of novel technologies may be affected by the cultural values of the individuals as described by their worldviews, that is, the attitude toward the world and the beliefs about society (Table 12.2). It has recently been shown that individuals with "hierarchical" and "communitarian" worldviews are less opposed to novel technologies applied to food, whereas individuals with "egalitarian" and "individualistic" worldviews are less keen for acceptance (in Canada, Yang, 2018). Although the test did not target nanotechnology in particular, the attitude toward innovation in food may be taken as representative of the issue of nanotechnology. Individualistic respondents did not approve of new technologies if they did not perceive a benefit for themselves, whereas communitarian respondents were more likely to embrace general advantages and benefits of the new technology. Hierarchists have a favorable attitude, presumably because of the trust in scientists and government. Specific testing identified three classes of consumers: supporters, doubters, and opponents. Supporters of nanotechnology in food were those who valued the "appearance" of food instead of "naturalness" or "origin": in other words, nanomaterials are perceived as nonnatural.

12.4 Rationale for rejection of nanotechnologies

Nanomaterials and nanotechnology cannot be detected or evaluated with human senses, and this is an important factor in the attitude of consumers. In the Eurobarometer (2010) survey of 2010, 54% of the Europeans had never heard about nanotechnologies, but the values in different countries ranged from 22% (Norway) to 79% (Portugal). Approximately 40% of respondents answered "don't know" to the question of the possible effects of nanotechnologies. In contrast only 20% responded the same for biotechnology. It is interesting that even with this relatively low level of knowledge, 50% of respondents felt that nanotechnology "could benefit some people but put others at risk." In addition, 42% agreed that nanotechnology is "fundamentally unnatural," 31% "felt uneasy," and 40% did not know if it is safe for health. It is well known that risk perception is greatly influenced by uncertainty. It is also clear that many will be sensitive to nanotechnology applications in food. The presence of nanomaterials in food should be clearly labeled as specified by EU regulations [Regulation (EU) 1169/2011], but according to several consumer associations, this is not done (Sodano, 2018).

As discussed by Frewer (2017), consumers may be inclined to reject new technologies, particularly in food production and processing, but this is not true for all technologies. In Brazil, "neophobia," rejection toward the new or unfamiliar, was shown to be higher for nanofoods and genetically modified foods in a consumer survey (Vidigal et al., 2015). The perception of benefits and risks is crucial to the behavior of consumers, but it is often difficult to predict reactions and behavior. One must remember that "consumers" as a group include many individuals with different expertise, preferences, knowledge, and personal histories. With regard to food, the recent occurrence of accidents and outbreaks have focused public attention on the role of science and technology in food production and processing. An increased interest in the "naturalness" of food is another factor that confounds acceptance of nanotechnologies, even when health benefits are recognized (Chuah et al., 2018; Cummings et al., 2018). Older age was another factor associated with rejection of nanotechnology in foods (Vidigal et al., 2015; Cummings et al., 2018). Trust in authority and perceived relative advantage are additional important factors in determining acceptance or rejection: for instance, food manufacturers could advertise the benefits of the nanofood to enhance consumer curiosity and willingness to try the product (Chang et al., 2017).

Understanding of the risks and impacts of nanotechnologies is far from complete, and the precautionary principle assumes the worst-case scenario. It is quite evident that people can form judgments and attitudes toward specific subjects without appropriate scientific information, basing the decision on "shortcuts" rooted in their own experience, predispositions, or from social media. When there is a perceived risk, scientific information may be not sufficient to change individual sentiment based on previous knowledge. And undoubtedly, the media coverage on food and technology trends toward the negative (Chuah et al., 2018).

12.5 Conclusion

Nanotechnology clearly offers the potential for significant benefit, although this continues to be a topic of wide debate (Yue et al., 2015). A consideration of the history associated with the introduction of transgenic organisms to agriculture can provide highly useful information and lessons learned that are relevant to introduction of nanotechnology in agriculture and food. The literature and surveys have demonstrated that many consumers have come to reject the application of transgenic organisms, and most particularly animals, in food production, whereas the applications to pharmaceuticals and health receive higher acceptance (Frewer, 2017). Data on US consumers (Yue et al., 2015) reveal that "modified" food is unpopular, with transgenic food being less acceptable than nanofood; any benefits for health and nutrition were more appreciated than improved taste or protection of environment. It is important to remember that the reaction of many experts to the rejection of genetically modified food by consumers was dismissive; it is clear now that there should have been an effort to understand the motivations and the rationale of the stakeholders. In the case of nanotechnology developers seem to be more considerate of these issues. Perception of the actual benefits is critical and can change the attitude of consumers that are skeptical about technologies (Giles et al., 2015). However, simply increasing the knowledge of nanotechnologies does not necessarily have significant effect on the attitude of consumers (van Giesen et al., 2018). Ethical considerations are also very important in acceptance of new technologies. If nanotechnologies could provide solutions to increasing food production, ensuring safety, and decreasing environmental impacts of agriculture, these factors would greatly increase the likelihood of acceptance.

Unfortunately, existing data and knowledge available on the effects of nanomaterials in humans, crop plants, and livestock are not yet sufficient to allow for a thorough evaluation of potential risk and safety. Evidence required for effective regulation and governance must be generated, and full knowledge exchange and a dissemination program with key stakeholders and end users, including the general public, must be implemented so as to avoid some of the mistakes which characterized the early introduction of transgenic crops.

Key points

Nanotechnology in the market products and the present state of the legislation to that related.

Rationale acceptance and rejection of nanotechnology applied to the everyday life, including agri-food.

Acknowledgments

NM and EM acknowledge the support of the project INTENSE, grant no. 652515. JCW acknowledges USDA NIFA AFRI 2011–67006-30181, USDA Hatch CONH00145, and USDA CONH00147. Authors want to thank Dr. Luca Pagano (University of Parma) for the editorial support.

References

Capon, A., Gillespie, J., Rolfe, M., Smith, W., 2015. Comparative analysis of the labelling of nanotechnologies across four stakeholder groups. J. Nanopart. Res. 17, 327. Available from: https://doi.org/10.1007/s11051-015-3129-8.

Chang, H.H., Huang, C.Y., Fu, C.S., Hsu, M.T., 2017. The effects of innovative, consumer and social characteristics on willingness to try nano-foods. Product uncertainty as a moderator. Inf. Technol. People 30 (3), 653–690. Available from: https://doi.org/10.1108/ITP-10-2015-0266.

Chuah, A.S.F., Leong, A.D., Cummings, C.L., Ho, S.S., 2018. Label it or ban it? Public perceptions of nano-food labels and propositions for banning nano-food applications. J. Nanopart. Res. 20, 36. Available from: https://doi.org/10.1007/s11051-018-4126-5.

Cummings, C.L., Chuah, A.S.F., Ho, S.S., 2018. Protection motivation and communication through nanofood labels: improving predictive capabilities of attitudes and purchase intentions toward nanofoods. Sci. Technol. Human Values 43 (5), 888–916. Available from: https://doi.org/10.1177/0162243917753991.

Dorbeck-Jung, B., Shelley-Egan, C., 2014. Meta-regulation and nanotechnologies: the challenge of responsibilisation within the European Commission's code of conduct for responsible nanosciences and nanotechnologies research. Nanoethics, 7, 55-68. Available from: http://dx.doi.org/10.1007/s11569-013-0172-8.

Eurobarometer, 2010. 73.1 Biotechnology. TNS Opinion & Social, Belgium.

European Commission, 2009. Commission recommendation on A code of conduct for responsible nanosciences and nanotechnologies research & Council conclusions on Responsible nanosciences and nanotechnologies research. Luxembourg: Office for Official Publications of the European Communities. ISBN 978-92-79-11605-6.

European Commission, 2012. Commission Staff Working Paper on Types and Uses of Nanomaterials, Including Safety Aspects Accompanying the Communication From the Commission to the European Parliament, the Council and the European Economic and Social Committee on the Second Regulatory Review on Nanomaterials. C (2012) 572 Final.

Frewer, L.J., 2017. Consumer acceptance and rejection of emerging agrifood technologies and their applications. Eur. Rev. Agric. Econ. 44 (4), 683–704. Available from: https://doi.org/10.1093/erae/jbx007.

Giles, E.L., Kuznesof, S., Clark, B., Hubbard, C., Frewer, L.J., 2015. Consumer acceptance of and willingness to pay for food nanotechnology: a systematic review. J. Nanopart. Res. 17, 467. Available from: https://doi.org/10.1007/s11051-015-3270-4.

Gupta, N., Fischer, A.R.H., Frewer, L.J., 2015. Ethics, risk and benefits associated with different applications of nanotechnology: a comparison of expert and consumer perceptions of drivers of societal acceptance. Nanoethics 9, 93–108. Available from: https://doi.org/10.1007/s11569-015-0222-5.

Ineichen, C., Biller-Andorno, N., Deplazes-Zemp, A., 2017. Image of synthetic biology and nanotechnology: a survey among university students. Front. Genet. 8, 122. Available from: https://doi.org/10.3389/fgene.2017.00122.

Rauscher, H., Rasmussen, K., Sokull-Kluttgen, B., 2017. Regulatory aspects of nanomaterials in the EU. Chem. Ing. Tech. 89 (3), 224–231. Available from: https://doi.org/10.1002/cite.201600076.

Rogers, E.M., 1983. Diffusion of Innovations, third ed. Free Press, New York.

Sodano, V., 2018. Nano-food regulatory issues in the European Union. In: AIP Conference Proceedings 1990, 020018. Available from: http://dx.doi.org/10.1063/1.5047772.

van Giesen, R.I., Fischer, A.R.H., van Trijp, H.C.M., 2018. Changes in the influence of affect and cognition over time on consumer attitude formation toward nanotechnology: a longitudinal survey study. Public Understanding Sci. 27 (2), 168–184. Available from: https://doi.org/10.1177/0963662516661292.

Vidigal, M.C.T.R., Minim, V.P.R., Simiqueli, A.A., Souza, P.H.P., Balbino, D.F., Minim, L.A., 2015. Food technology neophobia and consumer attitudes toward foods produced by new and conventional technologies: a case study in Brazil. LWT—Food Sci. Technol. 60, 832–840. Available from: https://doi.org/10.1016/j.lwt.2014.10.058.

Yang, Y., 2018. Deconstructing Public Perceptions of Novel Food Technologies: Human Values and Information Communication Strategies (Ph.D. thesis). University of Saskatchewan, Canada. Available from: <http://hdl.handle.net/10388/8506>.

Yue, C., Zhao, S., Kuzma, J., 2015. Heterogeneous consumer preferences for nanotechnology and genetic-modification technology in food products. J. Agric. Econ. 66 (2), 308–328. Available from: https://doi.org/10.1111/1477-9552.12090.

CHAPTER

Ethical issues of engineered nanomaterials applications and regulatory solutions

13

Elena Maestri[1,2], Nelson Marmiroli[1,3], Jing Song[4] and Jason C. White[5]

[1]*Department of Chemistry, Life Sciences, and Environmental Sustainability, University of Parma, Parma, Italy*

[2]*SITEIA.PARMA, University of Parma, Parma, Italy*

[3]*National Interuniversity Consortium for Environmental Sciences (CINSA), Parma, Italy*

[4]*Institute of Soil Science, Chinese Academy of Sciences, Nanjing, P.R. China*

[5]*Connecticut Agricultural Experiment Station, New Haven, CT, United States*

13.1 Introduction

Nanotechnology and nanoscience have developed rapidly since an initial discussion by Richard Feynman in 1959 (Ball, 2009). Applications of nanotechnologies and nanomaterials, both current and future, may well range in the thousands. Nonetheless, there is still a basic problem of definitions and terminology. However, nanotechnology clearly applies to several diverse materials and applications, as long as the scale is nano, and impacts will most certainly be felt in nearly every field of research and in everyday life (Arts et al., 2014). As has been described elsewhere in this book, it is widely thought that the unique properties of nanomaterials can provide benefits to human activities in many different fields, but at the same time, these materials may constitute a hazard, which is still difficult to quantify and predict (Roco et al., 2011).

Unlike nuclear technology and cloning technology, which came out earlier than nanotechnology and have gone through "development first and restriction afterwards" pattern, nanotechnology is the first technology in human history of which the ethical issues are considered and studied in the process of technology development. A prime example of the significant consideration given to ethical, legal, and social issues of particular technology [ethical, legal, and societal implications (ELSI) or ethical, legal, and societal issues (ELSA)] is with the Human Genome Project, which began in the 1990s (Walker and Morrissey, 2012). Nanotechnology, as one of the later developments, has inherited the issues previously raised by other technologies, many of which often generated problems and questions that remain unsolved. Another example can be found in the issues linked to genetic engineering and transgenic organisms (Frewer, 2017). A primary

Exposure to Engineered Nanomaterials in the Environment. DOI: https://doi.org/10.1016/B978-0-12-814835-8.00013-3

goal of many of the researchers contributing to this text is to incorporate the lessons learned from these prior endeavors so as to avoid some of the same pitfalls.

To harmonize the development of nanotechnology and nanoethics, it is necessary to establish a national nanoethics committee, promote the standardization of nanotechnologies, raise the awareness of nanoethics to practitioners in the nano industry, promote social studies on nanoethics, and formulate relevant laws to govern research on nanotechnologies (Chen and Xia, 2014).

In the European Union the main approach to ELSI is referred to as "responsible research and innovation" (RRI, Fig. 13.1): according to this, the actors must work together during the process of research and innovation so that the outcomes are in line with the values, needs, and expectations of the society (Ruggiu, 2014). The six dimensions of RRI (European Commission, 2012; https://www.rri-tools.eu/about-rri) are (1) public engagement, (2) gender equality, (3) science education, (4) open access, (5) ethics, and (6) governance.

Generally speaking, the main ethical issues of concern for the application of engineered nanomaterials (ENMs) are as follows (1) environmental impact, (2) equity, (3) privacy, and (4) safety/security. These aspects correspond to the principles of the "Georgetown paradigm" (Beauchamp and Childress, 1994) (1) nonmaleficence, do no harm; (2) beneficence, maximize possible benefits; (3) justice or equity, fairness in distribution; and (4) respect for persons, decisional autonomy. In the European Union the European Commission (EC) Code of Conduct for Responsible Nanosciences and Nanotechnologies Research (Commission of the European Communities, 2008) has been developed to regulate the applications of nanotechnology and to achieve responsible and active involvement of the main

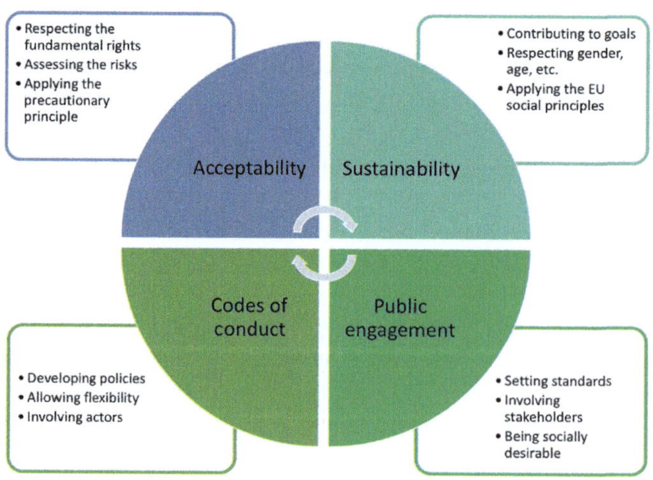

FIGURE 13.1

The main aspects of responsible research and innovation.

actors (Dorbeck-Jung and Shelley-Egan, 2014). This guidance is also meant to inform member states, as they develop their own regulations.

In the United States the ethical implications of nanoscience have been an essential component of the National Nanotechnology Initiative of the US Government, which began in 2000. An interesting analysis of the main steps in the development of nanotechnology and the associated regulatory approaches was reported by Hansen et al. (2008) and highlighted how the adoption of socially and economically responsive strategies has been hindered by large uncertainties. The application of ethical principles to the development of nanoscience and nanotechnology is obviously hindered by the same factors, and the implementation of regulatory schemes has been confounded. As highlighted by Mantovani et al. (2010), these uncertainties include (1) the variety of materials and applications; (2) the limited knowledge on toxic effects and fate in ecosystems; (3) proprietary issues; (4) lack of harmonization; and (5) potential inadequacy of authorities.

Nonetheless, in the past few years significant efforts have been made to include several aspects of nanotechnology applications into guidelines, standards, and regulations. Many of these activities have been reported in previous chapters, and a summary of the most relevant efforts is included here (Box 13.1). Their relevance will be briefly discussed in this chapter. Given the currently available information, the discussion will be primarily focused on positions in the European Union and the United States.

13.2 Ethical issues with medicine and human health (including cosmetics)

One document of reference for any application to humans and health is the Convention on Human Rights and Biomedicine (Council of Europe, 1997), also called Oviedo Convention, which protects individuals against exploitation. According to Article 1, "Parties to this Convention shall protect the dignity and identity of all human beings and guarantee everyone, without discrimination, respect for their integrity and other rights and fundamental freedoms with regard to the application of biology and medicine." In addition, the European Charter of Fundamental Rights (2012) ensures human dignity, equality, and solidarity. The use of nanotechnologies in medicine and related disciplines should therefore consider some fundamental questions, which might impact the rights of individuals.

The first issue is to establish the collateral benefits and damage/harmful effects of nanomedicine. The benefits can be estimated by considering current and future applications, and some chapters in this book have addressed these issues. The assessment of potential hazards is considered to be more of a work in progress and on a case-by-case basis. The EU Code of Conduct (Commission of the European Communities, 2008) is based on the Precautionary Principle, therefore having the goal of anticipating all possible harmful impacts of

Box 13.1 Regulatory solutions developed to address issues linked to nanotechnology and nanoscience applications. Each box briefly summarizes the role of the main institutions, indicating relevant recent documents and the existence of working groups or initiatives dedicated to nanotechnologies. Institutions are listed in alphabetical order.

Organization or institution	**EC**
Area of relevance	European Union
Website	ec.europa.eu/http://ec.europa.eu/research/industrial_technologies/nanoscience-and-technologies_en.html
Document	*Commission recommendation October 18, 2011 (2011/696/EU) on the definition of nanomaterial*
Year	2011 (under revision)
Citation and source	Official Journal of the European Union L275/38, 20.10.2011
Main message	"Nanomaterial" means a natural, incidental, or manufactured material containing particles, in an unbound state or as an aggregate or as an agglomerate and where, for 50% or more of the particles in the number size distribution, one or more external dimensions is in the size range 1–100 nm
Document	*Code of conduct for responsible nanoscience and nanotechnologies research*
Year	2008
Citation and source	http://ec.europa.eu/research/industrial_technologies/pdf/policy/nanocode-rec_pe0894c_en.pdf
Main message	Provides voluntary guidelines for responsible development of nanotechnologies and nanosciences, based on the precautionary principle
Groups, initiatives, etc.	*DG Environment GAARN*
Beginning	2012–13
Active	NO
Mission	Build a consensus on the best practice for assessing and managing the safety of nanomaterials under the REACH Regulation
Members	Experts from member states, EC, ECHA, industry
Website for outputs	non available
Organization or institution	**ECHA**
Area of relevance	European Union
Website	echa.europa.eu
Document	*REACH guidance for nanomaterials/best practice guide* *How to prepare registration dossiers that cover nanoforms: best practice*
Year	2017
Citation and source	Doi:10.2823/128306
Main message	Provide criteria for distinguishing between different nanoforms and to give a set of elements recommended to be reported on the characterization of nanoforms
Document	*Workplan on nanomaterials*
Year	2015
Citation and source	https://echa.europa.eu/documents/10162/21844190/mb_41_2015_workplan_nanomaterials_incl_annexes_en.pdf
Main message	Updating the document of 2011, the guidance has taken account of relevant new scientific studies that provide more insights to physicochemical properties, exposure assessment, and hazard characterization of nanomaterials
Groups, initiatives, etc.	*NMWG*

(Continued)

Box 13.1 Regulatory solutions developed to address issues linked to nanotechnology and nanoscience applications. Each box briefly summarizes the role of the main institutions, indicating relevant recent documents and the existence of working groups or initiatives dedicated to nanotechnologies. Institutions are listed in alphabetical order. (Continued)

Beginning	October 2012
Active	YES
Mission	Provides advice on scientific and technical issues regarding the implementation of REACH, CLP, and Biocidal Products Regulations in relation to nanomaterials
Members	Experts nominated from EU competent authorities, European Commission, EFSA, stakeholders
Website for outputs	https://echa.europa.eu/regulations/nanomaterials/nanomaterials-expert-group
Groups, initiatives, etc.	*EUON*
Beginning	2017
Active	YES
Mission	Information on the EUON offers interesting reading about the safety, innovation, research, and uses of nanomaterials
Members	non available
Website for outputs	euon.echa.europa.eu/
Organization or institution	**EFSA**
Area of relevance	European Union
Website	www.efsa.europa.eu
Document	*Guidance on assessing potential risks arising from applications of nanoscience and nanotechnologies to food, feed, and pesticides*
Year	2011
Citation and source	http://onlinelibrary.wiley.com/doi/10.2903/j.efsa.2011.2140/epdf
Main message	This scientific opinion offers practical guidance for the risk assessment of applications involving the use of ENMs in the food and feed chain (including food additives, enzymes, flavorings, food contact materials, novel foods, feed additives, and pesticides)
Document	*Guidance on risk assessment of the application of nanoscience and nanotechnologies in the food and feed chain: Part 1, human and animal health*
Year	2018
Citation and source	EFSA Journal 2018;16(7):5327, doi:10.2903/j.efsa.2018.5327
Main message	Updating the document of 2011, the guidance has taken account of relevant new scientific studies that provide more insights to physicochemical properties, exposure assessment, and hazard characterization of nanomaterials
Groups, initiatives, etc.	*Network for Risk Assessment of Nanotechnologies in Food and Feed (Nano Network)*
Beginning	2010
Active	YES
Mission	To strengthen the scientific cooperation for risk assessment of application of nanotechnologies and products thereof in the relevant areas within EFSA's remit
Members	Organizations appointed for 3 years, through EFSA Advisory Forum Members, one member from each EU member state, plus Norway and Iceland. One member per third country participating to EFSA's activities as observers
Website for outputs	https://www.efsa.europa.eu/en/cross-cutting-issues/networks

(*Continued*)

Box 13.1 Regulatory solutions developed to address issues linked to nanotechnology and nanoscience applications. Each box briefly summarizes the role of the main institutions, indicating relevant recent documents and the existence of working groups or initiatives dedicated to nanotechnologies. Institutions are listed in alphabetical order. (Continued)

Organization or institution	**EMA**
Area of relevance	European Union
Website	www.ema.europa.eu
Document	*Reflection paper on nanotechnology-based medicinal products for human use*
Year	2006
Citation and source	http://www.ema.europa.eu/ema/index.jsp?curl = pages/includes/document/document_detail.jsp?webContentId = WC500069728&mid = WC0b01ac058009a3dc
Main message	The scope of this document is to reflect the current thinking and the initiatives in view of recent developments in relation to nanotechnology-based medicinal products
Groups, initiatives, etc.	*Ad hoc CHMP expert group on nanomedicines*
Beginning	2009, Reinforced 2011
Active	YES
Mission	Providing specialist input on new scientific knowledge, help with the review of guidelines on nanomedicines
Members	Selected experts from academia and European regulatory network
Website for outputs	non available
Groups, initiatives, etc.	*International Regulators Working Group on Nanotechnology*
Beginning	2009
Active	YES
Mission	Nonconfidential information sharing, regulatory harmonization, or convergence focused on nanomedicines/nanomaterial in drug products and borderline and combination products
Members	EMA, US FDA, Japan MHLW, Health Canada, TGA Australia
Website for outputs	https://www.i-p-r-f.org/en/working-groups/nanomedicines-working-group/

Organization or institution	**EPA**
Area of relevance	United States
Website	www.epa.gov
Document	*Chemical substances when manufactured or processed as nanoscale materials: TSCA reporting and recordkeeping requirements*
Year	2017
Citation and source	https://www.regulations.gov/document?D = EPA-HQ-OPPT-2010-0572-0137 Federal Register/Vol. 82, No. 8/Thursday, January 12, 2017/Rules and Regulations
Main message	Establishes reporting and recordkeeping requirements for certain chemical substances when they are manufactured or processed at the nanoscale
Groups, initiatives, etc.	*Research on nanomaterials*
Beginning	non available
Active	YES
Mission	EPA scientists research the most prevalent nanomaterials that may have human and environmental health implications.
Members	non available
Website for outputs	https://www.epa.gov/chemical-research/research-nanomaterials

(*Continued*)

Box 13.1 Regulatory solutions developed to address issues linked to nanotechnology and nanoscience applications. Each box briefly summarizes the role of the main institutions, indicating relevant recent documents and the existence of working groups or initiatives dedicated to nanotechnologies. Institutions are listed in alphabetical order. (Continued)

Organization or institution	**European Agency for Safety and Health at Work (EU-OSHA)**
Area of relevance	European Union
Website	osha.europa.eu
Document	*Workplace exposure to nanoparticles*
Year	2009
Citation and source	https://osha.europa.eu/en/tools-and-publications/publications/literature_reviews/workplace_exposure_to_nanoparticles/view
Main message	To provide a broad overview, information from different sources such as scientific literature, policy documents, legislation, and work programs were collected
Organization or institution	**FAO of the United Nations**
Area of relevance	Worldwide
Website	www.fao.org
Document	*FAO/WHO paper state of the art on the initiatives and activities relevant to risk assessment and risk management of nanotechnologies in the food and agriculture sectors*
Year	2013
Citation and source	http://www.fao.org/docrep/018/i3281e/i3281e.pdf
Main message	Recommendations from international experts meetings, presents national, and international risk assessment and risk management approaches that identify and implement strategies to address potential hazards associated with the use of nanotechnology-related products or techniques
Groups, initiatives, etc.	*FAO/WHO joint meetings on nanotechnologies in food and agriculture*
Beginning	2009, 2012
Active	NO
Mission	Provide recommendations
Members	Experts
Website for outputs	http://www.fao.org/food/food-safety-quality/a-z-index/nano/en/
Organization or institution	**FDA**
Area of relevance	United States
Website	www.fda.gov
Document	*Nanotechnology guidance documents* *Final guidance* *Final guidance for industry—considering whether an FDA-regulated product involves the application of nanotechnology* *Final guidance for industry—safety of nanomaterials in cosmetic products* *Final guidance for industry—assessing the effects of significant manufacturing process changes, including emerging technologies, on the safety and regulatory status of food ingredients and food contact substances, including food ingredients that are color additives* *Final guidance for industry—use of nanomaterials in food for animals* *Draft guidance for industry—drug products, including biological products, that contain nanomaterials*
Year	2018
Citation and source	https://www.fda.gov/ScienceResearch/SpecialTopics/Nanotechnology/ucm602536.htm

(*Continued*)

Box 13.1 Regulatory solutions developed to address issues linked to nanotechnology and nanoscience applications. Each box briefly summarizes the role of the main institutions, indicating relevant recent documents and the existence of working groups or initiatives dedicated to nanotechnologies. Institutions are listed in alphabetical order. (Continued)

Main message	Guidance on application of nanotechnology in food, feed, drugs, and cosmetics
Groups, initiatives, etc.	*Nanotechnology Task Force*
Beginning	2006
Active	YES
Mission	Determining regulatory approaches that encourage the continued development of innovative, safe, and effective FDA-regulated products that use nanotechnology materials
Members	Internal
Website for outputs	https://www.fda.gov/ScienceResearch/SpecialTopics/Nanotechnology/ucm2006658.htm

Organization or institution	**OECD**
Area of relevance	Worldwide
Website	www.oecd.org
Document	*Series on the safety of manufactured nanomaterials no. 55. Harmonized tiered approach to measure and assess the potential exposure to airborne emissions of engineered nanoobjects and their agglomerates and aggregates at workplaces*
Year	2015
Citation and source	ENV/JM/MONO (2015)19. Paris
Main message	This document presents a harmonized tiered approach that is systematic, consistent, practical, and flexible for conducting field-based, real-time workplace release, and exposure measurement and assessment to airborne nanoobjects and off-line analyses of measurement samples
Groups, initiatives, etc.	*WPMN*
Beginning	2006
Active	YES
Mission	This program promotes international cooperation on the human health and environmental safety of manufactured nanomaterials and involves the safety testing and risk assessment of manufactured nanomaterials
Members	Physical and life sciences experts
Website for outputs	http://www.safenano.org/knowledgebase/standards/working-party-on-manufactured-nanomaterials/

Organization or institution	**UNITAR**
Area of relevance	Worldwide
Website	www.unitar.org
Groups, initiatives, etc.	*Regional workshops*
Beginning	2014
Active	YES
Mission	Raising awareness on nanosafety issues
Members	non available
Website for outputs	http://www.unitar.org/pillars/planet/nanotechnology

(*Continued*)

Box 13.1 Regulatory solutions developed to address issues linked to nanotechnology and nanoscience applications. Each box briefly summarizes the role of the main institutions, indicating relevant recent documents and the existence of working groups or initiatives dedicated to nanotechnologies. Institutions are listed in alphabetical order. (Continued)

Organization or institution	**US Federal Government**
Area of relevance	United States
Website	www.usa.org
Document	*National nanotechnology initiative strategic plan*
Year	2016
Citation and source	National Science and Technology Council, Washington, DC, http://www.nano.gov/sites/default/files/pub_resource/2016-nni-strategic-plan.pdf
Main message	Address how the agencies will collaborate to expand the ecosystem that supports fundamental discovery, fosters innovation, and promotes the transfer of nanotechnology discoveries from lab to market
Groups, initiatives, etc.	*National Nanotechnology Initiative*
Beginning	2001
Active	YES
Mission	Creating a framework for shared goals, priorities, and strategies that helps each participating federal agency to leverage the resources of all participating agencies
Members	Twenty federal agencies and departments
Website for outputs	www.nano.gov

Organization or institution	**WHO**
Area of relevance	Worldwide
Website	www.who.int
Document	*WHO guidelines on protecting workers from potential risks of manufactured nanomaterials*
Year	2017
Citation and source	http://www.who.int/occupational_health/publications/manufactured-nanomaterials/en/
Main message	The recommendations are intended to help policy-makers and professionals in the field of occupational health and safety in making decisions about the best protection against potential risks specific to MNMs in workplaces
Groups, initiatives, etc.	*WHO/NANOH GDG Guideline Development Group*
Beginning	2009
Active	NO
Mission	Development of the guidelines on workers' health
Members	Experts
Website for outputs	http://www.who.int/occupational_health/topics/nanotechnologies/en/

EC, *European Commission;* ECHA, *European Chemicals Agency;* EFSA, *European Food Safety Agency;* EUON, *European Union Observatory for Nanomaterials;* GAARN, *Group Assessing Already Registered Nanomaterials;* REACH, *registration, evaluation, authorization, and restriction of chemicals;* NMWG, *Nanomaterials Working Group;* EMA, *European Medicines Authority;* FDA, *Food and Drug Administration;* EPA, *Environmental Protection Agency;* FAO, *Food and Agriculture Organization;* WHO, *World Health Organization;* OECD, *Organisation for Economic Co-operation and Development;* WPMN, *Working Party on Manufactured Nanomaterials;* UNITAR, *United Nations Institute for Training and Research;* MNMs, *manufactured nanomaterials.*

nanotechnology, and striving to close the gaps in scientific knowledge. Thinking about application of nanomaterials in drug delivery or innovative sensors, a primary issue of concern is that knowledge on the effects in the human body is still incomplete.

Notably psychological and societal concerns and consequences must not be underestimated: any new technology, including nanotechnology, may be unsettling or perceived as hazardous. Informed consent by the patient could be difficult to obtain, especially if we consider that some information on the nanomedical approaches is not easily conveyed to the general public. The results obtained through nanosensors, for instance, could be difficult to interpret, and this will contribute to patient anxiety.

The most promising applications of nanomedicine are likely to be related to the approaches in personalized diagnosis and therapy. Combining nanosensors and nanoenabled pharmaceuticals with information and communication technology could provide the acquisition and transmission of personal data in real time. Under this approach the role of nanotechnology in precision medicine and personalized medicine will bring issues of privacy protection, data ownership, and patient communication to the forefront.

Another ethical issue concerns the primary goal of nanomedicine: does the approach stop with diagnosis and therapy, or can it be extended toward "enhancement" of properties and faculties? For example, could a therapy against neurodegenerative disorders be utilized as a means to enhance brain functions? A nanotherapy for improving bone and tissue repair could become a way to enhanced limbs functionality. Applications toward "illicit enhancement" are specifically mentioned in the EU Code of Conduct (Commission of the European Communities, 2008, art. 4.1.16).

One additional consequence of nanomedicine that may need consideration is the concept of distributive justice. There is potential unequal access to expensive new nanotechnologies for diagnosis or therapy, creating inequity and potentially discrimination among patients. Other common ethical issues, which could be touched by nanotechnologies are gender equality, ethnic equality, consideration for pluralism in lifestyles, and beliefs. These approaches have been evaluated by the European Medicines Authority for the European Union and by the Food and Drug Administration (FDA) for the United States (see Box 13.1). The FDA states in all the guidance documents that they do not consider nanomaterials and nanotechnologies to be intrinsically harmful or benign. All considerations are therefore product-specific and evaluated on a case-by-case basis. Concerning pharmaceuticals, a specific mention is made of the fact that nanomaterials might interact with biological systems in ways that depend on or vary with intrinsic factors such as gender or age and that this could introduce elements of inequality in drug treatment. In the European Union, approval of a drug does not extend to a nanoform of the same drug, for which a separate testing and authorization procedure is necessary.

In the European Union a specific directive covers nanomaterials in cosmetic products (European Commission, 2009), requiring labeling with an indication of the nanoformulation of the ingredient. Importantly, the nanomaterials may or may not be toxic; the labeling indication only has the purpose of allowing freedom of choice by the consumers. There is also the long-standing ethical problem of testing on animals, which are to be replaced with alternative assays [SCCS (Scientific Committee on Consumer Safety), 2012].

13.3 Ethical issues related to food

Nanomaterials in food have been recently included in the EU Directive on Novel Foods [Regulation (EU) 2015/2283; European Union, 2015], which also includes the currently valid definition of nanomaterials. Documents dealing with nanomaterials in food are mainly produced in the European Union by the European Food Safety Agency (EFSA, see Box 13.1). By defining the approaches to evaluate the hazards of nanomaterials in food and feed, EFSA works to protect the health of consumers and therefore performs a task which is in line with the EC Code of Conduct (Commission of the European Communities, 2008). The most recent document on risk assessment adopts the precautionary principle to protect the health of consumers, by assuming the worst-case scenario and applying uncertainty factors, when specific data are not available. Issues concerning nanomaterials are considered in several aspects of food and also feed, from additives, to food contact materials, flavorings, etc. It is explained that nanomaterials in food could be intentionally added or derived from production processes. The documents produced by EFSA do not explain the benefits derived from inclusion of nanomaterials in food production or processing. The documents mainly address the evaluation of potential hazards, due to the specific properties of the nanoscale, which could change the toxicokinetic behavior of substances.

Labeling should inform consumers about the presence of nanomaterials in food products, but it seems that currently no food product is correctly labeled (Sodano, 2018).

Perceived benefits of nanotechnologies can be envisaged in the development of sensors and measures for improving food safety.

FDA (see Box 13.1) has issued a guidance concerning nanomaterials in food ingredients, additives, and food contact substances. The presence of poisonous or deleterious substances makes the food unsafe, and adulterated, and it is a responsibility of the manufacturer and of the end user to ensure the safe use of food. Nanotechnology is seen as a possible source of changed properties in food products, requiring specific testing to assess the safety. Also in this case, FDA does not consider nanoengineered food to be intrinsically benign or harmful. Labeling requirements are not discussed.

Food and Agriculture Organization (FAO) of the United Nations and World Health Organization (WHO) together have issued a document addressing all issues of nanomaterials in food and agriculture [see Box 13.1, FAO/WHO (Food and Agriculture Organization of the United Nations/World Health Organization), 2013]. They report that mandatory labeling would bring to transparency and enable the freedom of choice in the consumers; the European Union is the only one with mandatory labeling of nanomaterials in food. However, this is felt as a barrier to trade, in case a nanolabeled food is rejected by consumers. In fact, FAO and WHO have also stressed the possibility of challenges to the developing countries, both in accessing nanotechnologies and performing safety tests. This becomes relevant for nanopesticides or nanofertilizers applied to agriculture.

13.4 Ethical issues related to occupational health and worker safety

Working environments are the places where the risk of exposure to high concentrations of ENMs is the greatest. The precautionary principle mandates controls in the work environment, and when there is a lack of adequate information about nanotechnology hazards and risks, employers are to implement additional control measures. This situation may result in difficult choices for employers. As noted earlier, the risk assessment for nanotechnologies is performed on a case-by-case basis. There are the added difficulties of effective risk communication, as well as how to obtain acceptance by workers: there are difficulties in obtaining the timeliness of communication, and there are requirements to include workers in the decision-making.

A common problem encountered in work environments is that workers often have little control over their working conditions, due to pressures from economic and social conditions.

Regarding the need for medical surveillance and screening of workers, the usual ethical problems also apply to nanotechnologies: how to stimulate voluntary participation of workers, how to plan actions in case of positive results, and how to protect the privacy of results (Schulte and Salamanca-Buentello, 2007).

A final consideration on possible consequences involves the economics of nanotechnology production: the development of new production approaches based on nanoscience could displace traditional markets, affecting the employment of workers in traditional sectors, and potentially enhancing the North–South divide (Sodano, 2018).

On the subject of worker safety, the WHO has issued a set of guidelines (see Box 13.1), which are based on the precautionary principle and advocate the involvement of workers in all phases of risk assessment and control. The Organisation for Economic Co-operation and Development has also issued documents on nanomaterial exposure in workplaces (see Box 13.1). Documents prepared by EU-OSHA (see Box 13.1) advocate the use of "control banding" procedures for risk assessment and management to assist in minimizing the exposure of workers and provide employers with accessible tools to ensure compliance and safety. However, these documents do not provide guidance relative to situations in which "susceptible" workers might be exposed. However, a reference is made to asthmagenic substances, which could specifically impact vulnerable workers.

13.5 Ethical issues related to environmental impacts

The Parma Declaration on Environment and Health issued by the Fifth Ministerial Conference on Environment and Health (2010) considered the implications of nanotechnology to be a crucial concern, equivalent to bioaccumulating and endocrine disrupting chemicals. Ultrafine particles were specifically mentioned as pollutants responsible for respiratory diseases, whereas the presence of nanoparticles and nanomaterials in products was considered to be a topic that requires additional research. The European Code of Conduct (Commission of the European Communities, 2008) states that nanotechnology research should contribute to the Millennium (Sustainable) Development Goals and should do no harm to animals, plants, and the environment. A portion of this book has dealt with analysis of risks and impacts of nanotechnologies. The full potential benefits that nanotechnologies could bring with them in regard to reaching the goals are still unclear, and similarly, it remains difficult to estimate all possible impacts (EEA, 2015). One possible positive impact can be seen in the nanoformulations of pesticides and fertilizers: if these materials were more effective than conventional counterparts, this would decrease the amount of chemicals used and reduce the risks to workers.

The detectability and traceability of nanomaterials in the environment, including air, water, and soil, is a critical issue that needs significant additional research. Accurate and detailed exposure assessment is still elusive in many instances, and concern is great where children and other vulnerable subgroups are involved (WHO, 2013). It is also highly likely that exposure will not be equally distributed for all countries, populations, or individuals; this creates obvious concern with regard to the principles of equity.

In the European Union the European Chemicals Agency (ECHA) has worked toward the adaptation of REACH regulation (registration, evaluation, authorization, and restriction of chemicals) to nanomaterials (see Box 13.1), implementing procedures for registration and disseminating information on use. The website of ECHA is the reference point for all those seeking information on nanomaterials. From the perspective of toxicity relevant to environmental impact, the European Union approach is to "group" nanomaterials based on similarities in physicochemical, toxicological, ecotoxicological, and environmental fate properties [Arts et al., 2014; ECHA (European Chemicals Agency), 2016]. A specific procedure for assessment of environmental impacts should likely be required for nanomedicines or nanopharmaceuticals, which can reach water and soil after transfer through or metabolism in the human body.

The US Environmental Protection Agency (Box 13.1) has issued a "science in ACTION" leaflet (https://www.epa.gov/sites/production/files/2013-12/documents/nanotechnology-fact-sheet.pdf) to explain to the lay community the research being done on nanomaterials and nanotechnologies in order to protect the environment and human health. Applications of nanomaterials to environmental problems, such as pollutant remediation, are described to highlight the benefits of these new technologies. The website (https://www.epa.gov/chemical-research/research-nanomaterials) provides additional information to citizens about environmental impacts of nanotechnology.

13.5.1 Role of media and nongovernmental organizations on nanoethics

To protect the public's right to know, media needs to take the initiative to report technology with a responsible attitude. Media practitioners need to improve their scientific knowledge so that news on nanoethics can be readily understood by a wider readership. Last but not the least, scientists also have obligation to disseminate the outcome of their nanoethics research to the general public.

In the good governance of nanotechnology, nongovernmental organizations (NGOs) play a huge role as a third force besides the government and the market. NGOs should conduct objective evaluation of governments' policy on good governance of nanotechnology, supervise the market's economic behavior on the promotion of nanoproducts, and disseminate nanotechnology and nanoethics information to the general public. In so doing, NGOs contribute to the in-depth study of nanoethics and promotes the sustainable development of nanotechnology (Zhang and Wang, 2014).

13.6 Conclusion

The ethical aspects of nanotechnologies and nanoscience have been addressed previously in several documents and working groups formed by the main institutions involved in their development and regulation. Some undefined areas still exist, and there is a need for a more targeted effort, particularly to address eventual inequalities in approaches and unfairness toward categories of users and/or workers.

Key point

- Ethical issues related to different contexts with regards to health (food, medicine, occupational safety) and environment risks.

Acknowledgments

NM and EM acknowledge the support of the project INTENSE, grant no. 652515. JCW acknowledges USDA NIFA AFRI 2011-67006-30181, USDA Hatch CONH00145, and USDA CONH00147. Authors want to thank Dr. Luca Pagano (University of Parma) for the editorial support.

References

Arts, J.H.E., Hadi, M., Keene, A.M., Kreiling, R., Lyon, D., Maier, M., et al., 2014. A critical appraisal of existing concepts for the grouping of nanomaterials. Regul. Toxicol. Pharmacol. 70, 492–506. Available from: https://doi.org/10.1016/j.yrtph.2014.07.025.

Ball, P., 2009. Feynman's fancy. Chem. World 2009, 58–62.

Beauchamp, T.L., Childress, J.I., 1994. Principles of Biomedical Ethics, fourth ed. Oxford University Press, New York.

Charter of Fundamental Rights of the European Union, 2012. Off. J. Eur. Union C326/391.

Chen, S.Z., Xia, B.H., 2014. The collaborative construction of nanotechnology and ethics. J. Huazhong Univ. Sci. Technol. 28 (2), 132–136.

Commission of the European Communities, 2008. Commission Recommendation of 07/02/2008 on a Code of Conduct for Responsible Nanosciences and Nanotechnologies Research. C (2008) 424 final.

Council of Europe, 1997. Convention for the Protection of Human Rights and Dignity of the Human Being With Regard To the Application of Biology and Medicine: Convention on Human Rights and Biomedicine. Available from: <https://www.coe.int/en/web/conventions/full-list/-/conventions/treaty/164>.

European Commission, 2009. Council regulation (EC) 1223/2009 of the European Parliament and of the Council of 30 November 2009 on cosmetic products (recast). Off. J. Eur. Union. L342/59.

Dorbeck-Jung, B., Shelley-Egan, C., 2014. Meta-regulation and nanotechnologies: the challenge of responsibilisation within the European Commission's code of conduct for responsible nanosciences and nanotechnologies research. Nanoethics 7, 55–68. Available from: https://doi.org/10.1007/s11569-013-0172-8.

ECHA (European Chemicals Agency), 2016. Usage of (Eco)toxicological Data for Bridging Data Gaps Between and Grouping of Nanoforms of the Same Substance. Elements to Consider. ECHA. Available from: http://dx.doi.org/10.2823/982046.

EEA, 2015. The European Environment—State and Outlook 2015: Synthesis Report. European Environment Agency, Copenhagen.

European Commission, 2012. Responsible Research and Innovation: Europe's Ability to Respond to Societal Challenges.

European Union, 2015. Regulation (EU) 2015/2283 of the European Parliament and of the Council of 25 November 2015 on novel foods, amending Regulation (EU) no 1169/2011 of the European Parliament and of the Council and repealing Regulation (EC) no 258/97 of the European Parliament and of the Council and Commission Regulation (EC) no 1852/2001. Off. J. Eur. Union. L327/1-22.

FAO/WHO (Food and Agriculture Organization of the United Nations/World Health Organization), 2013. FAO/WHO Paper: State of the Art on the Initiatives and

Activities Relevant to Risk Assessment and Risk Management of Nanotechnologies in the Food and Agriculture Sectors. Geneva.

Fifth Ministerial Conference on Environment and Health, 2010. Parma declaration on environment and health. In: EUR/55934/5.1 Rev.2. Available from: <http://www.euro.who.int/__data/assets/pdf_file/0011/78608/E93618.pdf>.

Frewer, L.J., 2017. Consumer acceptance and rejection of emerging agrifood technologies and their applications. Eur. Rev. Agric. Econ. 44 (4), 683–704. Available from: https://doi.org/10.1093/erae/jbx007.

Hansen, S.F., Maynard, A., Baun, A., Tickner, J.A., 2008. Late lessons from early warnings for nanotechnology. Nat. Nanotechnol. 3, 444–447. Available from: https://doi.org/10.1038/nnano.2008.198.

Mantovani, E., Porcari, A., Morrison, M.J., Geertsma, R.E., 2010. Developments in Nanotechnologies Regulation and Standards 2010—Report of the Observatory Nano. Available from: <www.observatorynano.eu>.

Roco, M.C., Harthorn, B., Guston, D., Shapira, P., 2011. Innovative and responsible governance of nanotechnology for societal development. J. Nanopart. Res. 13 (9), 3557–3590. Available from: https://doi.org/10.1007/s11051-011-0454-4.

Ruggiu, D., 2014. Responsibilisation phenomena: the EC code of conduct for responsible nanosciences and nanotechnologies research. Eur. J. Law Technol. 5, 3.

SCCS (Scientific Committee on Consumer Safety), 2012. Guidance on the safety assessment of nanomaterials in cosmetics. In: SCCS/1481/12.

Schulte, P.A., Salamanca-Buentello, F., 2007. Ethical and scientific issues of nanotechnology in the workplace. Environ. Health Perspect. 115, 5–12. Available from: https://doi.org/10.1289/ehp.9456.

Sodano, V., 2018. Nano-food regulatory issues in the European Union. In: AIP Conference Proceedings 1990, 020018. <https://doi.org/10.1063/1.5047772>.

Walker, R.L., Morrissey, C., 2012. Charting ELSI's future course: lessons from the recent past. Genet. Med. 4 (2), 259–267. Available from: https://doi.org/10.1038/gim.2011.60.

WHO, 2013. Nanotechnology and human health: scientific evidence and risk governance. In: Report of the WHO Expert Meeting 10–11 December 2012, Bonn, Germany. WHO Regional Office for Europe, Copenhagen.

Zhang, Z.G., Wang, A.Q., 2014. The third force in the good governance of nanotechnologies—the role of NGOs in discussions on nano-ethics. J. Henan Sci. Technol. 8, 225–227.

Index

Note: Page numbers followed by "*f*" and "*t*" refer to figures and tables, respectively.

D

E

R

S

T

CPI Antony Rowe
Eastbourne, UK
June 18, 2019